Studies in Logic

Volume 116

Logic

A Primer

Volume 106
New Directions in Term Logic
George Englebretsen, ed.

Volume 107
Non-commutative Algebras. Pseudo-BCK Algebreas versus m-pseudo-BCK Algebras
Afrodita Iorgulescu

Volume 108
Semitopology: decentralised collaborative action via topology, algebra, and logic
Murdoch J. Gabbay

Volume 109
The Cognitive Dimension of Social Argumentation. Proceedings of the 4th European Conference on Argumentation, Volume I. Fabio Paglieri, Alessandro Ansani and Marco Marini, eds.

Volume 110
The Cognitive Dimension of Social Argumentation. Proceedings of the 4th European Conference on Argumentation, Volume II. Fabio Paglieri, Alessandro Ansani and Marco Marini, eds.

Volume 111
The Cognitive Dimension of Social Argumentation. Proceedings of the 4th European Conference on Argumentation, Volume III. Fabio Paglieri, Alessandro Ansani and Marco Marini, eds.

Volume 112
Implicative-groups vs. Groups and Generalizations. Second Edition
Afrodita Iorgulescu

Volume 113
Meaning as a Set-theoretic Object. A Gentle Introduction to the Ideas Behind Formal Semantics
Jaroslav Peregrin

Volume 114
Logic and Language: New Frontiers in Analysis and Interpretation
José David García-Cruz, ed.

Volume 115
Four Functors. The 'ancestor' of Mundici's equivalence between unital commutative *l*-groups and MV algebras
Afrodita Iorgulescu

Volume 116
Logic: A Primer
Neil Tennant

Studies in Logic Series Editor
Dov Gabbay dov.gabbay@kcl.ac.uk

Logic

A Primer

Neil Tennant

ISBN 978-1-84890-506-1

College Publications
Scientific Director: Dov Gabbay
Managing Director: Jane Spurr

http://www.collegepublications.co.uk

Cover prepared by Laraine Welch

Dedication

To my dear wife Janel and our beloved sons Henry and Hugo,
who are my necessary and sufficient conditions
for a life of goodness, truth, and beauty.

Logic must look after itself.

Ludwig Wittgenstein
Tractatus Logico-Philosophicus, 5.473

Contents

Preface

I was born exactly two months after the publication of Quine's *Methods of Logic*. Since then Quine's primer, and other introductory books in its mold, have had (in my humble opinion) an adverse effect on the teaching of formal logic. They have entrenched Fitch-style proofs in Logic courses taught in Philosophy Departments. This is a sub-optimal situation, given the rapid development of more sophisticated proof theory over the last several decades. Fitch-style proofs cannot hold a candle to the tree-structured natural deductions of Gentzen's proof theory, which have been available since 1934-5. Gentzen-style natural deductions render to our primate brains the tree-structure of logical reasoning in a perspicuous and immediate way. By being introduced to formal logical proofs Fitch-style, and not Gentzen-style, students have been deprived of the all-important initial exposure that is required for mastery of the basics of modern proof theory. They have been given relatively impoverished resources for proving results at the metalevel *about* proofs at the object level. This primer is intended to be an antidote to Quine's and Fitch's no doubt unintended but eventually untoward influence.

We cover here the material that one would expect to find in a university-level Introduction to Symbolic Logic course. The emphasis will be on working *within* a logical system, and on why and how primitive rules of inference in natural deduction capture the meanings of the usual logical operators. We shall provide the reader (whether an instructor or a beginning student) with plenty of worked examples and appropriately timed exercises. We shall reserve treatment of the main *meta*-logical results (except for some easy ones) for a sequel volume, tentatively titled (something like) *Logic: The Sequel.* It will pick up where this primer ends, but with all the foundational concepts, symbolism for their expression, and rules governing them already in place.

Acknowledgments

In preparing this Primer I have had the benefit of Jane Spurr's remarkable efficiency in answering all manner of important queries. It has been a pleasure to work with College Publications.

This Primer and its intended sequel form a happy culmination of fifty years' teaching of Logic at all levels, from *Introduction to Informal Logic* through to *Advanced Symbolic Logic*; and of various research projects during those fifty years, to which that teaching gave rise. I suspect that I have learned a great deal more from my students than they have from me. So it gives me great pleasure to thank, however anonymously, past and present students at the University of Edinburgh, the University of Stirling, the Australian National University, and The Ohio State University. I hope that none of these students, on re-visiting any of the material included herein, suffers any discomforting flashbacks to their earlier attempts to learn Logic.

I have two great and lifelong intellectual debts to two of my teachers, for their formative influences in the very early 1970s.

First, to Timothy Smiley. Tim was my first logic teacher. He instilled by his fine example a taste for economy and elegance of expression in logic. No doubt my efforts here will have fallen short of his Platonic prescriptions. Still, I owe to him my original grounding and inspiration in logic, and the lesson that one should focus on inference in general, rather than on theoremhood. Tim's lectures on Aristotle's method of perfecting or completing proofs, and his continual insistence on the primacy of consequence over logical truth, and of deducibility over theoremhood, have had an enduring influence on my subsequent thinking about logic.

Second, to Ian Hacking. Ian's lectures conveyed an infectious enthusiasm for the philosophical insights to be had from Gentzenian proof theory. They helped to shape my subsequent methodological approach to logic.

A third lifelong intellectual debt, incurred just a few years later, is to Dag Prawitz, for his work on natural deduction and its relevance to the debate between realists and anti-realists on the question of which logic is the right logic. Dag was an inspiration on his 1976 visit to Edinburgh and St. Andrews when I was a rookie lecturer at Edinburgh. I took my bike on the train from Waverley to Leuchars and rode it through the mists and drizzle to Edgecliff, for each seminar lecture he gave in a week-long series. What I learned from Dag was foundational for a great deal of my own subsequent work.

Two later intellectual debts have been incurred to my colleagues Harvey Friedman and Stewart Shapiro at *The* Ohio State University.

Harvey has been unstinting in sharing his immense wealth of *a priori* and self-generated knowledge. Like Tim, he has always placed great value on philosophical relevance and general intellectual interest in whatever technical results one pursues.

Stu has been a resourceful sounding board for ideas and predilections at odds with his own. He is a structuralist; I am a Platonist. He is a classicist and realist; I am a constructivist and anti-realist. He prioritizes model theory; I prioritize proof theory. He is a second-orderist about mathematics; I am a first-orderist. He is a pluralist and anti-exceptionalist about logic; I am a monist and *a priorist*, despite being a naturalist. I would like to think that the joint *polemos* has imparted to our students a certain degree of hybrid vigor.

Last, but not least, I wish to record my thanks in the production of this text to Owain Griffin and to Charles Crumpler.

Owain, as my Teaching Assistant in Introduction to Symbolic Logic, rendered sterling help to students working through the material of Chapter 21, 'Proofs Galore'. It resulted in the current ordering of those deductive problems by their 'levels of difficulty'.

Charles, with his eagle-eyed reading of a near-final draft, has helped me to produce a book that might be worthy of a reader's attention. I owe to him corrections of many typos; improvements in exposition for the benefit of the beginner reader; and many valuable suggestions as to what should be included in the Index. All remaining defects, by way of commission or omission, are my own fault. *Vae mihi!*

Chapter 1

Introductory Considerations

> ... I shall not stop to explain this in more detail, because I should deprive you of the pleasure of mastering it yourself, as well as of the advantage of training your mind by working over it, which is in my opinion the principal benefit to be derived from this science.
>
> René Descartes, *La Geometrie*, 1637.

1.1 Target audience(s); and advice for the reader

This text is written in the belief that Logic now lives under four roofs: Philosophy, Linguistics, Mathematics, and Computing Science. Any student in any of these areas should study the other three. Logic draws its insights and methods from, and enjoys deep applications within, all four of these areas.

The text is aimed at a reader with some background in philosophy, linguistics, mathematics, or computing. But it should also be accessible to the autodidact with no background in any of these fields. For, if anything can be learned on one's own, from scratch, given only a sufficiently sustaining love of the life of the mind, it is Logic. The text provides the beginner with ample exercises, in order both to perfect and to streamline one's logical techniques, and to develop a deductive grasp of the main concepts used in Metalogic.

*Meta*logic might be a new term for the reader. The prefix 'meta' comes from ancient Greek, meaning 'after' and entered the academic vernacular owing to Aristotle's *Metaphysics*. It is not known whether Aristotle wrote his *Metaphysics after* his *Physics*. But his *Metaphysics* came to occupy a later spot in the curriculum than his *Physics*, and has come to be regarded as a set of philosophical reflections on the latter. In the scholarly vocabularies of contemporary mathematicians, philosophers, and logicians, 'meta' now means, roughly, 'at one level up'. Metalogic talks about logical systems. Metamathematics talks about mathematical theories. Metaphilosophy discusses how the discipline of Philosophy should itself develop and proceed.

As the material presented here for the student reader unfolds, we shall rely on the context of introduction of new technical terms and formal symbols to impart a good grasp of what they mean. The previous paragraph is an illustrative case in point.

This book will serve even an absolute beginner to the discipline of Symbolic Logic—who, once equipped with the concepts and skills developed here will then be able to build further on these in more advanced courses in mathematical or philosophical logic. On occasion in this introductory work we shall be referring to publications in the research literature; but this will only be to orient the reader and to make promissory notes. In a planned sequel to this work, we intend to cover all the usual metatheorems in mathematical logic, about the expressive and deductive powers of first-order logics. These include The Strong Completeness Theorem, The Compactness Theorem, The Countable Models Theorem, The Noncompossibility Theorem, The Cut Admissibility Theorem, The Implicit Definability Theorem, and The Interpolation Theorem. Also to be treated are the usual metatheorems about first-order *theories*. Among such theories are the theory of natural numbers, the theory of rational numbers, the theory of real numbers, and set theory. It should be a seamless progression (from this primer to its sequel) for the student who reads this work carefully to the end. All the basic metalogical concepts that one needs to deploy on the more advanced material—those aforementioned theories that seek to characterize certain *infinite* structures—can be acquired at this introductory stage by considering only *finite* ones.

The present author is a firm believer in the centrality, fundamentality, and *a priori* status of Logic as of critical importance for the pursuit of knowledge in both mathematics and the natural sciences. A beginner who approaches Logic with no preconceptions at all, with a willingness to learn, and with an open mind should be able to master fully the basic concepts and technicalities of the discipline through careful study of this primer. They should also be able thereby to acquire confidence in the prospect that the formal systems developed here will be able to fulfill the original foundational aim of the discipline of Logic. This aim is the formalizability, in principle, of deductive reasoning—both constructive and classical—in mathematics and the natural sciences.

The author asks the beginning reader to persevere. Try to 'stay the course'. Read carefully, and be prepared to undertake some rote learning. Logic can at times look deceptively simple, while in fact revealing something profound. It may also appear very complicated when the matters under discussion are quite simple. The appearance of complication is frequently the price to be paid in order to achieve the necessary level of rigor, generality, and attention to expository details.

Some ways of setting things out (such as Rules of Inference) may at the outset strike the beginner as rather complex. That is where the reader's forbearance and resilience is required. Formal rules of inference are central to logic. They are the means by which one constructs *proofs* of conclusions from premises. In the author's experience, students benefit from *learning* things like rules of inference 'off by heart'—but, to be sure, only after receiving a principled explanation of what is involved. The author might be old-fashioned in this regard; but he is aware of no easy way to proficiency in Logic without committing at least *some* slightly complex patterns to memory.

In this text we shall be selecting our rules of inference very carefully indeed. They are to govern the so-called *logical operators*. These are words or phrases such as 'it is not the case that ...'; 'both ... and ...'; 'either ... or ...'; 'if ... then ...'; 'every'; and 'some'. The *governing* of these operators by their rules is a *normative* matter. If I wish to assert 'P and Q' to you, I am *obliged* to be able both to assert P and to assert Q. When you hear my assertion 'P and Q', you are *entitled* both to infer that P and to infer that Q. These normativities are expressed by the inferential rules

$$\wedge\text{-Introduction}\quad \frac{P \quad Q}{P \wedge Q} \qquad\qquad \wedge\text{-Elimination}\quad \frac{P \wedge Q}{P} \quad \frac{P \wedge Q}{Q}$$

where of course we are using the symbol $\wedge$ as shorthand for 'and'.

It is an enduring philosophical problem whether one can explain such normativities if one's overall background theory of the world is a merely *descriptive* one. A guiding aim of ours in this text is not to engage in any attempt to solve this problem of 'locating the normative within the descriptive'; but only (or at least) to furnish some formal-symbolic food for thought for the philosophically minded reader who might wish to pursue a deeper philosophical understanding of what Wittgenstein once called *the hardness of the logical 'must'*. This is the metalogical analog, perhaps, of the parent or moral tutor who conveys to the learner of values such normativities as the Golden Rule and the Silver Rule; but who leaves to the learner the further *philosophical* problem of trying to grasp how it is that moral values can arise within a world of brute physical facts.

The discipline of Logic is not entirely theoretical. Like musicianship, or the martial arts, it requires practice.[1] The beginning logician needs to acquire a certain sort of practiced dexterity, an intellectual nimbleness, that is based on rote repetition but which eventually affords creative combinations. Therefore, when the text advises the reader to pause and commit certain material to memory, it is meant seriously and with sincere concern for the reader's continuation of the journey. The reader is now advised to commit to rote memory the above rules of Introduction and Elimination for conjunction.

1.2 Starting on a philosophical note

We begin with one of the classic moves in philosophizing—that of raising a 'problem of reflexivity'.

The study of logic is itself logical. And any explanation, to a beginner, of what logical systems are, and what their basic properties are, will presuppose at least some ability on the beginner's part to think logically. Thus we have a problem of potential circularity. Can the reflective study of logic be free of any contamination by what one might regard as unexamined preconceptions about logic, or unexamined logical presuppositions? How can one convey to beginners the basic theoretical facts about logical systems if the beginners really are innocent of logic at the

[1] See Leonard [1992].

outset? And how does one negotiate a route, free of presuppositional obstacles, to the point where one can lay out a study of *object-logical* systems designed to furnish the means for laying things out in fully rigorous deductive detail, yet do so in compliance with the immediately prevailing *metalogical* demand for fully rigorous deductive detail? How can the novice be guided down a path to that endpoint?

Let us for the time being assume a sufficiently sophisticated and shared level of knowledge of logic and of foundational matters. (The beginner can therefore return to re-read this section after reading the rest of the text.) In order to theorize about even the simplest logical systems, namely systems of propositional logic, one needs to be able to refer to *symbols* (such as $\wedge$); to *sentences*—which are finite strings or sequences of symbols, such as $P \wedge Q$; and to *proofs*—which are finite tree-like arrays of sentence-occurrences, such as

$$\dfrac{\dfrac{P \wedge Q}{P} \qquad \dfrac{R \wedge S}{S}}{P \wedge S}\ .$$

One needs to be able to identify and distinguish such things, and assert and prove generalizations about them, both existential and universal. Example: *Every valid argument has a proof.* The logical form of this claim is at least something along the lines of 'For every x, if x is a valid argument, then there is some y such that y is a proof of x'. And here we are generalizing both universally ('for every') and existentially ('there is some').

So theorizing about logical systems involves at least the logical refinement and power of quantificational (or first-order) logic. It requires, that is, a language powerful enough for the expression of so-called *combinatorial* or *discrete* mathematics. Not only do we need the language, however; we also need to be able to reason within it logically, when we prove results within it about logical systems. That is, one needs a commitment to a certain logical system (for our quantificational metalanguage) before one can even begin laying the logical foundations for the study of logical systems themselves. Let us call this the 'core *logico-linguistic* commitment'.

Discrete mathematics is the study of finite 'discrete' objects such as strings and trees, and other combinations of basic symbols. Moreover, theorizing about logical systems involves *espousing and applying* the discrete mathematics in question. That is, one needs a theoretical commitment to a certain fragment of contemporary mathematics before one can even begin studying the *logical foundations* of that very fragment. Let us call this the 'core *mathematical* commitment'.

Clearly it is desirable to minimize both kinds of commitment. By doing so, we render more certain the results we obtain about logical systems. Moreover, the weaker we can make these core commitments at the outset, the greater will be the theoretical gains that we are able consequently to garner.

1.3 Our emphasis will be on proof theory

The plan of exposition of logical material in this book is unusual, but it is informed by these reflections on minimizing one's mathematical commitments. We shall set out an unusual—if not to say unprecedented—amount of *proof-theoretic* material, or material in the so-called 'theory of syntax', *before* we go on to cover formal semantics and the completeness theorem. We do this because all the proof-theoretic theorizing can be accomplished within a relatively weak mathematical theory. Formal sentences and formal proofs involving them are inductively definable objects. The study we undertake of rules of inference, proofs, and normalization of proofs is relatively *weak*, mathematically speaking, even if at times a little taxing because of the unfamiliarity of the material and the occasional complexities of the patterns involved. For the reader who perseveres, there is the prospect of richer understanding within a framework of more modest resources.

1.4 The main logical system in focus

This primer focuses on *free first-order logic with identity* as the main system of logic. It is the logical system that is needed for the regimentation of deductive reasoning in mathematics. That is how modern logic was born, in Frege's 1879 masterpiece *Begriffschrift: eine der arithmetischen nachgebildete Formelsprache des reinen Denkens* (in English: *Conceptual notation: a formulaic language for pure thought modeled on that of arithmetic*—henceforth, *Begriffsschrift*). It was this work that first illuminated and made precise the workings of the two quantifiers 'some' (or 'there exists') and 'all' in the hands of mathematicians. These quantifiers are rendered today by the respective symbols $\exists$ and $\forall$.

They are called 'first order' quantifiers because they 'range over' the *individuals* in the *domain of discourse.* The individuals are also called the objects, or things, or entities, or particulars, being talked about. They may be, for example, in the mathematical case, *abstract* objects: the natural numbers; or the rational numbers; or the real numbers; or the complex numbers; or, most ambitiously, the *pure sets* in the so-called cumulative hierarchy of sets, which—according to some—is the unifying and all-encompassing ontology for mathematics, in which all the more familiar mathematical objects can be represented by 'set surrogates'. The individuals in the domain of discourse may also be *concrete* objects, not abstract ones, at any scale studied by natural science, ranging from galaxies to sub-atomic particles.

The logical system is called *free* because in one important and fundamental respect the formal language involved differs from Frege's. He made the unfortunate decision to demand that every singular term of his formal language should be secured a denotation. But this meant that he could not treat properly the many complex logico-grammatical terms that *do not*—indeed, *cannot*—denote objects. Examples of such terms are the following:

$$1 \div 0 \ ; \qquad \{x|x = x\} \ ; \qquad \{x|x \notin x\} \ ; \qquad \sum_{n>0} \frac{1}{n}$$

In free logic we free ourselves from this disastrous choice on Frege's part—which, ironically, he made in pursuit of what he called a 'logically perfect' language. We shall be paying the necessary rigorous attention to one's existential presuppositions in deductive reasoning. For our overarching aim is to produce a formal language, governed by a logical system consisting of rules of inference, that will serve all the deductive needs of the mathematician.

And that brings us to the final part of the label 'free first-order logic with identity'—which is the logician's sign '=' for *identity*, and which the mathematician often calls 'equality'. Everyone knows that mathematics is full of *equations*. These are *identity statements* involving singular terms. Examples are

$$2+3=5 \ ; \quad (x+1)\times(x-1)=x^2-1 \ ; \quad \int_1^e \frac{1}{x}dx=1 \ ; \quad \sum_{n>0}\frac{1}{2^n}=1.$$

Every *kind* of object, abstract or concrete, will have its own *criteria of identity.* And the binary identity predicate '=' has the same meaning, and is governed by the same logical laws, whatever the kind of things being talked about. Thus the identity predicate in the statement '$2+3=5$' is the same predicate as the one in the statement 'The Morning Star = The Evening Star'. As Quine once put it: 'No entity without identity.'

Quine also called first-order languages with identity the 'Grade A idiom' for science. We agree with that. In this text we set out to furnish exactly the logical rules that are *needed* for this Grade A idiom to be employed. Our logical rules will also *suffice* for the formalization of deductive reasoning in all scientific investigations—in pure mathematics, in applied mathematics, in scientific prediction and explanation, and in empirical confirmation or refutation of scientific hypotheses.

1.5 Logic and philosophical analysis

It is no accident that Logic is rooted in the work of philosophers. For philosophers are concerned with fundamental questions about truth, meaning, existence, possibility and necessity, rationality and reason. They are also fascinated by the connection between language and thought; and by the nature of the communicative and contractual ties that bind individuals into a community.

In analytic philosophy, moreover, as it has developed from the early twentieth century, there has been a focus on the analysis of central concepts such as meaning and knowledge. This has engaged the two kinds of logical ability discussed above: the ability to construct proofs (that is, to engage in chains of logically rigorous reasoning); and the ability to think up counterexamples when one suspects that there are no proofs in the offing (for whatever argument is in question). A famous recent instance of *counterexemplification* in analytical philosophy is that of so-called 'Gettier cases' in epistemology.[2] These are imagined situations

[2]See Gettier [1963].

that serve as counterexamples to the argument

> S believes that p;
> p is true;
> S is justified in believing that p;
> *therefore*, S knows that p

Contemporary epistemologists now accept that Gettier-style counterexamples have rendered untenable Plato's famous analysis of knowledge as justified true belief. Much of so-called reliabilist or externalist epistemology is trying to make good the logical gap that these Gettier-style counterexamples have exposed.

Two other well-known counterexemplifications in recent philosophy are those offered by Saul Kripke to infirm the arguments

> p is necessarily true;
> *therefore*, p is knowable *a priori*

and

> p is knowable *a priori*;
> *therefore*, p is necessarily true

According to Kripke, 'Water is H_2O' is metaphysically necessary, but knowable only *a posteriori*. Thus it serves as a counterexample to the first of these two arguments. And 'The standard meter rod is one meter long' is knowable *a priori*, but is metaphysically contingent (that particular rod might have been longer, or shorter, than it happens to be). So this serves as a counterexample to the second argument.[3]

The philosophical literature is full of ingeniously constructed counterexamples to various suggested arguments. The arguments typically arise from the project of furnishing necessary and sufficient conditions for the application of an important concept (such as the concept of knowledge, *à la* Plato). All claimed analytic implications, from some ingeniously specified set of antecedent conditions to some target *analysandum* supposedly consequent upon them, is hostage to the misfortune of even more ingenious counterexamples thought up by fertile philosophical minds. Such thought-experiments often invite the consideration of

[3]See Kripke [1980].

extremely bizarre possibilities, in which concepts might be stretched to the very limits of their intelligible application. These possibilities are designed to establish that there is a logical gap between the antecedent conditions and the supposedly consequent one. That is, they are designed to counterexemplify the argumentative inference from the former (as premises) to the latter (as conclusion).

It is perhaps less usual to find, in philosophical writing, sustained examples of the complementary logical activity—that of providing a rigorous, step-by-step proof, for a valid argument. Even philosophy of the most rigorously argumentative kind tends to be ruminative, in a way that makes it difficult to reconstruct a precise direction of logical reasoning. In most cases of the inevitable exceptions to this rule, one finds that the premises and the conclusion of the diagnosed 'proof' within the text form an argument-pattern that is not terribly demanding, from a logical point of view. Much critical exegesis of philosophical 'argument' consists in identifying or isolating more clearly than the original author did those claims that happen to serve as premises for whatever 'argument' the author provided for his or her 'conclusion'. The secondary literature is filled with such diagnoses of argumentative intent, accompanied by identification of the inevitable logical lacunae, and proposals for filling them.

It would seem that Philosophy, as a rigorous discipline, attracts that cast of mind that is preoccupied with finding logical fault in attempted argumentation. It is the 'But what if ...?'-mentality at work. To be sure, there is a perverse kind of intellectual joy to be had from the highly imaginative counterexamples that philosophers at times think up ('OK, so there's this Martian who .../homunculus that .../zombie that .../Twin Earth where .../brain-in-a-vat that .../molecule-by-molecule *Doppelgänger* that .../universe with just two identical metal spheres where .../doughnut-shaped space-time where .../infinitely divisible gunk that .../telepathic evil android that .../clairvoyant dice-thrower who .../disembodied Cartesian soul who .../color-blind but neuroscientifically omniscient sheila called Mary who ...')—but at the same time there is a relative dearth of sustained, deep, deductive forging of necessary connections between premises and conclusion. Philosophy, that is to say, is short on *proofs*, but full of *counterexamples*. For the best examples of the complementary activity of *proving conclusions from premises* one does better to turn to that other great area of highly abstract human intellectual endeavor—mathematics.

1.6 Glimpses aside

This is a primer on formal logic, which would naturally lead to more advanced studies of deep results comprising what is known as Mathematical Logic. We have already mentioned some of staples, on p. 2. But we did not mention there the famous *limitative* theorems: Alonzo Church's Undecidability Theorem for first-order logic, and the two Incompleteness Theorems of Kurt Gödel: (I) the first-order truths in the language of arithmetic cannot be axiomatized; and (II) no consistent and sufficiently strong theory of arithmetic can establish its own consistency.[4] These limitative theorems bear on the philosophical questions of whether mathematical thinking could be mechanized, and whether our minds could be modeled as computational systems.[5]

In Quine's introductory text[6] his brief discussion of more advanced material was titled 'Glimpses Beyond'. But, as the title of this section intimates, there are other ways that a core training in formal logic can lead to further developments within logic that are pertinent for understanding, discussing, and potentially solving certain problems in philosophy. We shall briefly describe some of the main further developments here.

Philosophical logic is a branch of philosophical and logical inquiry, typically involving extensions of the standard, extensional logic of connectives ('not', 'and', 'or', and 'only if'), quantifiers ('some' and 'all'), identity ('=') and definite descriptions ('the ...'). Philosophical logicians study a variety of extensions of standard logic, each one typically focusing on one or two chosen concepts of great philosophical importance.

In *modal logic*, for example, one deals with the concepts of *necessity* ('truth in *all* possible worlds') and *possibility* ('truth in *some possible* world, even if not in the actual world').[7] One extends the formal language by adding the so-called *modal operators* $\Box$ and $\Diamond$, which are

[4]See Church [1936] and Gödel [1931].
[5]See Lucas [1961], Benacerraf [1967], Penrose [1989], and Tennant [2001].
[6]Quine [1950].
[7]The *locus classicus* for possible-worlds semantics for modal logic is Kripke [1963].

one-place connectives:

$$\Box\varphi : \text{It is necessarily the case that } \varphi;$$
$$\Diamond\varphi : \text{It is possibly the case that } \varphi.$$

In *epistemic* and *doxastic* logic, one uses a new kind of operator, an expression of a so-called 'propositional attitude', which forms a sentence from a singular term and a sentence:

$$aK\varphi : a \text{ knows that } \varphi;$$
$$aB\varphi : a \text{ believes that } \varphi.$$

Modal, epistemic, and doxastic logics are *not extensional.* This is because of the failure, in those logics, of the principle of *Substitutivity of Identicals*:

$$\frac{\psi(t) \qquad t=u}{\psi(u)}\ .$$

When such substitution (of u for t) cannot be relied upon to preserve truth from the two premises displayed to the conclusion, the context $\psi(\ )$ is said to be *referentially opaque.* By contrast, mathematical language and its logic (the very birthplace of modern logic) is *referentially transparent.*

In *tense logic* one deals with the concepts of the past, present and future. On one approach, this involves adopting three new one-place connectives:

$$P\varphi : \text{It was the case that } \varphi;$$
$$N\varphi : \text{It is now the case that } \varphi;$$
$$F\varphi : \text{It will be the case that } \varphi.$$

while on another approach one takes every statement to be time-indexed, and one quantifies over temporal instants t presumed to be subject to some ordering $<$. Among these instants is a distinguished one called 'now' (n):

$$\exists t(t < n \wedge \varphi(t)) : \text{It was the case that } \varphi;$$
$$\varphi(n) : \text{It is now the case that } \varphi;$$

$\exists t(n < t \wedge \varphi(t))$: It will be the case that φ.[8]

Philosophers interested in the analysis of causation often make use of the *counterfactual* conditional, either in the form 'if φ had been the case, then ψ would have been the case' or in the form 'if φ were the case, then ψ would be the case'. This is usually regimented as $\varphi \,\Box\!\!\rightarrow \psi$.[9]

These are just a few examples of important notions whose logical behavior philosophical logicians have sought to characterize. With each of these notions, there is typically a great variety of competing systems of logic, differing in their account of the inferential powers of sentences involving the new operators.

Each notion calls for a new kind of *formal semantics* to model the essential features of the world that are germane to the interpretation of the new operator(s) in question. With modal operators, for example, the formal semantics provides a plurality of 'possible worlds', among which is the actual world. Each such world is made up of a different range of contingent facts. From each such world, some (but not necessarily all) of the others are 'accessible'. The truth-condition for $\Box\varphi$ in a given world w is that in *every* world accessible from w, φ is true. And the truth-condition for $\Diamond\varphi$ in a given world w is that in *some* world accessible from w, φ is true. Different modal logics will be obtained by specifying different conditions on the accessibility relation among worlds (conditions such as reflexivity, symmetry, transitivity, ...). In the intended sequel to this primer we shall be giving a proof-theoretic treatment of various systems of modal logic that is strongly informed by the conditions on the accessibility relation in the possible-worlds semantics for the system in question.

In some cases a workable logic can be achieved only by idealizing the notion being treated. For example, it is often objected that people's actual knowledge is never closed under logical deducibility; yet it is difficult, if not impossible, to lay down normative rules of inference accommodating logical myopia on the part of the agents concerned. One tends therefore to idealize the epistemic claim $aK\varphi$ so that it reads '[*a*

[8] The beginning reader who is interested in learning more about tense logic would do well to study Massey [1969] and Burgess [1979].

[9] The beginning reader who is interested in learning more about counterfactuals and the various logics governing them should consult Lewis [1973].

knows or] it follows logically from what a knows that φ'. But even if the knower were a logical saint, she would still be afflicted by the quirks of intensionality (i.e. failures of extensionality) with which the logic of propositional attitudes is mainly preoccupied. For example, logical omniscience on Alberto's part notwithstanding, the following inference scheme is not valid:

$$\frac{\begin{array}{l} aK\psi(t) \\ t = u \end{array}}{aK\psi(u)} \ .$$

An obviously invalid instance of this scheme is:

$$\frac{\begin{array}{l} \text{Alberto knows that Tiny was at the crime scene} \\ \text{Tiny is the undercover agent} \end{array}}{\text{Alberto knows that the undercover agent was at the crime scene}} \ .$$

This phenomenon, which we have described as *referential opacity* (manifested in the above failure of substitutivity of identicals) is the most serious problem to be addressed by any epistemic or doxastic logic.[10] (It arises also in modal logic, but there the solutions are easier to formulate.)

Corresponding to the innovations called for in the formal semantics for any of the 'new' notions mentioned above (by 'new' here we mean 'not central enough to be addressed in this work'), one will have new rules of inference in one's proof system, which will be an extension of the proof system for standard extensional logic.

Let us return now to the basics of standard extensional logic.

1.7 The main concepts to be explained

Beginning students should not be overwhelmed by finding themselves unacquainted with many (indeed, nearly all) of the items on the following 'laundry list'. They can skip ahead to the next section, which covers in more detail the rudimentary skills to be acquired. They can return to this section when they have completed their study of this primer, so

[10] For detailed attempts to construct a logic governing verbs of propositional attitude, in which distinctions can be drawn between referentially transparent and referentially opaque contexts, the reader could study Hintikka [1964] and Tennant [1977].

as to experience a sense of satisfaction and accomplishment from their diligent study.

Syntactic concepts

Syntactic concepts are ones that apply to *bits of notation*, or *signs*, or *symbols*. Items like these in Logic include the following.

name; variable; function symbol; predicate symbol; variable-binding term-forming operators (definite description operator; set abstraction operator; number abstraction operator); connectives (negation; conjunction; disjunction; implication/conditional); quantifiers (universal; existential); atomic sentence; absurdity constant; term; well-formed formula; subformula; free v. bound variable occurrences in a term or formula; scope of an occurrence of an expression; dominant occurrence; disjunctive normal form; conjunctive normal form; prenex normal form; substitution instance; instantiation; generalization; axiom; rules of inference (introduction rules; elimination rules; strictly classical rules for negation); reduction procedure; proof; reductio ad absurdum (refutation); discharge of assumptions; application of a rule; deducibility ($\vdash$)*; theorem (of logic); consistency v. inconsistency; deductive closure; normal form of proof; mechanical checkability*

Semantic concepts

Semantic concepts are ones that have to do with the relationship between syntactic expressions and their 'subject matter', or 'interpretation'. Among such concepts in Logic are the following.

truth tables for the connectives; truth-value assignment to atomic sentences; reference; denotation; extension; mapping; individual; domain; model; satisfaction; truth; verification; falsification; truth-conditions; logical consequence ($\models$)*; satisfiability v. incoherence; isomorphism of models; categorical description of a model*

Metalogical concepts

Metalogical concepts are concepts that are employed in theorizing (at the so-called 'meta-level') *about* proof systems, languages, and theories expressed within those languages at the so-called 'object level'. Metalogical theorizing is conducted in the so-called 'metalanguage' (usually written this way, without a hyphen). The metalanguage is typically more expressively powerful than the object language. Some theorists (such as Alfred Tarski, who first introduced the object language/metalanguage distinction in the 1930s)[11] do not allow the object language to express concepts concerning its own relationship to its subject matter. One such concept is that of *truth*. If one allows oneself to speak *within the object language itself* of the truth or falsity of sentences of the object language, then paradox can and will ensue. The most famous example, of great vintage since the ancients, is the 'Liar Paradox'. This takes the form 'This sentence is not true'.

Among the metalogical concepts deployed in Logic are the following.

truth of a sentence under an interpretation of the appropriate kind; soundness of a proof system; completeness of a proof system; theory-completeness; categoricity of a theory; axiomatizability of a theory; decidability of a theory; definability; submodel; elementary extension of a model; non-standard model.

1.8 The skills to be developed by this primer

The beginner should be able to develop the following skills by reading the material carefully and doing the exercises provided. The beginner should be able to re-visit this list to check on their progress in acquiring the promised skills to be imparted. They are arranged in the rough order of expected acquisition. Chapter 22 provides a longer and more detailed list of these skills, describing them in terms that will be explained more fully in the intervening chapters.

1. checking terms and formulae for well-formedness

[11] Tarski [1935. First published, in Polish, in 1933]. The standard reference for the English translation is Tarski [1956].

2. constructing terms and formulae
3. finding formulae to represent logical forms of English sentences
4. determining truth values of (formal) sentences in models
5. describing (finite) models exactly
6. checking proofs for formal correctness
7. checking counterexamples to invalid arguments
8. finding proofs of valid arguments
9. finding counterexamples to invalid arguments
10. appraising arguments in formal notation
11. appraising arguments in English

Note that it is easier to *check* a proof (or a counterexample) than it is to find one. This holds even if you are told that there is indeed a proof of the argument, and are asked to find such a proof. Likewise it holds even if you are told that there is indeed a counterexample to the argument, and are asked to find such a counterexample. Finding or constructing a complicated logical object is always more difficult than checking (once you have found it, or been given it) that it is indeed an object of that kind.

When it comes to *appraising* arguments, the difficulty is even greater. One has to decide whether the argument is valid or invalid. If it is valid, one has to find a proof. But if it is invalid, one has to find a counterexample. It is much easier to check what you find; and much harder to find something (that you think is correct) for anyone to check.

Moreover, when appraising arguments that are expressed in English, the task is at its most difficult, and it requires all the skills mentioned above. If you think the argument is valid, then you have to translate its premises and conclusion into formal notation, so as to reveal enough logical form to enable you then to construct a proof of the (formal) conclusion from the (formal) premises, using only basic rules of inference. If, on the other hand, you think the argument is invalid, then you have to translate its premises and conclusion into formal notation in such a way that every possibly relevant aspect of logical structure is revealed; then you have to construct a counterexample—let us call it M (for model)—and determine that each (formal) premise is true in M while the (formal) conclusion is false in M.

Exercise 1. Try to work out why the Liar Paradox poses a serious problem for any theory of language and logic.

Exercise 2. Why do you think it is easier to check constructs for correctness than it is to find them?

Exercise 3. What do you think are the main differences between formal and natural languages? Are they so great that formal languages will not be of much use for one who is interested in deductive reasoning in natural language?

Chapter 2

Sentences

2.1 Declarative sentences

Logic studies the laws governing transmission of truth among declarative sentences. Declarative *sentences* are used to 'tell one how things are'. They are 'truth apt'. When a sentence is true (under a particular interpretation) we say that it *has the truth value T*. When it is false we say that it *has the truth value F*. A sentence has the truth value it does when it is interpreted as a statement or claim about some subject matter. The subject matter may be abstract or concrete, natural or social. But in logic, we do not care about the *content* of declarative sentences insofar as such content concerns their particular subject matter; rather, we care only about their *logical forms*, and the *logical connections* among them.

Henceforth by 'sentence' we shall mean 'declarative sentence'. But on occasion we shall reinstall the adjective 'declarative'.

Here are two questions that generate heated philosophical debate:

1. Does every declarative sentence have one or other of the two truth values? (Or can there be truth-value *gaps*?)

2. Can any declarative sentence have *both* of the truth values? (Can there be truth-value *gluts*?)

One of the main skills that the beginner in Logic needs to acquire is that of translating English sentences into formal logical notation. One can think of the formal sentence that results as providing the *logical form* of the English sentence in question. The converse skill is also important: given a formal sentence, find an example of a fluent English sentence that has that particular formal sentence as its logical form.

A lot depends, of course, on how complicated, or expressively powerful, the English (or 'mathematese') sentences are allowed to be, for the purposes of assigning logical forms to them. A little bit of history might be helpful here.

2.2 The expressive needs of mathematics

The single most powerful motivation, in the late 1800s, to provide a formal language for the expression of thought was to be able to handle the sentences of mathematical discourse. Mathematics is the discipline where deductive precision and correctness is at a premium. In mathematics, rigorous proofs are the be-all and end-all of the whole endeavor. So it behooves the logician to examine carefully the language of mathematics.

The logician needs to identify which logical operators need to be treated. The logician also needs to specify what modes of logico-grammatical combination permit the correct formation of mathematical expressions, including declarative sentences, in general.

There are special constants (names), such as 0 and 1. There is the ubiquitous identity predicate (the 'equals' sign) $=$. There are various kinds of complex singular terms for referring to mathematical entities. *Singular* terms are expressions which stand for a unique individual or object, if they succeed at all in connecting with the subject matter of the discourse. Consider, for example, complex singular terms in mathematical discourse such as

$$a+b\ ;\ \ a\times b\ ;\ \ a^b\ ;\ \ \frac{d}{dx}f(x)\ ;\ \ \int_0^{\frac{\pi}{2}}\sin(x)dx\ ;\ \ \sum_{n=1}^{\infty}\frac{1}{n^2}\ ;\ \ \{x|x\subseteq\omega\}.$$

There are also the familiar mathematical relations such as $<$ ('is less than'), $\leq$ ('is less than or equal to'), and $\in$ ('is a member of'), to name but a few.

The most important expressions in mathematics, from the logician's point of view, are the so-called *quantifiers*—the universal quantifier $\forall x$ ('for all x') and the existential quantifier $\exists x$ ('there exists x such that

. . . '). During the mid- to late-1800s, mainly as a result of the works of Augustin Cauchy and Karl Weierstraß, mathematicians became increasingly sophisticated and (informally) rigorous in their manipulations of these expressions in their definitions of, and proofs about, the notions of limit, least upper bound, continuity, uniform continuity, derivatives, and integrals. But they still had no precise *logical grammar* for the quantifying expressions.

The great breakthrough in this regard came in Gottlob Frege's 1879 masterpiece *Begriffsschrift*. In that seminal work, Frege 'cracked the grammatical code' of the quantifying expressions, and laid bare their logical workings by inventing the 'bound variable' notation.

We shall reprise the relevant Fregean ideas in due course. We should take the opportunity, at this juncture, to advise the reader that we shall try to avoid the usual introductory-textbook approach of dealing only with propositional logic (the logic of the *connectives*—negation $\neg$, conjunction $\wedge$, disjunction $\vee$, and the conditional $\rightarrow$) at the outset, and only later embarking on a treatment of the quantifiers $\exists$ and $\forall$. That conventional approach conveys the mistaken impression that the logical behavior of the quantifiers is much more difficult to grasp than the behavior of the connectives. It is much more interesting and motivating, however, to have quantifying expressions occurring in one's examples from the very beginning. One can make logical ideas accessible right away for a formal language that contains *both* the connectives *and* the quantifiers. Indeed, certain very central logical ideas can be introduced much more vividly and accessibly if one has the quantifying expressions in one's logical toolkit at the very outset. See the treatment in §4.4, for example, of the notions of numerosity, inconsistency, and logical consequence.

We have now mentioned all the main logical operators, with their logical symbols and their English names. The latter names are used also to designate formulae created by using the logical symbols. If φ and ψ are formulae, we speak of their *disjunction* $(\varphi \vee \psi)$, their *conjunction* $(\varphi \wedge \psi)$ and the *conditional* $(\varphi \rightarrow \psi)$ with *antecedent* φ and *consequent* ψ. The free occurrences of variables in these compound formulae are those that are free either in φ or in ψ. The formulae φ and ψ are, respectively, the left and right *disjuncts* of $(\varphi \vee \psi)$ and the left and right *conjuncts* of $(\varphi \wedge \psi)$. As already mentioned, the formula φ is the *antecedent* of the conditional $(\varphi \to \psi)$, and ψ is its *consequent.* If ψ is a formula with a free occurrence of the variable x, then we shall speak of the *existential* $\exists x\psi$ and of the *universal* $\forall x\psi$. In each of these the variable x will no longer be free. It is *bound* by the quantifier prefix in question.

We shall include the quantifiers as important logical expressions in our early exercises in *translating* English sentences into logical notation and in rendering sentences in the formal language as fluent English sentences. Only later will we focus more narrowly on just the connectives—before moving on to include the quantifiers—in order to have a simpler context within which to introduce the reader to certain metalogical concepts and to certain metaproofs 'by induction' on the internal complexity of the syntactic objects being talked about. An historical irony, given modern logicians' proclivity to deal first with the connectives and only later with the quantifiers, is that the founder of logic, Aristotle, over two millenia ago, dealt in his syllogistic with the *quantifier* expressions 'All As are B', 'Some As are B', 'No As are B', and 'Some As are not B'; he did not deal at all with the connectives in their own right.

There are some metalogical ideas and techniques, however, that are easier to introduce in the slightly simpler setting of propositional logic; and, as already indicated, we shall take advantage of these at certain expository junctures.

We have mentioned that Frege made a monumental breakthrough in 1879 with his characterization of the quantifying expressions 'all/every' and 'some/there exists' as variable-binding operators. Here we explain how his insights can be achieved, and exactly how variables come to be bound.

We prepare the ground by looking at how atomic sentences of English can enjoy so-called 'predicate-argument' structure. Then we shall move on to quantifying expressions.

2.3 Predicates and names

As the reader will already be aware, the basics of Logic can be explained by means of examples of sentences of a natural language, such as English, or sentences of a formal language, such as that of mathematics. And this is because Logic applies to both kinds of discourse, in a topic-indifferent way. It also means, happily, that one can explain logical (and grammatical) considerations that may have had their source in mathematics by using simple examples from a preferred natural language—which in this case is English. So, the reader is invited to journey to an understanding of Frege's bound-variable notation for the quantifiers—for him, at the time of his breakthrough, a pressing conundrum in *mathematical* discourse—by reflecting on some simple examples in everyday discourse in English.

One-place predicates

Take a simple singular sentence like 'John is friendly'. It contains the *one-place predicate* '...is friendly', which is completed by the name 'John' (a basic singular term) to form the sentence. (One-place predicates are also called *monadic*, or *unary*.) Logicians use an uppercase letter with a parenthetically enclosed gap—say,

$$F(\ )$$

—to symbolize the predicate, and use a lowercase letter—say, j—as the name that completes the predicate by filling the gap:

$$F(j).$$

This 'atomic sentence' of the logician's formal language *symbolizes* or *regiments* the English sentence, by bringing out its salient logical structure.

For what purpose is this logical structure *salient*?—why, for the discernment of *logical relationships* among sentences, such as the *logical implication* of 'Someone is friendly' by 'John is friendly', or that of 'John is friendly' by 'Everyone is friendly':

$$\frac{\text{John is friendly}}{\text{Someone is friendly}} \qquad \frac{\text{Everyone is friendly}}{\text{John is friendly}}$$

These are *valid arguments*. In each case the premise (above the inference stroke) logically implies the conclusion (below the inference stroke). That is, any interpretation of the non-logical expressions—the name 'John' and the predicate 'is friendly'—that makes the premise true makes the conclusion true also.

Two-place predicates

With *two-place*, or *relational* predicates, the procedure is much the same as for one-place predicates.

Two-place predicates in ordinary discourse

The sentence 'John loves Mary' involves the two-place predicate

$$L(\ ,\ )$$

whose gaps are filled by the names j and m—*in that order*—to symbolize the sentence:

$$L(j, m).$$

(Two-place predicates are also called *dyadic*, or *binary*.) In general, the order of the 'arguments' (here, j and m) after the predicate is important. The sentence 'Mary loves John' would be symbolized as

$$L(m, j).$$

As a little wordly experience teaches one, each of these two sentences could be true while the other one is false. All too frequently, Cupid's arrows are one-way arrows.

Two-place predicates in mathematics

Two-place predicates are common in mathematics. Among them are the ordering relation $<$ ('less than') in the theory of real numbers, and the relation $\in$ ('is a member of') in set theory. Indeed, the most frequently used predicate in mathematics is the two-place *identity predicate* '$=$', which is used to form *equations*.

Infix v. prefix notation for predications

An expression-forming symbol is *infixed* within what it forms when it is not the first (leftmost) symbol in the latter. For example, the symbol $\wedge$ is infixed within any conjunctive sentence $A \wedge B$ that it forms. If, however, the expression-forming symbol is *prefixed*, then it would be placed leftmost. That is what happens in so-called 'left Polish' notation, where the conjunction symbol would be used to form the conjunctive sentence $\wedge AB$.

It is worth noting that mathematicians almost always prefer to *infix* their expression-forming symbols. Thus they place their two-place predicate-letters *in between* the singular terms involved. They write '$0 < 1$' rather than '$< (0, 1)$', and they write '$\emptyset \in \{\emptyset\}$' rather than '$\in (\emptyset, \{\emptyset\})$'.

Logicians, however, tend to prefer *prefixing*—they place their predicate-letters leftmost. The only exception that proves the rule is the identity predicate, which is *always* infixed. Thus, in regimenting the claim that two plus two is four, one writes the equation '2+2=4' rather than '=(2+2,4)'—let alone: '$= (+(2, 2), 4)$'. (Mathematicians also prefer to infix their two-place *function signs*—but we are getting ahead of ourselves).

Three- and four-place predicates

A good example of a three-place predicate in ordinary discourse is 'x gave y to z'; and in scientific discourse: 'x is at location y at time t'.

In Euclidean geometry, there is use for a *three*-place predicate of 'betweenness': point x is between points y and z.

In two-dimensional projective geometry, there is the *four*-place relation of 'harmonic conjugacy' among collinear points (and, dually, among concurrent lines).

In principle, a predicate could involve any finite number of such *argument-places*: $P(t_1, \ldots, t_n)$; but natural language hardly ever provides for more than four argument-places for any (primitive) predicate.

2.4 Quantifying phrases are not names

Beguiled by the workings of natural language, one might entertain the naive belief that *quantifying phrases* like 'someone' and 'everyone' behave in the same way as *names* like 'John' and 'Mary'. After all, it seems as though in any grammatical sentence of English containing the name 'John' (say), one could substitute the quantifying phrase 'someone' (say), and the result would be grammatical. 'Everyone loves John' is grammatical; and so too is 'Everyone loves someone'. And perhaps also (one might think) substitutions in the other direction can always be carried out, again *salva congruitate* (that is, without loss of grammaticality).

This naive belief, however, does not withstand closer scrutiny. The sentence 'Everyone loves someone else' is grammatical; but 'Everyone loves John else' is not. Many similar examples can be found. Names and quantifying phrases in English *cannot* always be substituted for one another *salva congruitate.* The same holds for other natural languages. The differences between quantifying phrases and names run deep. No natural language can make these differences disappear.

Moreover, if quantifying phrases were to be treated as names, then they would need to behave *logically* like names in every respect. But they do not.

2.5 Inferences that names must obey

If an expression n is a *name*, then the following two inference patterns involving n (in English) must be valid:

$$\text{(I)} \qquad \frac{n \text{ is } (F \text{ or } G)}{(n \text{ is } F) \text{ or } (n \text{ is } G)}$$

$$\text{(II)} \qquad \frac{(n \text{ is } F) \text{ and } (n \text{ is } G)}{n \text{ is } (F \text{ and } G)}$$

Moreover, if expressions n_1 and n_2 are names, then the following 'passivization' inference pattern involving n_1 and n_2 (in English) must be valid:

$$\text{(III)} \qquad \frac{n_1\, R\text{'s}\, n_2}{n_2 \text{ is } R\text{'ed by } n_1}$$

2.6 'Everyone' and 'someone' are not names

The so-called quantifier expressions 'everyone' and 'someone' do not pass these logical tests for namehood.

'Everyone' fails test (I).

'Someone' fails test (II).

Together, 'everyone' and 'someone' fail test (III).

To appreciate these failures, note first the *invalidity* of the following instances of (I), (II) and (III) respectively:

$$\text{(I)} \qquad \frac{\text{Everyone is } (F \text{ or } G)}{(\text{Everyone is } F) \text{ or } (\text{everyone is } G)}$$

$$\text{(II)} \qquad \frac{(\text{Someone is } F) \text{ and } (\text{someone is } G)}{\text{Someone is } (F \text{ and } G)}$$

$$\text{(III)} \qquad \frac{\text{Everyone loves someone}}{\text{Someone is loved by everyone}}$$

Counterexample to the invalid instances of (I) and (II)

In the following situation, called M, the domain of discourse is represented by the outer box. In the domain there are exactly two individuals (the bullets), α and β. As indicated by the inner boxes for the predicate-extensions, the individual α has the property F but not the property G; while the individual β has the property G but not the property F.

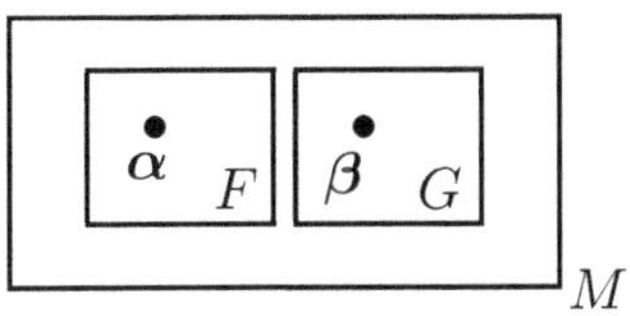

In this little world M we have the following evaluations of the premises and conclusions of our arguments that are instances of (I) and of (II):

$$\text{(I)} \qquad \frac{\text{Everyone is } (F \text{ or } G)}{(\text{Everyone is } F) \text{ or } (\text{everyone is } G)} \qquad \begin{array}{l} \text{True in } M \\ \text{False in } M \end{array}$$

(II)	(Someone is F) and (someone is G)	True in M
	Someone is (F and G)	False in M

Thus M is a counterexample to each of those two arguments.

But what our argument-instance of (III)? A little extra thought produces the following counterexample to it.

Counterexample to the invalid instance of (III)

In the following situation, called N, there are again exactly two individuals, α and β. The 'atomic facts' are that α and β love each other, but not themselves.

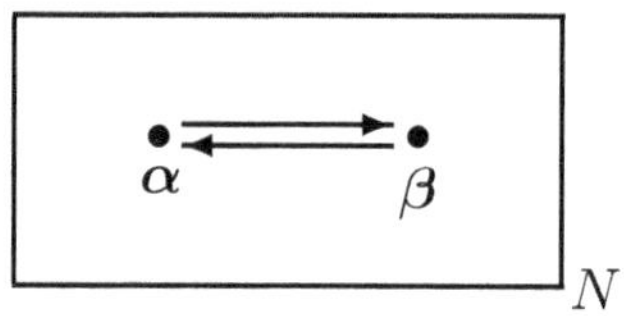

We have

(III)	Everyone loves someone	True in N
	Someone is loved by everyone	False in N

2.7 Quantified sentences are not of predicate-argument form

Let us use the symbol $\forall$ to mean 'everyone' and the symbol $\exists$ to mean 'someone'. Let $F(\ )$ be any predicate, with the gap ready to be filled by a name to form a sentence of predicate-argument form. Thus 'John is F' is a properly formed sentence with the predicate-argument form

$$F(j).$$

The foregoing considerations about the invalidity of (I), (II), and (III) prevent us, however, from assigning the following logical forms:

Everyone is F	:	$F(\forall)$	[Wrong!]
Someone is F	:	$F(\exists)$	[Wrong!]

Only names can be put into the gap of a predicate; and we have just seen that $\forall$ and $\exists$ are not names. They cannot *go into* gaps of predicates.

How, then, does a quantifier relate to the gap in a predicate with which it combines, so as to make up a sentence with no gaps in it? Put another way: how do $\forall$ and $\exists$ actually 'work' (logico-grammatically), in simple sentences like 'Everyone is F' and 'Someone is F'? This is the profoundly important logico-grammatical question to which Frege provided an ingenious and enduring answer.

2.8 Quantifiers are *second-order* predicates

Frege's answer to this question is that the quantifier-expressions $\forall$ and $\exists$ have their own, 'bigger' gaps, into which the predicate $F(\;)$ itself can be placed, *thereby sealing the gap in* $F(\;)$ *itself.* We shall indicate the gap in the quantifier by using square brackets—thus $\forall[\quad]$ and $\exists[\quad]$.

The picture of forming the sentence 'Everyone is F' is then like this:

$$\forall[F(\;)]$$

where the linkage shows that the quantifier $\forall$ seals the gap in the predicate $F(\;)$. This makes the sentence 'Everyone is F' grammatically complete, no longer containing any open gaps that can still receive names. The same account works for the quantifier $\exists$.

2.9 The role of bound variables

In the interests of having a *linear* logical notation, we indicate the linkage simply by *marking its ends.* From the markers of its ends, the linkage itself can be recovered. The markers in question are occurrences of some same letter, called a *variable.*

Nowadays these letters are, conventionally, taken from the end of the Latin alphabet: x, y, z, w. Thus the modern way of rendering the last logical form above is

$$\forall x F(x),$$

in which we also drop the square brackets visibly enclosing the 'bigger gap' of the quantifier.

In modern notation, the initial placement of the quantifier prefix involving x means that the whole *formula* $F(x)$ that follows it is enclosed in that (now invisible, 'bigger') gap of the quantifier. The occurrences of the variable x that fill the gaps of the predicate $F(\ )$ itself are now said to be *bound.* And we say that the formula $F(x)$, in which the variable x is *free*, is the *scope* of that occurrence of the quantifier $\forall x$.

Multiple occurrences of a variable

Note that one and the same variable can have more than one occurrence, to take care of examples where some single *name* has enjoyed more than one occurrence. Take, for example, the sentence 'John loves John' (more felicitously: 'John loves himself'). Its regimentation as a formal sentence is

$$L(j, j).$$

Here, the 'gap for a name' (i.e., for the single name j, for 'John') is really twofold. Removing the name j from $L(j, j)$ yields the 'single-gap' predicate

$$L(\_\,,\_),$$

which can now be quantified thus:

$$\exists x L(x, x).$$

The underlying pattern of 'binding', i.e., of 'sealing the gaps against any possible further insertion of names', is the following:

$$\exists[L(\ ,\ )]\,.$$

Bound variables can make distinctions of scope

The use of bound variables enables one to make *distinctions of scope* when quantifiers are used with relational predicates. For example the sentence

Everyone loves someone : $\forall x \exists y L(x, y)$

is implied by, but does not imply—recall (III) above—the sentence

Someone is loved by everyone : $\exists y \forall x L(x, y)$

The invalid inference

$$\frac{\forall x \exists y L(x,y)}{\exists y \forall x L(x,y)}$$

is called the *Quantifier Switch Fallacy*, for the obvious reason: it involves switching the order of the two quantifiers. We have already seen a counterexample to this fallacy—look back at model N on p. 28.

On logical form

Logical form arises from the manner of combination of logical words with descriptive words, and the pattern of repetitions of those descriptive words, rather than what those descriptive words actually mean, or refer to, themselves. For example, the following two sentences have the same logical form:

> If you are happy, then you will not cry if someone else is happy

> If you are hungry, then you will not chew if someone else is hungry

Using the initial letters involved, we can say that the common logical form is

> If you are H, then you will not C if someone else is H.

Note that the repetition of H contributes to the sentence's logical form; but what H actually means is irrelevant to the logical form. On the other hand, what the logical words 'if ...then...', 'not', 'some', and 'else' actually mean, and the pattern of their occurrences within the sentence, are crucial to the sentence's logical form.

2.10 Formulae, logicians' English, and logical forms

If, instead of just writing out sequences of expressions in a formal language, one reads them out loud, for one's audience to hear, one will be tempted to embroider them as phrases of ordinary language, rather than as symbols of a formal language. Thus, for example, the formal sentence

$$\exists x \forall y \; x = y$$

would, if read exactly as written down, be rendered as

ee x ay y x equals y.[1]

Usually, however, the expert user will lapse into so-called 'logicians' English', and say out loud

there exists x such that, for all y, x is identical to y.

The temptation is close at hand to anglicize what is said even further, and read the original formula more fluently as

There is some thing to which every thing is identical.

We have, then, a spectrum of expressive possibilities in rendering the formal sentence. These range from an utterance 'at the formal end' that reads off a formal sentence symbol-by-symbol, in linear sequence; through to an utterance, 'at the natural end', of a stylistically attractive English sentence that contains no hint of the logician's formal devices. But the *truth-conditions* of the English sentence, one wishes to say, are nevertheless 'captured' by the formal sentence in question. The formal sentence provides the *logical form* of the English sentence.

The primitive symbols of the formal language—as we have just seen from the example $\exists x \forall y \; x = y$—may be used to form complex expressions called *formulae*, among which are the *sentences* of the formal language. (Sentences are formulae in which all variable-occurrences are bound.) One needs, of course, an unlimited supply of parentheses. These help to indicate the internal structure of complex formulae when two-place connectives are infixed. One needs also an unlimited supply of *(individual) variables*: $x, y, z, \ldots$ (with or without numerical subscripts) in order to be able to form all possible sentences involving the quantifier-

[1]Recall that '$\exists$' is a backward 'E', while $\forall$ is an upside-down 'A'. Hence the phonetic readings suggested.

expressions. Bear in mind that there is no upper limit, in principle, to how many embeddings a formal sentence may contain of one quantifier-expression within the scope of another.[2] That is one reason why we need to furnish ourselves, in advance, with an unlimited supply of lexicographically distinct variables.

2.11 Multiple quantification

Our last example, $\exists x \forall y\ x = y$, illustrates the expressive power of quantifiers when they are analyzed as variable-binding expressions. In this example, (the occurrence of) $\forall$ is *within the scope of* (the occurrence of) $\exists$. The formal sentence $\exists x \forall y\ x = y$ says that there is exactly one thing. This is a very powerful (and usually false) claim. If we reverse the order of the quantifiers, however—putting (the occurrence of) $\exists$ within the scope of (the occurrence of) $\forall$—we obtain $\forall y \exists x\ x = y$. This is the (trivially and) *logically true* claim that everything is identical to something (i.e., itself). The order of the quantifiers matters.

Nor is it just the special features of the *identity* predicate '=' that make for differences upon re-ordering of quantifiers. Take any descriptive two-place predicate, such as 'x loves y'. The following multiply quantified sentences are (pairwise) grammatically distinct. Moreover, it is only the doubly-existentially and doubly-universally quantified sentences for which the ordering of the quantifiers does not make a difference to truth-conditions. Note that when two sentences φ and ψ have the same truth-conditions, they are logically equivalent—that is, each logically implies the other. Any situation or interpretation that makes φ true makes ψ true; and conversely.

$\exists x \exists y Lxy$	is logically equivalent to	$\exists y \exists x Lxy$;
$\exists x \forall y Lxy$	is *not* logically equivalent to	$\forall y \exists x Lxy$;
$\forall x \exists y Lxy$	is *not* logically equivalent to	$\exists y \forall x Lxy$;
$\forall x \forall y Lxy$	is logically equivalent to	$\forall y \forall x Lxy$.

Fluent English readings of these formal sentences are as follows. Logically equivalent pairs are braced together. The result is *six* logically distinct claims.

[2]See §2.11 and §4.5 for more detailed explanation of this notion of scope.

$$\left\{\begin{array}{ll} \exists x\exists yLxy & \text{Someone loves someone} \\ \exists y\exists xLxy & \text{Someone is loved by someone} \end{array}\right\}$$

$$\begin{array}{ll} \exists x\forall yLxy & \text{Someone loves everyone} \\ \forall x\exists yLxy & \text{Everyone loves someone} \\ \forall y\exists xLxy & \text{Everyone is loved by someone} \\ \exists y\forall xLxy & \text{Someone is loved by everyone} \end{array}$$

$$\left\{\begin{array}{ll} \forall x\forall yLxy & \text{Everyone loves everyone} \\ \forall y\forall xLxy & \text{Everyone is loved by everyone} \end{array}\right\}$$

Not only does the order of the quantifiers matter; so too does the order of the bound variables after the predicate-letter L:

$\exists x\forall yLxy$ is *not* logically equivalent to $\exists x\forall yLyx$;
$\forall x\exists yLxy$ is *not* logically equivalent to $\forall x\exists yLyx$.

Note here that the sentences on the right now have 'Lyx' in place of 'Lxy'. But we can make the same point, using the latter ordering of the variables after the predicate 'L', as follows:

$\exists x\forall yLxy$ is *not* logically equivalent to $\exists y\forall xLxy$;
$\forall x\exists yLxy$ is *not* logically equivalent to $\forall y\exists xLxy$.

The proper analysis of the truth-conditions of multiply quantified claims was Frege's great breakthrough in logical analysis. It afforded mathematicians a logical way to make rigorous the distinction between, for example, the mere continuity of a function over an interval, as opposed to its *uniform* continuity over the interval.

It is not necessary to employ multiple quantification in order to form a sentence from a relational predicate. One may, for example, quantify into *two* (or more) argument places with a single quantifier. Thus, for example, one can form the sentences

$\exists xLxx$ Someone loves himself (or herself);
$\forall xLxx$ Everyone loves himself (or herself).

In order fully to appreciate the extraordinary extent of the logical distinctions that can be made with just the two-place predicate L and the two quantifiers $\exists$ and $\forall$, the reader can look forward to reading in due course the jaw-dropping details in Chapter 12. Or they can sneak a peek right now at p. 233.

2.12 Restricted quantification

In the standard language of first-order logic it is customary to use an appropriate *two-place connective* in order to construct a logical form, or formal translation, for an English sentence like 'All Fs are Gs' or 'Some F is G'. For the former, we use the conditional connective $\rightarrow$, which can be read as '... only if __' (or as 'if ..., then __'); for the latter, we use the connective $\wedge$ for conjunction. We then have the following logical forms, with their 'logician's English' renderings:

Every F is G	$\forall x(Fx \rightarrow Gx)$	For all x, if Fx, then Gx
Some F is G	$\exists x(Fx \wedge Gx)$	For some x, Fx and Gx

The quantifiers $\forall$ and $\exists$ are interpreted as having their variables 'range over' *all* the individuals that the speaker may take to exist. These individuals form the *domain of discourse.* But frequently, in ordinary language, we restrict our generalizations to particular kinds of individuals. We may speak of all (or some) *persons*, or of all (or some) *animals*, or of all (or some) *numbers*; and when we do so, we are implicitly allowing for the existence of things other than persons, animals, and numbers, respectively.

We have displayed above the standard way of translating into formal notation claims in which quantifications (whether universal or existential) are thus restricted. The intuitive pictures to keep in mind are those of *overlap* and *inclusion* between the extensions of the predicates F and G. The following Venn-Euler diagrams illustrate the difference:

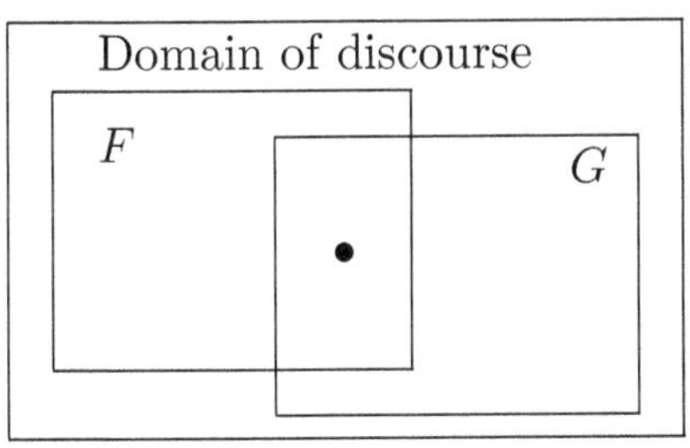

Some F is G

$\exists x(Fx \wedge Gx)$

The extension of F *overlaps* the extension of G

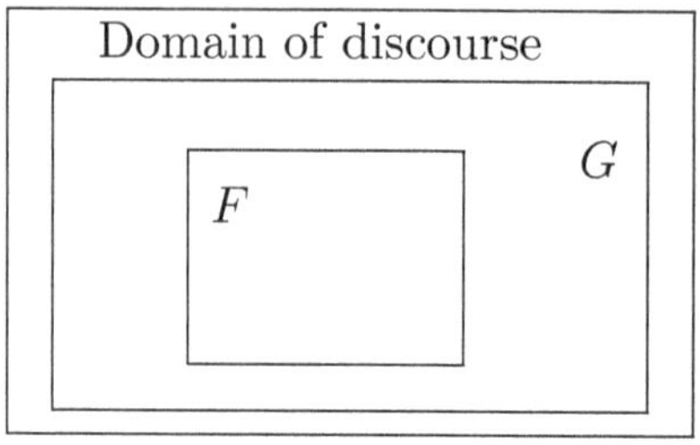

Every F is G

$\forall x(Fx \rightarrow Gx)$

The extension of F is *included in* the extension of G

Beginners should try to avoid the rather frequent error of assigning the incorrect logical form $\exists x(Fx \rightarrow Gx)$ to 'Some F is G'. $\exists x(Fx \rightarrow Gx)$ is true if the universe of discourse contains a non-F individual. It is also true if the universe contains a G-individual.

One *could*, if one wished, provide more directly for restricted quantification, by using quantifiers that bind variables in two predicates—the restricting predicate, and the main predicate. Thus one could write 'Some Fs are Gs' in formal notation as

$$\exists x[Fx]Gx$$

and one could write 'All Fs are Gs' as

$$\forall x[Fx]Gx.$$

Note the absence of any connectives in these two logical forms. They employ only (new) *binary* quantifiers. And these logical forms using binary quantifiers are arguably more natural renditions of the English sentences themselves. For there is no occurrence of 'and' in 'Some Fs are Gs'; and likewise there is no occurrence of 'if ... then ...' (or of 'only if') in 'All Fs are Gs'.

These new binary quantifiers would then have to be furnished with their own inference rules—which would be quite easy to devise. If one did adopt such binary quantifiers as primitives, however, one would not thereby enjoy any gain in expressive power. For, in the resulting system one would still need to be able to prove the equivalence of $\exists x[Fx]Gx$ (using the new restricted quantifier) with $\exists x(Fx \wedge Gx)$ (using the standard unrestricted quantifier). Similarly, one would have to be able to prove the equivalence of $\forall x[Fx]Gx$ with $\forall x(Fx \rightarrow Gx)$. For that reason, logicians simply opt for the unrestricted quantifiers, and employ the two forms of expression $\exists x(Fx \wedge Gx)$ and $\forall x(Fx \rightarrow Gx)$ in order to capture the respective restricted quantifications. These means of regimenting the logical forms of mathematical claims are sufficient for the mathematician's needs. They come only at the slight cost, for the beginner, of having to learn that $\exists$ typically calls for $\wedge$, and $\forall$ typically calls for $\rightarrow$, when one is translating restricted quantifications (in English, or mathematese) into the usual logical notation that employs only the *unary* quantifiers.

If, however, one wished to formalize the so-called *plurality* quantifiers *few, most, many, several* ...etc., then one would be well-advised to resort to binary forms of these quantifiers as logical primitives.[3] For one can show that with such plurality quantifiers Q, no 'unrestricted' form $Qx\Phi(F, G)x$ is logically equivalent to the restricted form $Qx[Fx]Gx$. This is so for any formula $\Phi(F, G)$ formed from F and G by the use of connectives, and with the variable x interpreted as ranging over all individuals in the domain. What *counts as* few, most, many or several Fs very much depends on *how many Fs there are*; and this number can in general be much smaller than the number of individuals in the domain of discourse.

Fortunately, 'some' and 'all' are not thus *essentially restricted.* Their 'restricted' forms can be analyzed by means of the unrestricted existential and universal quantifiers, respectively. So we do not need to introduce binary versions of these quantifiers either in place of or alongside their standard, unrestricted forms.

2.13 First-order v. second-order quantification

It is because the quantifier-expressions $\forall$ and $\exists$ are understood as expressing generalizations about *things* (rather than about sets of things, or properties of things) that a formal language containing only those two quantifiers is called a *first-order* language. A *second-order* language, by contrast, would allow one to generalize about all *properties* of things. A good example to clarify the distinction is this: 'Everyone but John loves Mary' is a first-order claim, since it generalizes only about individuals; whereas 'John is everything but a lover to Mary' is a second-order claim, since it generalizes about relationships in which individuals can stand.

[3]See Altham [1971] and Altham and Tennant [1975].

By adopting a first-order language we deliberately eschew the means to say things that can only be said at second-order. Whether this is an insuperable handicap in the case of mathematics is debatable. For, in the language of first-order *set theory* (about which, more in due course), there are ways to circumvent the apparent expressive difficulty incurred by staying at first order. Moreover, as we shall subsequently learn, there is a principled trade-off to be made between *expressive power*, on the one hand, and *deductive power*, on the other. By staying at first order, one achieves full deductive power, albeit at the cost of some expressive power. But if one were to ascend to second order in order to increase one's expressive power, the price would be that one could no longer attain full deductive power. There will be more to say about this methodological dilemma in the sequel to this primer.[4]

We have explained the meanings of the two-place connectives $\wedge$ (for conjunctions) and $\rightarrow$ (for conditionals). In addition to those there is the two-place connective $\vee$ for disjunctions, and the one-place connective $\neg$ for negations. A formal sentence of the form $A \vee B$ is to be read as 'either A or B (or both)'. This is called *inclusive disjunction.* The symbol $\vee$ comes from the Latin word 'vel', which means just that. A formal sentence of the form $\neg A$ is to be read as 'It is not the case that A'.

The formal language whose primitive vocabulary consists of just the expressions $\neg, \vee, \wedge, \rightarrow, \exists, \forall$, and $=$, whose meanings have now been explained—along with variables and parentheses—is known as the *pure first-order language of identity.* In §4.4 we study the pure first-order language of identity in some detail, in order to introduce further useful concepts, such as the syntactic notions of *scope* and *dominant occurrence.* Then we move on to consider *descriptive vocabulary* (that is, names, function signs, and predicates other than the identity predicate) and formulae of descriptive first-order languages in general.

Try now to do these exercises, even though we have not yet covered the details of the formal grammar of first-order languages. Wrestling with subtleties of logical form at this stage is excellent preparation for the enlightenment that will come when you encounter the precise rules of logico-grammatical formation in Chapter 4.

---

[4]The interested reader can also pursue this problem of expressive vs. deductive power by consulting Tennant [2000a].

Exercise 4. *Translate the following English sentence-forms into logical notation. Then supply English predicates as replacements for 'F' and 'G', and use them to give more fluent English sentences enjoying the forms in question.*

1. *All Fs are non-Gs*
2. *Some non-Fs are Gs*
3. *No Fs are Gs*
4. *Not all non-Fs are Gs*
5. *All Fs are Gs*
6. *All Gs are non-Fs*

Exercise 5. *Give fluent English sentences with these logical forms:*

1. $\neg(\exists xFx \vee \exists yGy)$
2. $\neg\exists xFx \vee \exists yGy$
3. $\neg\exists x(Fx \wedge \forall yLxy)$
4. $\neg\exists xFx \wedge \exists x\forall yLxy$
5. $\forall x(Fx \rightarrow \exists yLxy)$
6. $\forall xFx \rightarrow \exists y\forall xLxy$
7. $\forall x\forall y(Lxy \rightarrow Fx)$
8. $\forall x(\exists yLxy \rightarrow Fx)$
9. $\forall x((Fx \vee Gx) \rightarrow Lxx)$
10. $\exists x((\neg Fx \wedge \neg Gx) \wedge \neg Lxx)$

Exercise 6. *Using Lxy to mean 'x loves y', translate the following sentences into logical notation:*

1. *Everyone loves someone*
2. *Everyone loves everyone*
3. *Someone loves someone*
4. *Someone loves everyone*
5. *Everyone is loved by someone*
6. *Everyone is loved by everyone*
7. *Someone is loved by someone*
8. *Someone is loved by everyone*
9. *Everyone loves themselves*
10. *Someone loves themselves*
11. *No one loves themselves*

12. *No one loves everyone*
13. *No one loves anyone*
14. *Someone loves no one*
15. *No one is loved by anyone*
16. *No one is loved by everyone*
17. *Everyone is loved by no one*
18. *Someone is loved by no one*
19. *Everyone loves no one*
20. *Not everyone loves someone*
21. *Anyone who is loved by someone loves themselves*
22. *Anyone who is loved by everyone loves themselves*
23. *Anyone who is loved by anyone loves themselves*
24. *Someone who loves someone is loved by everyone*
25. *Anyone who loves someone loves themselves*

Exercise 7. *Show that each of the sentences*

$$\exists x \forall y Lxy,\ \forall y \exists x Lxy,\ \forall x \exists y Lxy,\ \exists y \forall x Lxy,\ \exists x \exists y Lxy,\ \forall x \forall y Lxy$$

is not logically equivalent to any of the others. That is, for any two distinct sentences φ and ψ on this list, show that either φ does not logically imply ψ, or ψ does not logically imply φ. (In order to show that a sentence φ does not logically imply another sentence ψ, one needs to describe a possible situation in which φ would be true but ψ would be false.)

Exercise 8.

1. *Show that $\exists x \exists y Lxy$ does not logically imply $\exists x Lxx$.*
2. *Show that $\forall x Lxx$ does not logically imply $\forall x \forall y Lxy$.*

Chapter 3

Arguments

An *argument* consists of finitely many *premises*, an *inference marker*, and a *conclusion*. Premises and conclusions are *sentences*. An argument whose premises are $P_1, \ldots, P_n$ and whose conclusion is Q can be displayed schematically in any of the following ways:

$$P_1, \ldots, P_n; \textit{ergo } Q$$

$$P_1, \ldots, P_n; \text{therefore } Q$$

$$P_1, \ldots, P_n; \text{so, } Q$$

$$P_1, \ldots, P_n \therefore Q$$

$$P_1, \ldots, P_n / Q$$

$$\frac{P_1 \ \ldots \ P_n}{Q}$$

$$\begin{array}{c} P_1 \\ \vdots \\ P_n \\ \hline Q \end{array}$$

$$P_1, \ldots, P_n : Q$$

The final form just given— $P_1, \ldots, P_n : Q$ —is called a *sequent*. This form will be used in what is known as the *sequent calculus*, which is a proof system that will be developed in due course. It will be introduced in §10.3 for propositional logic (dealing with the connectives) and developed further in Chapter 14 (to deal with the quantifiers).

The premises $P_1, \ldots, P_n$ are the 'starting points' of the argument; its conclusion Q is its 'endpoint'. Metaphorically speaking, the endpoint can be far removed from the (collection of the) starting points. This is often the case in mathematics, where the premises of a stated argument are 'self-evident' or obviously true *axioms* or *postulates* of the branch of mathematics in question, and the conclusion can be a deep conjecture whose truth values is not yet known.

This is the case, for example, with the argument whose premises are the axioms of complex number theory, and whose conclusion is the famous Riemann Hypothesis (that every solution z in the complex plane of the equation $\sum_{i=1}^{\infty} \frac{1}{n^z} = 0$ has $\frac{1}{2}$ as the real part of z, if that real part is not a negative even integer).

So too with the argument whose premises are the axioms of arithmetic (the theory of the natural numbers 0, 1, 2, ...) and whose conclusion is Goldbach's Conjecture (that every even number greater than 2 is the sum of two prime numbers). The Riemann Hypothesis and Goldbach's Conjecture have yet to be determined as true or as false of their respective subject matters.

The 'far removal' of the conclusion from the premises of an argument is the case also with the argument whose premises are the axioms of Euclidean geometry and whose conclusion is the famous Theorem of Pythagoras that the area of the square on the hypotenuse of any right-angled triangle is the sum of the areas of the squares on the other two sides—a statement whose known truth we owe to the ancient Greeks. The statement of the latter *argument*

Axioms of Euclid : Theorem of Pythagoras

gives no hint as to *how* the logical journey from left to right is to be undertaken. That is, it gives no hint as to how the argument is to be *proved*.

Note that in stating an argument, one is not making out that its conclusion is *known to follow* from its premises; rather, one is simply *referring to* a would-be transition (from premises to conclusion) without necessarily knowing that the transition is legitimate (or valid). One can *refer to* the argument $P_1, \ldots, P_n : Q$ even in cases where, as it happens, Q does *not* follow logically from $P_1, \ldots, P_n$. In fact, one *must* be able so to refer, for otherwise one would not be able to say that the argument in question is *invalid* (hence, not provable). Moreover, one can (and would need to) do so even in cases where it is the *negation* of Q that follows logically from $P_1, \ldots, P_n$.

The logical journey from the premises to the conclusion (the precise inferential 'itinerary') is not part of the statement of the argument. That journey can be long and complex; and there could be more than one route to be taken. Stating an argument merely tells one's listener 'where to start' and 'where to finish' that logical journey, if it can be made at all. It gives no advice on the actual step-by-step route to be taken.

3.1 Logic is concerned with ***validity*** of arguments

To repeat: $P_1, \ldots, P_n$ are the *premises*, and Q is the *conclusion*. The horizontal lines and the colon in the last three displays on p. 41 are the inference markers that indicate the argumentative transition. Note that in actual applications, the sentences $P_1, \ldots, P_n$ and Q will in general be logically *complex*, containing both logical words and descriptive words in grammatically correct combinations.

In one of the earliest studies of deductive logic and methods of proof of arguments, Aristotle, in his *Prior Analytics*, treated *syllogisms*, which are two-premise arguments, both of whose premises, and whose conclusions, are logically complex. Famous examples are his syllogisms called *Barbara* and *Celarent*.

$$\textit{Barbara} \qquad \begin{array}{l} \text{Every M is P} \\ \text{Every S is M} \\ \hline \text{Every S is P} \end{array}$$

$$\textit{Celarent} \qquad \begin{array}{l} \text{No M is P} \\ \text{Every S is M} \\ \hline \text{No S is P} \end{array}$$

In logic, we aim (as Aristotle did, with arguments of these relatively simple forms) to identify the *valid* arguments, that is, the ones that are *guaranteed to preserve truth from their premises to their conclusions*:

> Under every possible interpretation of
> the descriptive vocabulary in the argument:
> *if* all the premises of the argument are true,
> *then* the conclusion of the argument is true also.

In logic we aim to understand why, and how, truth is transmitted from premises to conclusions of valid arguments. See Chapter 20 for a modern account of why, and how, truth is transmitted from the premises to conclusions of Aristotelian syllogisms.

Logic is utterly general and universally applicable. The notion of *logical validity of argument* is fundamental to every branch of investigation and inquiry. It transcends all disciplinary boundaries. It would have to be employed by *any* rational thinker, even *non-human* ones; even *extra-terrestial* intelligences; even *artificial* intelligences. It can be made the subject of rigorous study, employing symbolic methods. One can think of Logic as 'reasoning gone self-conscious'.

3.2 Areas in which deductive argumentation is important

Deductive argumentation is used with varying degrees of precision in widely varying domains.

The most impressive example is *mathematics*, where one is concerned solely with valid arguments whose premises are *axioms* (that is, simple and self-evident claims about the mathematical domain in question). Deep conclusions deduced from the axioms are called mathematical *theorems*. There are also results that mathematicians call *lemmas*. These labels are subjectively applied. A lemma is typically 'not as deep' as a theorem, but is also widely applicable as a fresh deductive starting point on the way to yet other theorems. And theorems have special cases called *corollaries*. From the logician's point of view, all of these deduced results—lemmas, theorems, corollaries—are simply conclusions (our generic Q's) validly drawn from the mathematicians' ultimate starting points, namely the *axioms* (our generic P's) of the branch of mathematics in question. What the mathematician does is supply the needed *proofs* of Q from $P_1, \ldots, P_n$, thereby *establishing the validity* of the argument $P_1, \ldots, P_n : Q$. The important thing about proofs is that they can be effectively checked for formal correctness. Proofs are finite; and they are put together in accordance with rules of inference . Every step in a proof can be checked for formal correctness, since there are only finitely many rules in our canon.

Another important arena for deductive argumentation is the method of so-called *hypothetico-deductive explanation* in the natural sciences. Here, the arguments' *premises* include scientific *hypotheses* and statements of initial and boundary conditions; and their *conclusions* will be *predictions* of observable and/or measurable fact, which may or may not be borne out by observation. When they are *not* borne out by observation, one has a potential *refutation* of the scientific hypotheses that generated the faulty prediction. Overall, the 'disproof tree' in such a case has the following form:

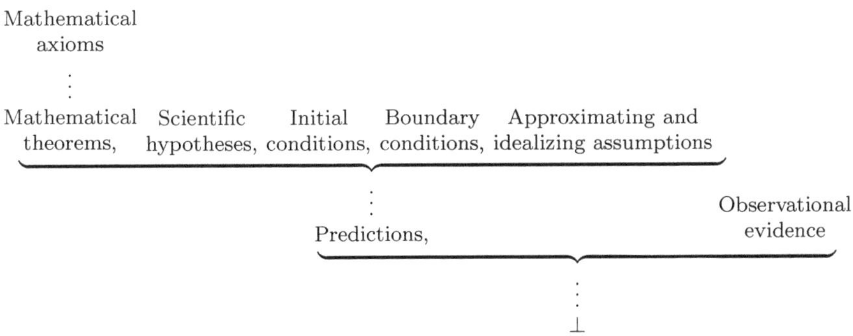

where the symbol $\perp$ represents *absurdity*, or *contradiction.*

The vertical ellipses represent the passages of detailed deductive reasoning that would have to be filled in. The deductive direction is *downward*—in this case, *from* the premises at the tips of the branches (namely, Scientific hypotheses, Initial conditions, Boundary conditions, Approximating and idealizing assumptions, and Observational evidence) *to* the conclusion at the bottom (here, $\perp$).

Yet another important arena for deductive argumentation that is worth mentioning is the *law*, where a verdict, properly arrived at, would be the conclusion of an argument whose premises would be statements of tried fact and of statute and precedent. *This much* of the overall legal reasoning would be deductive—even if the earlier reasoning undertaken to establish the 'tried facts' might have been probabilistic, or inductive, or by way of 'inference to the best explanation'.

3.3 Two simple examples of valid arguments

Here are two simple examples of arguments, each with two premises. The reader will recognize the second one as an instance of Aristotle's syllogism *Barbara*.

$$\frac{\begin{array}{l}\text{All men are mortal}\\\text{Socrates is a man}\end{array}}{\text{Socrates is mortal}} \qquad \frac{\begin{array}{l}\text{Every Armenian is a Bolshevik}\\\text{Every Bolshevik is a Communist}\end{array}}{\text{Every Armenian is a Communist}}$$

When an argument is seriously *put forward*. or *endorsed*, and not simply stated, or referred to, or mentioned, or gestured at—we understand the arguer to be claiming that *if* all the stated premises are true, *then* the conclusion is true also. That is, the arguer's claim is that the argument is (logically) *valid*. The two examples above are both valid. But while both premises of the first argument are true, it is highly unlikely that both premises of the second argument are true. Nevertheless, the second argument is valid. And this, as with all valid arguments, is because of its *logical form*, which can be illustrated in 'logician's English' as follows:

$$\frac{\begin{array}{l}\text{All } A\text{s are } B\text{s}\\\text{All } B\text{s are } C\text{s}\end{array}}{\text{All } A\text{s are } C\text{s}}\ .$$

It is this form that is represented *yet more formally* by sentences of a formal language, when the latter is brought to bear on the problem of appraising arguments for validity. The fully formal 'regimentation' of this argument, for a contemporary logician, using logical operators that were introduced and explained in Chapter 2, is

$$\frac{\begin{array}{l}\forall x(Ax \to Bx)\\\forall x(Bx \to Cx)\end{array}}{\forall x(Ax \to Cx)}\ .$$

3.4 A fallacy and a counterexample to it

Invalid arguments are also called *fallacies.* Here is a famous fallacy, committed by Saint Thomas Aquinas:

$$\frac{\text{Every event has a prior cause}}{\text{Some first cause caused every subsequent event}}$$

A *counterexample* to an invalid argument $P_1, \ldots, P_n/Q$ is an interpretation of the descriptive vocabulary that makes the premises $P_1, \ldots, P_n$ *true*, while making the conclusion Q *false.*

The existence of a counterexample shows that truth is *not* in general preserved from the premises of the argument to its conclusion—so the argument is *in*valid.

Here is a counterexample to the famous fallacy above. Interpret 'event' as 'positive or negative integer', interpret 'prior' by the mathematical 'less than' relation $<$, and interpret 'x caused y' as '$x + 1 = y$'. Then the premise is true but the conclusion is false.

3.5 The techniques of proof and counterexample

Two of the main techniques to be acquired from a course in logic are:

- *proving that an argument is valid, if it is valid*; and
- *counterexemplifying it, if it is invalid.*

The role of deductively valid arguments in rational thought

Note that the arguer's claim *in putting an argument forward* is not just that if all the premises $P_1, \ldots, P_n$ are true, then the conclusion Q *might* be true, or that if all the premises $P_1, \ldots, P_n$ are true, then Q would be a *reasonably reliable* (though fallible) belief to hold. The claim is much stronger than that.

The claim is that the argument is *deductively valid.* A deductively valid argument *guarantees* the transmission of truth from its premises $P_1, \ldots, P_n$ to its conclusion Q. If you believe the premises $P_1, \ldots, P_n$, *and* you know that the argument is valid, then you *ought* to believe its conclusion Q.

This is a *requirement of rationality.* It arises from the very constitution of thought and language. The transmission of truth is not under our control; nor is it accidental, or intermittent. It is a matter of inviolable, eternal, logical and conceptual necessity.

That said, it is important to remark explicitly on some perhaps unexpected properties that a valid argument might have.

Valid arguments can have false premises

Valid arguments can have (obviously) *false premises*:

$$\begin{array}{l} \text{All women are saints} \\ \text{Socrates is a woman} \\ \hline \text{Socrates is a saint} \end{array}$$

The arguer's claim, in putting the argument forward, is *not* that the premises are true. The claim is rather that *if* the premises are true, *then* the conclusion is true also. It is a claim of *logical dedendency.*

Invalid arguments can have true conclusions

An *in*valid argument can have a(n obviously) *true conclusion*:

$$\begin{array}{l} \text{If human life is precious, then God exists} \\ \text{God exists} \\ \hline \text{Human life is precious} \end{array}$$

Despite the good intentions of anyone who might argue in this way, the argument is fallacious. It is of the form

$$\frac{\text{if } A \text{ then } B \qquad B}{A}$$

This is known as *the fallacy of 'affirming the consequent'.* It uses the 'if ... then ... ' premise in the wrong direction. The correct direction would be

$$\frac{\text{if } A \text{ then } B \qquad A}{B}$$

This is known as *modus ponens.*

Proofs are perfected arguments

A valid argument would, ideally, admit of *proof.* A *proof* is a *perfected* argument—one with all its steps filled in:

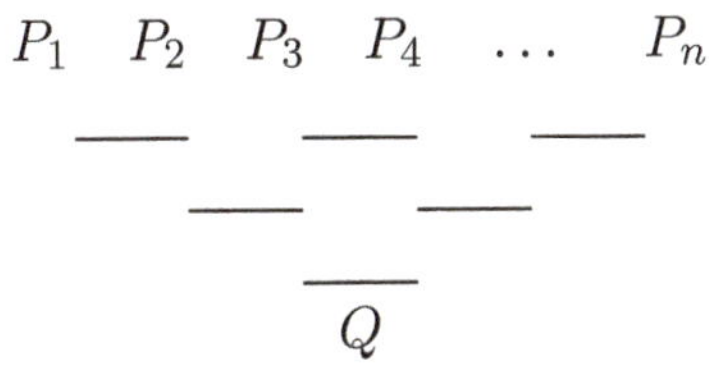

The steps in a proof would be *logically primitive.* Logically primitive steps do not need any further justification. They are obviously valid. They are so because they are based only on the *meanings of the logical words* that they involve, and the *patterns of occurrence* of the other words.

Logically primitive steps

For example, any step of the form

$$\frac{A \qquad B}{A \text{ and } B}$$

(where of course A and B are sentences) is obviously valid. By the very meaning of 'and', one is entitled to infer 'A and B' as soon as one has inferred or assumed A and inferred or assumed B.

Likewise, the form of inference we called *modus ponens* above:

$$\frac{\text{if } A \text{ then } B \qquad A}{B}$$

is obviously valid. By the very meaning of 'if ... then ...', one is entitled to infer B as soon as one has inferred or assumed the claim A and the conditional claim 'if A then B'.

3.6 Introduction and elimination rules

Logically primitive steps are taken in accordance with rules of inference like the two just mentioned. Note that the rule of inference

$$\frac{A \qquad B}{A \text{ and } B}$$

brings an occurrence of 'and' into the *conclusion of the step*, *below* the line. When this happens, we call the rule an *introduction* rule.

By contrast, the rule of inference

$$\frac{\text{if } A \text{ then } B \qquad A}{B}$$

involves an occurrence of 'if ... then ...' in a *premise* of the step, *above* the line. When this happens, we call the rule an *elimination* rule. And the premise whose dominant logical operator is being eliminated is called the 'major premise' for the elimination.

Every logical operator has both an introduction rule and an elimination rule. The introduction rule tells one how to get the operator into a conclusion (in so-called *dominant position*); the elimination rule tells one what one can do with it (in dominant position) in a premise. The full story about the logical operator *and*, the logical operator *if ... then ...*, and the logical operator *every* is told by the following pairs of introduction and elimination rules.

Rules for 'and', 'if... then...' and 'for every'

	INTRODUCTION	ELIMINATION
and	$\dfrac{A \qquad B}{A \text{ and } B}$	$\dfrac{A \text{ and } B}{A} \qquad \dfrac{A \text{ and } B}{B}$
if ... then ...	$\dfrac{\begin{matrix} \overline{\;A\;}^{(i)} \\ \vdots \\ B \end{matrix}}{\text{if } A \text{ then } B}{\scriptstyle (i)}$	$\dfrac{\text{if } A \text{ then } B \qquad A}{B}$

Comment on the Introduction Rule for 'if... then...'

In order to prove a conclusion of the form 'if A then B', one *assumes* A 'for the sake of argument', and proves (or *deduces*, or *derives*) B from that assumption. This shows the conditional dependence of B on A.

When we summarize that dependence in the conditional claim 'if A then B' in the conclusion of the rule, that conclusion no longer depends on the assumption A. So we are entitled to *discharge* the assumption A. This operation of discharge is indicated by the line placed over the assumption A, annotated with a numeral (i) which is attached also to the inference stroke below, in order to show which step was responsible for that discharge.

	INTRODUCTION	ELIMINATION
every	$\dfrac{\begin{array}{c}\vdots\\ F(a)\end{array}}{\text{for every } x\, F(x)}$ where the parameter a does not occur in any assumptions on which $F(a)$ depends	$\dfrac{\text{for every } x\, F(x)}{F(t)}$

Comment on the Introduction Rule for 'every'

In order to prove a conclusion of the form 'for every $x\,F(x)$', one first proves a syntactic instantiation $F(a)$, where the name-like expression a (which is called a *parameter*) is new in the context, and can therefore be thought of as standing for 'an arbitrary individual'. One makes no assumptions about this individual, and proceeds to prove $F(a)$ from assumptions not involving 'a'. This ensures the requisite degree of generality in the result; and we can accordingly infer the conclusion 'for every $x\,F(x)$'.

In such a context a is called *the parameter for* the step of 'universal introduction' that is taking place. We shall take care to keep parameters separate from genuine names, because their logical functions are so different. Names denote individuals; parameters do not. Parameters simply allow one to maintain an important 'smack of generality' that is needed when proving universal generalizations as conclusions (or using existential generalizations as premises—of which more anon).

Comment on the Elimination Rule for 'every'

In the other direction, the elimination rule for *every* allows one to conclude from a universal generalization 'for every $x\,F(x)$' any instance $F(t)$. The only background assumption needed here is of course that if the term t is parameter-free then there should be such a thing as t. Put another way: the term t should denote.

3.7 Bringing in logical symbols

We now re-state the foregoing rules for 'and', 'if... then...' and 'for every', using the special logical symbols $\wedge$, $\rightarrow$, and $\forall$ respectively.

We can now understand these rules as conferring upon the new symbols $\wedge$, $\rightarrow$ and $\forall$ the meanings 'and', 'if... then...' and 'for every', respectively. For it is the permitted inferential patterns of introduction (in a conclusion) and elimination (from a premise) that determine the meaning of a logical operator.

Rules for $\wedge$, $\rightarrow$, and $\forall$

INTRODUCTION ELIMINATION

$\wedge$

Introduction:

$$\frac{A \qquad B}{A \wedge B}$$

Elimination:

$$\frac{A \wedge B}{A} \qquad \frac{A \wedge B}{B}$$

$\rightarrow$

Introduction:

$$\begin{array}{c} \overline{\;A\;}^{\,(i)} \\ \vdots \\ \dfrac{B}{A \rightarrow B}\,(i) \end{array}$$

Elimination:

$$\frac{A \rightarrow B \qquad A}{B}$$

$\forall$

Introduction:

$$\begin{array}{c} \vdots \\ \dfrac{F(a)}{\forall x\, F(x)} \end{array}$$

where the parameter a does not occur in any assumptions on which $F(a)$ depends

Elimination:

$$\frac{\forall x\, F(x)}{F(t)}$$

Rules of introduction and elimination for logical operators make up what is known as a system of *natural deduction* for them.

We hasten to point out that these rules are stated here exactly as they were in the two great 'scriptures' of natural deduction—Gentzen [1934, 1935] and Prawitz [1965]. And the elimination rules for $\wedge$, $\rightarrow$, and $\forall$ are written in the *serial* form that Gentzen pioneered and that Prawitz adopted. The following snippets are included here both for scholarly accuracy and with reverential admiration.

The first snippet shows the rules as stated in Gentzen [1934, 1935], at p. 186.

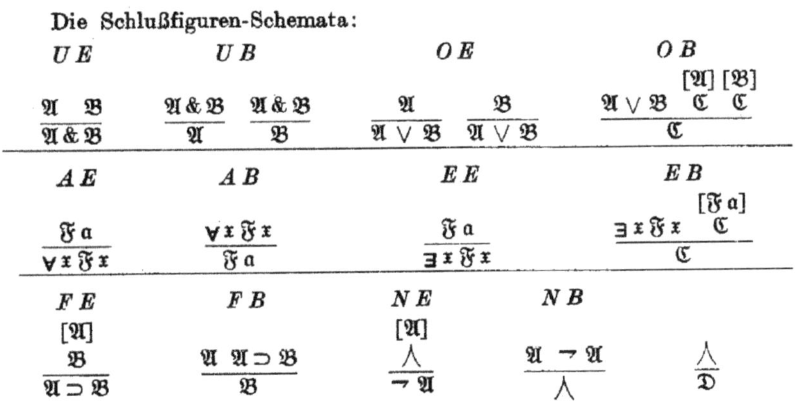

Die Schlußfiguren-Schemata:

UE: $\dfrac{\mathfrak{A} \quad \mathfrak{B}}{\mathfrak{A} \,\&\, \mathfrak{B}}$ *UB*: $\dfrac{\mathfrak{A} \,\&\, \mathfrak{B}}{\mathfrak{A}} \quad \dfrac{\mathfrak{A} \,\&\, \mathfrak{B}}{\mathfrak{B}}$ *OE*: $\dfrac{\mathfrak{A}}{\mathfrak{A} \vee \mathfrak{B}} \quad \dfrac{\mathfrak{B}}{\mathfrak{A} \vee \mathfrak{B}}$ *OB*: $\dfrac{\mathfrak{A} \vee \mathfrak{B} \quad \overset{[\mathfrak{A}]}{\mathfrak{C}} \quad \overset{[\mathfrak{B}]}{\mathfrak{C}}}{\mathfrak{C}}$

AE: $\dfrac{\mathfrak{F}a}{\forall \mathfrak{x} \mathfrak{F}\mathfrak{x}}$ *AB*: $\dfrac{\forall \mathfrak{x} \mathfrak{F}\mathfrak{x}}{\mathfrak{F}a}$ *EE*: $\dfrac{\mathfrak{F}a}{\exists \mathfrak{x} \mathfrak{F}\mathfrak{x}}$ *EB*: $\dfrac{\exists \mathfrak{x} \mathfrak{F}\mathfrak{x} \quad \overset{[\mathfrak{F}a]}{\mathfrak{C}}}{\mathfrak{C}}$

FE: $\dfrac{\overset{[\mathfrak{A}]}{\mathfrak{B}}}{\mathfrak{A} \supset \mathfrak{B}}$ *FB*: $\dfrac{\mathfrak{A} \quad \mathfrak{A} \supset \mathfrak{B}}{\mathfrak{B}}$ *NE*: $\dfrac{\overset{[\mathfrak{A}]}{\curlywedge}}{\neg \mathfrak{A}}$ *NB*: $\dfrac{\mathfrak{A} \quad \neg \mathfrak{A}}{\curlywedge}$ $\dfrac{\curlywedge}{\mathfrak{D}}$

In the labels for his rules the first letters indicate the operator involved:

	German	Gentzen's Symbol	Modern Symbol
U	Und	&	$\wedge$
O	Oder	$\vee$	$\vee$
A	All(e)	$\forall$	$\forall$
E	(es) Existiert	$\exists$	$\exists$
F	Falls	$\supset$	$\rightarrow$
N	es ist Nicht der Fall daß	$\neg$	$\neg$
	Absurdität	$\curlywedge$	$\bot$

The second letters in Gentzen's rule-labels are '*E*' for *Einführung* (*Introduction*) or '*B*' for *Beseitigung* (*Elimination*). The only unlabeled rule is lurking at bottom right. It has been called the Absurdity Rule, and is also known by the acronym EFQ, for *Ex Falso Quodlibet*. The rules in this snippet are those in Gentzen's system of natural deduction for Intuitionistic Logic. On p. 190 (*loc. cit.*) he points out that for Classical Logic one could add either the Law of Excluded Middle:

$$\mathfrak{A} \vee \neg \mathfrak{A}$$

or the rule of Double Negation Elimination:

$$\frac{\neg\neg \mathfrak{A}}{\mathfrak{A}} .$$

The second snippet shows how Gentzen's rules were reprised in the seminal monograph Prawitz [1965]. Prawitz's strictly classical rule is the one at bottom right, labeled '$\curlywedge_C$)'. But because Prawitz allowed the discharge indicated in his classical rule to be *vacuous*, it yields the rule $\curlywedge_I$ as well (that is, the Absurdity Rule).

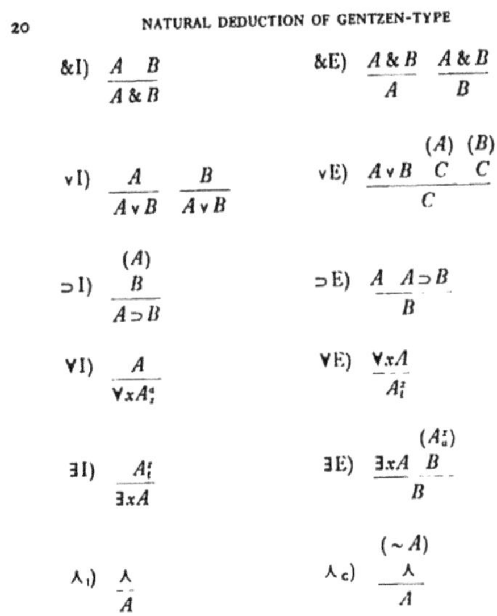
20 NATURAL DEDUCTION OF GENTZEN-TYPE

&I) $\dfrac{A \quad B}{A \,\&\, B}$ &E) $\dfrac{A \,\&\, B}{A} \quad \dfrac{A \,\&\, B}{B}$

∨I) $\dfrac{A}{A \vee B} \quad \dfrac{B}{A \vee B}$ ∨E) $\dfrac{A \vee B \quad \overset{(A)}{C} \quad \overset{(B)}{C}}{C}$

⊃I) $\dfrac{\overset{(A)}{B}}{A \supset B}$ ⊃E) $\dfrac{A \quad A \supset B}{B}$

∀I) $\dfrac{A}{\forall x A^a_x}$ ∀E) $\dfrac{\forall x A}{A^x_t}$

∃I) $\dfrac{A^x_t}{\exists x A}$ ∃E) $\dfrac{\exists x A \quad \overset{(A^x_a)}{B}}{B}$

$\curlywedge_I$) $\dfrac{\curlywedge}{A}$ $\curlywedge_c$) $\dfrac{\overset{(\sim A)}{\curlywedge}}{A}$

Let us return to our foregoing thread of exposition about the quantifier rules ∃-E and ∀-I, with their respective uses of a parameter (usually 'a' when it is the only one featuring.) Here now is an immediate example of a proof in which a parameter a occurs in the way explained on p. 52, for the rule of ∀-Introduction. It is the contemporary logician's proof of Aristotle's syllogism *Barbara*, as that argument was formalized in contemporary logical notation at the end of §3.3. We take the liberty here of jumping ahead to a fully formalized proof of a multi-premise argument, with no trace of English left in it. Note how by the stage in the following proof at which one draws the parametric conditional subconclusion $Aa \rightarrow Ca$ of the step labeled (1)—which is an application of the rule (→-I)—the assumption Aa (the antecedent of the conditional) that had been made 'for the sake of argument' at top left *has been discharged.* This ensures that a does not occur in any of the remaining assumptions on which $Aa \rightarrow Ca$ depends. That is why one can 'universally generalize' at the final step, drawing the conclusion $\forall x(Ax \rightarrow Cx)$ by applying the rule of ∀-Introduction to the subconclusion $Aa \rightarrow Ca$

'for arbitrary a'.

$$\cfrac{\cfrac{\cfrac{\cfrac{\overset{(1)}{\overline{Aa}} \quad \cfrac{\forall x(Ax \to Bx)}{Aa \to Ba}}{Ba} \qquad \cfrac{\forall x(Bx \to Cx)}{Ba \to Ca}}{Ca}\,{\scriptstyle(1)}}{Aa \to Ca}}{\forall x(Ax \to Cx)}$$

There is an important aspect of the elimination rule for 'every' (i.e., $\forall$) which has been used (twice over) in this proof. In its general form, as we have seen, this elimination rule is stated as follows:

$$\frac{\forall x \varphi(x)}{\varphi(t)},$$

where it is understood that the singular term t (which in the proof just displayed was the parameter a) is substituted for every so-called 'free occurrence' of the variable x in $\varphi(x)$. (In the context of the foregoing proof it is obvious what those 'free' occurrences of x are. The notion of a free occurrence of a variable in formulae quite generally will be rigorously defined in §4.1.)

The important aspect that we are referring to is that this simple and original form of the elimination rule for $\forall$ is known as *serial* elimination. It has enjoyed currency ever since being stated this way in Gentzen [1934, 1935]. (See the first snippet above.) It is the form which, for the time being, the student should master in solving deductive problems involving $\forall$. Later on in this work the student will be shown an interestingly different form of $\forall$-Elimination, which we shall call the *parallelized* form. The two forms (serial and parallelized) are constructively equivalent—each can be derived by using the other. (See §18.2.) But the parallelized form has certain theoretical virtues that we shall be emphasizing in due course. (The elimination rules for both $\wedge$ and $\to$ can and will be parallelized in similar fashion, with the same theoretical motivations. See §7.4.)

3.8 Regimenting (or formalizing) arguments

Before we can construct proofs, we need to *regiment* or *formalize* the English sentences involved as premises and conclusions of the arguments concerned. The reader has already seen how this was done with Aristotle's syllogism *Barbara*.

This is a process of making the logical forms clear by a judicious use of symbols (for logical operators) and so-called *place-holders*, which are single letters doing duty for descriptive expressions such as names, functions symbols, and predicates. Thus one uses, say, j for 'John', $f(\ldots)$ for 'the father of ...', $F(\ldots)$ for '... is friendly', and so on. (The convention among logicians is that names and function symbols are lowercase, and predicates are uppercase.) This enables us to detect the logical forms of the sentences more easily. It also helps us to work out the order in which we should apply rules of inference, as we seek to negotiate a step-by-step inferential transition from the premises to the conclusion.

Learning how to translate English sentences into the formal notation of logic is another main technique to be mastered in a course in logic. Indeed, it is a prerequisite for then going on to *prove* the argument (if it is valid) or *counterexemplifying* it (if it is invalid).

3.9 Constructing proofs

Before seeking a proof of a valid argument, it is worth regimenting its premises and conclusion. Recall, for example, the other argument alongside *Barbara* in the pair on p. 46:

All men are mortal
<u>Socrates is a man</u>
Socrates is mortal

Here is a slightly more regimented rendering of this argument in so-called 'logician's English':

for every x(if x is a man, then x is mortal)
<u>Socrates is a man</u>
Socrates is mortal

Proofs are constructed as tree-like arrays of steps taken in accordance with rules of inference like the ones we have been discussing. For the argument in logician's English we can provide the following two-step proof.

$$\dfrac{\dfrac{\text{for every } x(\text{if } x \text{ is a man, then } x \text{ is mortal})}{\text{if Socrates is a man, then Socrates is mortal}} \qquad \text{Socrates is a man}}{\text{Socrates is mortal}}$$

The first step in this proof eliminated the logical operator 'for every'; the second step eliminated the logical operator 'if ... then ...'.

The same proof can be made easier to read by using the more succinct symbolism introduced above, and completely regimenting the sentences involved:

$$\dfrac{\dfrac{\forall x(Mx \to Dx)}{Ms \to Ds}{\scriptstyle \forall\text{-}E} \qquad Ms}{Ds}{\scriptstyle \to E}$$

The first step in the proof eliminated $\forall$; the second step eliminated $\to$.

It is time to remark on a general feature in all formal proofs, no matter how long they are, or how complex are their premises and conclusions. The individual steps, or 'micro-moves', of inference within formal proofs deal with *one logical operator-occurrence* at a time. Moreover, the operator-occurrence in question is alway in *dominant position.* That is to say, it is 'applied last' in building up the sentence in question in accordance with the strict rules of logical grammar that govern its formation. (These rules will be given in rigorous detail in Chapter 4.)

Here we have used the translation manual

$M(\ldots)$	... is a man
$D(\ldots)$	... is mortal (i.e., will <u>d</u>ie)
s	Socrates

Note that we have labeled each of the two steps of the little proof above with the name of the rule of inference being applied. It is not necessary to have these labels count as integral to the proof. Rather, they are supplied as a helpful afterthought for the learner, to enable them to appreciate why the step is licit. As soon as one is thoroughly familiar with the graphic forms of the rules of inference , one will be able to check the correctness of any proof without any need for labels on its steps.

3.10 Constructing counterexamples

The reader will recall the simple model M that was given on p. 27:

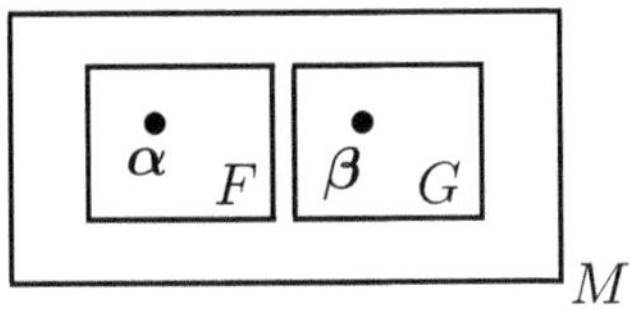

and which served as a counterexample to each of the invalid arguments that we called (I) and (II) thus:

(I)	Everyone is (F or G)	True in M
	(Everyone is F) or (everyone is G)	False in M
(II)	(Someone is F) and (someone is G)	True in M
	Someone is (F and G)	False in M

We propose to spell out here in greater justificatory detail the diagram-based *evaluations* of the sentences involved—the evaluations of the premises as true (in M) and of the conclusions as false (in M). Note that these evaluations will be 'M-relative'. This means that the verifications and falsifications are constructed relative to the atomic facts that make up M. The reader should get the intuitive idea quite easily from the examples that we can provide at this stage; and will find their more detailed, rigorous treatment in Chapter 11.

It will make matters easier when dealing with these two arguments if we express them in our formal language:

$$\text{(I)} \quad \frac{\forall x(Fx \vee Gx)}{\forall x Fx \vee \forall x Gx} \quad \begin{array}{l}\text{True in } M\\ \text{False in } M\end{array}$$

$$\text{(II)} \quad \frac{\exists x Fx \wedge \exists x Gx}{\exists x(Fx \wedge Gx)} \quad \begin{array}{l}\text{True in } M\\ \text{False in } M\end{array}$$

With an eye to the diagram above of the model M, one can easily see that the attributions of truth to the respective premises of arguments (I) and (II), and the attributions of falsity to their respective conclusions, are correct. We now intend to leverage off the ease of those insights to usher in two new notions that 'encode' them. These are the interdependent notions of an *M-relative verification* and an *M-relative falsification.* They are the two species of what we call *M-relative evaluations.* They will be very simple constructions in the context of these two simple arguments and the small, simple model M. But such evaluations are of great theoretical importance, and, especially with more complex sentences and large, more complicated models, they are of great practical utility.

Let us illustrate, now, by constructing the four evaluations that are needed. We shall construct them 'bottom up'. To make the process of construction as easy as possible to follow, we shall repeat the depiction of the model M with each evaluation.

We begin with the task of verifying the premise of argument (I).

Verification of $\forall x(Fx \vee Gx)$

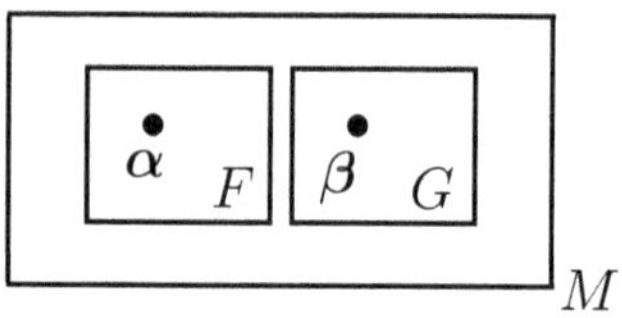

The sought M-relative verification will *end* with a step of what we shall call 'M-relative $\forall$-Verification', since $\forall$ is the dominant operator in the sentence to be verified. The relativity to the model M is incurred because *every instance* of the generalization needs to be verified. Thus we shall need to verify both the α-instance and the β-instance of the disjunction; and that will suffice, since α and β are *all the individuals there are* in the domain of M. Thus the nether part of the verification will look like this

$$\frac{\begin{matrix}\vdots & & \vdots \\ F\alpha \vee G\alpha & & F\beta \vee G\beta\end{matrix}}{\forall x(Fx \vee Gx)}{\scriptstyle M\,\forall\mathcal{V}}$$

where the final step is labeled as one of *M-relative universal verifica-*

tion. The two premises for this step of $\forall\mathcal{V}$ will be verified, as is now evident, by appropriate steps of '$\vee$-Verification', because now it is the operator $\vee$ that needs to be dealt with. The latter steps' premises, in turn, will be verified by appeal to strategically well-chosen 'positive factoids' of M. This yields the following completed M-relative verification of $\forall x(Fx \vee Gx)$:

$$\dfrac{\dfrac{\dfrac{}{F\alpha}\,M}{F\alpha \vee G\alpha}\,\vee\mathcal{V} \qquad \dfrac{\dfrac{}{G\beta}\,M}{F\beta \vee G\beta}\,\vee\mathcal{V}}{\forall x(Fx \vee Gx)}\,M\ \forall\mathcal{V}$$

Extended comment; As the reader will realize on reflection, *this* verification is *the only one possible*, for *this* sentence $\forall x(Fx \vee Gx)$, with respect to *this* model M. At no stage in our 'upward' construction of the verification did we exercise any choice in what could be done. But that is not always the case. In general, for some same sentence φ that is true in some same model N, there will be a plurality of *pairwise distinct* N-verifications of φ. We shall encounter such examples in due course. (The reader should already be able to imagine how such cases might arise: disjunctions, conditionals, and existential claims are particularly susceptible to having multiple verifications; while conjunction, negation, and universal claims are susceptible to having multiple falsifications, *within one and the same model.*) We can already say: the truth of a sentence φ in a model N *consists in the existence of at least one N-relative verification of φ.* There are, in general, many different ways that a sentence φ can be true in a given model N; and to each such way there corresponds an N-relative verification of φ. We turn now to the task of falsifying the conclusion of argument (I).

Falsification of $\forall xFx \vee \forall xGx$

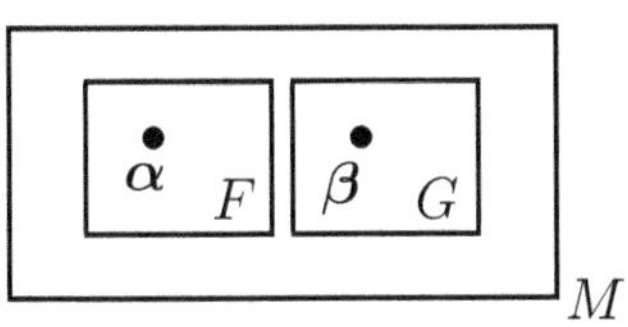

Again, we construct our sought evaluation upwards. The disjunction operator $\vee$ is dominant. To falsify a disjunction one needs to falsify each disjunct. So the bottommost step of our falsification will look like this:

$$\dfrac{\forall xFx \vee \forall xGx \qquad \begin{matrix} {}^{(1)}\overline{\quad\;\;}\\ \forall xFx \\ \vdots \\ \bot \end{matrix} \qquad \begin{matrix} {}^{(1)}\overline{\quad\;\;}\\ \forall xGx \\ \vdots \\ \bot \end{matrix}}{\bot}\,{\scriptstyle (1)\,\vee\mathcal{F}}$$

To falsify each of the two universals now exposed, it will suffice to furnish, for each of them, a falsifying instance. Note that in M, the individual $\boldsymbol{\alpha}$ is not in the extension of G; and the individual $\boldsymbol{\beta}$ is not in the extension of F. Thus we can use these 'negative M-factoids' as follows:

$$\dfrac{\forall xFx \vee \forall xGx \qquad \dfrac{\begin{matrix} {}^{(1)}\overline{\quad\;\;}\\ \forall xFx \end{matrix} \quad \dfrac{\overline{\;F\boldsymbol{\beta}\;}^{\,(2)}}{\bot}\,{\scriptstyle M}}{\bot}\,{\scriptstyle (2)\,\forall\mathcal{F}} \qquad \dfrac{\begin{matrix} {}^{(1)}\overline{\quad\;\;}\\ \forall xGx \end{matrix} \quad \dfrac{\overline{\;G\boldsymbol{\alpha}\;}^{\,(2)}}{\bot}\,{\scriptstyle M}}{\bot}\,{\scriptstyle (2)\,\forall\mathcal{F}}}{\bot}\,{\scriptstyle (1)\,\vee\mathcal{F}}$$

Once again, we had no choices to exercise in the upward construction of this evaluation. What is displayed here is the only available M-relative falsification of $\forall xFx \vee \forall xGx$.

We have now completed our counterexemplification of argument (I).

We turn now to the counterexemplification of argument (II). First we verify its premise $\exists xFx \wedge \exists xGx$; then we shall falsify its conclusion $\exists xFx \wedge \exists xGx$.

Verification of $\exists xFx \wedge \exists xGx$

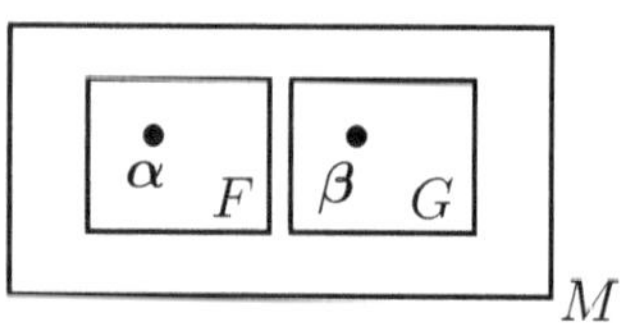

Once again, in building the sought verification upwards, we note the dominant logical operator that we have to deal with: it is the conjunction operator $\wedge$. To verify a conjunction we need to verify each conjunct. So the bottommost step of the verification will be

$$\dfrac{\begin{matrix}\vdots \\ \exists xFx\end{matrix} \qquad \begin{matrix}\vdots \\ \exists xGx\end{matrix}}{\exists xFx \wedge \exists xGx}\wedge\mathcal{V}$$

To verify an existential, one needs but one 'witnessing individual' from the domain. We see that α is the only F in the domain of M, and likewise β is the only G. So once again we have no choice; the final moves of construction are forced upon us:

$$\dfrac{\dfrac{\dfrac{}{F\alpha}M}{\exists xFx}\exists\mathcal{V} \qquad \dfrac{\dfrac{}{F\beta}M}{\exists xGx}\exists\mathcal{V}}{\exists xFx \wedge \exists xGx}\wedge\mathcal{V}$$

Falsification of $\exists x(Fx \wedge Gx)$

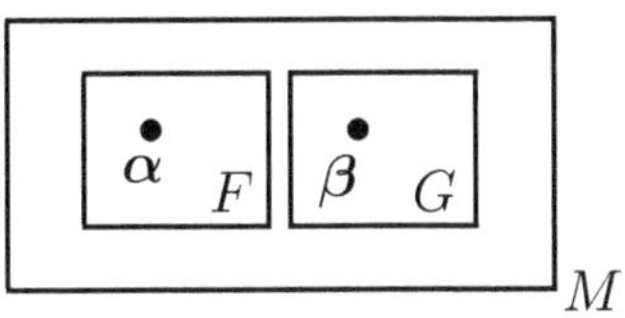

To falsify an existential, one needs to falsify its every possible instance in the domain. Thus we need to falsify both $F\alpha \wedge G\alpha$ and $F\beta \wedge G\beta$. The bottommost part of our falsification will therefore be of the following form. Note how the final step is labeled as one of *M-relative existential falsification*:

$$\dfrac{\exists x(Fx \wedge Gx) \qquad \begin{matrix}{}^{(1)}\overline{F\alpha \wedge G\alpha} \\ \vdots \\ \bot\end{matrix} \qquad \begin{matrix}{}^{(1)}\overline{F\beta \wedge G\beta} \\ \vdots \\ \bot\end{matrix}}{\bot}{\scriptstyle (1)\, M\, \exists\mathcal{F}}$$

We have now exposed two conjunctions as in need of falsification. To falsify a conjunction one need only falsify one of its conjuncts. And indeed there is only one way to do this, for each of the displayed conjunctions. The model M reveals both that α is not in the extension of the predicate G, and that β is not in the extension of the predicate F. Thus we can insert appeal to these two 'negative M-factoids' as follows:

$$\dfrac{\exists x(Fx \wedge Gx) \qquad \dfrac{\overline{F\alpha \wedge G\alpha}^{(1)} \qquad \dfrac{\overline{G\alpha}^{(2)}}{\bot}{\scriptstyle M}}{\bot}{\scriptstyle (2)\,\wedge\mathcal{F}} \qquad \dfrac{\overline{F\beta \wedge G\beta}^{(1)} \qquad \dfrac{\overline{F\beta}^{(2)}}{\bot}{\scriptstyle M}}{\bot}{\scriptstyle (2)\,\wedge\mathcal{F}}}{\bot}{\scriptstyle (1)\,M\,\exists\mathcal{F}}$$

Notice once again how at every stage of construction of this falsification our hand was forced. We had no choices to make. This may be the reason why with arguments (I) and (II) it was so easy to tell, by looking at the diagram M and looking at each sentence involved, what the sentence's truth value would be in the diagram.

It has not been otiose, however, to have spelled out in detail how these truth values arise, by constructing the simple evaluations that we have displayed. These ones have provided a manageable setting for our explanation of what the basic moves would be, in any detailed verification or falsification of a sentence in a model. We see that for each logical operator @, there is a verification rule (which we call @$\mathcal{V}$) and a falsification rule (which we call @$\mathcal{F}$). These evaluation rules for the connectives will be set out in fuller detail in Chapter 6, which will also deal with negation and the conditional (neither of whose evaluation rules have been seen at work in the foregoing evaluations). And when @ is a quantifier, the applications of $\forall\mathcal{V}$ and of $\exists\mathcal{F}$ will be *sensitive to the domain of the model.* They will call for immediate sub-evaluations for each and every individual in the domain. Thus, if M has an infinite domain—as, say, with the model $\mathbb{N}$ of the natural numbers—there will be infinitely many sub-evaluations 'above the line' for applications of $\forall\mathcal{V}$ and of $\exists\mathcal{F}$. (The $\mathbb{N}$-relative rule of universal verification was called, by Carnap, the *ω-rule.*) Verification and falsification rules for the quantifiers, as we mentioned earlier, will be set out in full detail in Chapter 11.

Thus: *evaluations* (verifications and falsifications) *can be infinitary.* But *deductive proofs* (and *disproofs*) must be *finite.* We therefore need to make a principled distinction between rules governing steps of model-relative evaluations, and rules governing steps of deductive inference. The latter rules are *model-invariant.* Each logical operator @ will have its deductive Introduction rule @-I and Elimination rule @-E, in whose formulations there will be no mention of models. By contrast, each logical operator @ will have its own model-relative verification rule $@\mathcal{V}$ and falsification rule $@\mathcal{F}$. In due course we shall be seeing how the $\mathcal{V}$- and $\mathcal{F}$-rules 'morph over' into the I- and E-rules; but that exposition will first require a more thorough acquaintance with all the rules involved.

We return now to speaking of deduction, not evaluation.

3.11 Deductive progress

There is such a thing as *deductive progress.* Proofs can be *accumulated.* Suppose we first deduce Q from premises $P_1, \ldots, P_n$ by means of the proof Π:

$$\begin{array}{c}\underbrace{P_1, \ldots, P_n}\\ \Pi \\ Q\end{array}$$

and suppose that thereafter we deduce R from premises $Q, Q_1, \ldots, Q_m$ by means of the proof Σ:

$$\begin{array}{c}\underbrace{Q, Q_1, \ldots, Q_m}\\ \Sigma \\ R\end{array}$$

Note that Q is the *conclusion* of the proof Π, but is a *premise* of the proof Σ. This means that if the combined premises $P_1, \ldots, P_n, Q_1, \ldots, Q_m$ are true, then so too will be the conclusion R:

$$\begin{array}{c}\underbrace{\begin{array}{c}\underbrace{P_1, \ldots, P_n}\\ \Pi \\ [Q]\end{array}, \quad Q_1 \ , \ldots, \ Q_m}\\ \Sigma \\ R\end{array} .$$

We have bracketed Q in this proof-schema to stress the fact that the downward flow of truth will 'pass through' Q. The overall reasoning of the two proofs Π and Σ taken together does not use Q as a starting-point, and does not have Q as its endpoint. Instead, Q is a 'halfway house' for the deductive reasoning.

Logicians call Q an *interpolant* for the argument

$$\frac{P_1, \ldots, P_n, Q_1, \ldots, Q_m}{R} .$$

It mediates between the premises $P_1, \ldots, P_n, Q_1, \ldots, Q_m$ and the conclusion R, in the way just depicted.

The ingenuity of mathematicians consists at least in part in their being able to find intelligent interpolants to mediate between their mathematical *axioms* and the mathematical *theorems* that they are trying to prove from those axioms. Mathematicians call such interpolants *lemmas*. Think of Q in the foregoing as a lemma. Think of $P_1, \ldots, P_n, Q_1, \ldots, Q_m$ as drawn from the mathematician's set of axioms, or 'first principles'. And think of R as some 'deep theorem' that the mathematician has succeeded in establishing, by means of the two proofs Π and Σ, as a consequence of the axioms. By finding the right lemmas (such as Q), one can drastically reduce the time and memory space needed in a search for a proof of a given conclusion from given premises.

Proving an argument *via* interpolants enables us to come to know that truth is guaranteed to be transmitted from the premises of the argument to its conclusion. Every step in a chunk of proof that is interpolant-free can be methodically checked for correctness; we can all reach agreement that such a chunk of proof is properly put together. This means that the validity of an argument is a property that we can know about—provided, of course, that every valid argument does indeed admit of proof in this fashion.

In the case of mathematical reasoning, there is a further important point to note. In the notation of the foregoing general example, the mathematician is morally certain that the axioms $P_1, \ldots, P_n, Q_1, \ldots, Q_m$ are *jointly satisfiable*. That moral certainty amounts to this:

> *there is some abstract structure of mathematical entities, and an interpretation of the mathematical constants, function signs, and predicates therein, under which those axioms are all true.*

So, provided only that the two proofs Π (*to* the lemma Q) and Σ (*from* the lemma Q) are truth-preserving under all interpretations, the overall conclusion R (the theorem) will be known to be true in any structure in which the axioms $P_1, \ldots, P_n, Q_1, \ldots, Q_m$ are true.

That is how mathematical truths come to be known. Ironically, one does not have to be able actually to *put together*, or *graft one on top of the other*, the two formal proofs Π and Σ so as to make a yet bigger, licitly constructible, formal proof. All that is required is that each of the proofs Π and Σ be truth-preserving. This much is tacitly acknowledged in the standard practice of giving 'informally rigorous' proofs in mathematics. The deductive passages are broken down by invoking (i.e., interpolating) various lemmas on the deductive journey from the axioms to the sought theorem. The actual passages of deductive reasoning *to* and *from* the lemmas are set out explicitly and separately, under the headings '*Proof: ...*', and with the end-of-proof indicator '*QED*' (for 'Quod erat demonstrandum'—'that which had to be proved') or (in modern typesetting) the box-symbol □. And the resulting proofs do not actually end up being grafted together explicitly. That they will nevertheless, collectively, guarantee the transmission of truth—from the axioms, via one's lemmas, to the theorem thereby established—is *tacitly presupposed.*

The logician, therefore, who is aiming to furnish a system of formal proof that provides adequately for the rigorous regimentation of mathematical reasoning, is relieved of the need to characterize formal proofs as objects that can always be 'grafted' one on top of another so as to form yet another formal proof. It is enough just to have the separate 'stage-wise' proofs (like Π and Σ above) and the assurance that those proofs are truth-preserving. There are no further logical demands than these, for the logical attainment of mathematical knowledge (i.e., of the truth of mathematical theorems) on the basis of axioms taken to be true of some particular mathematical structure that one is thinking about.

We shall investigate this matter more deeply in the sequel.

3.12 Soundness and completeness

A system $\mathcal{S}$ of rules of inference that furnishes proofs *only* for valid arguments is said to be *sound*:

$$\left(\text{there is an } \mathcal{S}\text{-proof of } \frac{P_1, \ldots, P_n}{Q}\right) \Rightarrow \left(\frac{P_1, \ldots, P_n}{Q} \text{ is valid}\right)$$

A system $\mathcal{S}$ of rules of inference that furnishes proofs for *all* valid arguments is said to be *complete*:

$$\left(\frac{P_1, \ldots, P_n}{Q} \text{ is valid}\right) \Rightarrow \left(\text{there is an } \mathcal{S}\text{-proof of } \frac{P_1, \ldots, P_n}{Q}\right)$$

(The double arrow $\Rightarrow$ here regiments the conditional connective in our metalanguage.) The proof systems for *propositional* logic and *first-order* logic that we study in this book are indeed *both* sound *and* complete. *Propositional logic* deals with the logical notions *not* ($\neg$), *and* ($\wedge$), *or* ($\vee$), and *if... then...* ($\rightarrow$). *First-order logic* deals further with the logical notions '*... is identical to...*' ($=$), *some* ($\exists$), and *every* ($\forall$). First-order languages are powerful enough to express all of classical mathematics. It is a major aim of the intended sequel to this work to establish the soundness and completeness of first-order classical logic.

3.13 Sequents and subsequents

As already mentioned, another technical term often used for entities of the form

$$P_1, \ldots, P_n : Q$$

(or, more generally, of the form $\Delta : \varphi$, where Δ is a finite set of sentences, and φ is a sentence) is *sequent.* The premises are said to be on the left side of the sequent, and the conclusion on its right side. We treat here only of *single-conclusion* sequents $P_1, \ldots, P_n : Q$ rather than *multiple-conclusion* sequents of the more general form

$$P_1, \ldots, P_n : Q_1, \ldots, Q_m.$$

Multiple-conclusion sequents allow one, however, to make some points in a nicely symmetrical way, so we shall dwell on them a little longer.

The validity of a multiple-conclusion sequent is to be understood as follows: there is no interpretation that makes $P_1, \ldots, P_n$ all true but makes $Q_1, \ldots, Q_m$ all false. Hence the *in*validity of such a sequent can be understood as consisting in the existence of an interpretation (M, say) that dichotomizes the language **L** into the true sentences and the false ones, and thereby 'drives a wedge' between the sequent's premises and its conclusions. An interpretation M that does this, as illustrated by the following diagram, is said to be a *counterexample* to the sequent $P_1, \ldots, P_n : Q_1, \ldots, Q_m$.

L	True-in-M	False-in-M
	$P_1, \ldots, P_n$	$Q_1, \ldots, Q_m$

One says that a set Δ is a subset of a set Γ just in case every member of Δ is a member of Γ. In formal symbols: $\Delta \subseteq \Gamma$. Such containment of Δ in Γ is said to be *proper* just in case (in addition) some member of Γ is not a member of Δ. In that case one writes $\Delta \subset \Gamma$.

Subsequents are obtained by 'subsetting down' on either the left or the right (or both). Thus, in general, $\Delta' : \Gamma'$ is a subsequent of $\Delta : \Gamma$ just in case both $\Delta' \subseteq \Delta$ and $\Gamma' \subseteq \Gamma$. And $\Delta' : \Gamma'$ is a *proper* subsequent of $\Delta : \Gamma$ just in case at least one of these containments is proper.

For those readers not yet familiar with the notation of set theory, a quick word of explanation is in order. Δ is a *subset* of Γ—in symbols, $\Delta \subseteq \Gamma$—just in case every member of Δ is a member of Γ. And a subset Δ of Γ is a *proper* subset of Γ if moreover some member of Γ is not in Δ.

A *single*-conclusion sequent $\Delta : \varphi$ can be thought of as a special case of a multiple-conclusion sequent, one having a singleton on its right—

$$\Delta : \{\varphi\}.$$

But consider the case where one has the empty set $\emptyset$ on the right, instead of a singleton. $\Delta : \emptyset$ is a proper subsequent of $\Delta : \{\varphi\}$. And on the construal of validity just given, the sequent $\Delta : \emptyset$ is valid just in case there is no interpretation that makes all of Δ true. So $\Delta : \emptyset$ is valid just in case $\Delta : \bot$ is valid, where $\bot$ ('absurdity') is the logical constant that is always false. Hence $\Delta : \bot$ can be treated as though it were a (proper) subsequent of $\Delta : \{\varphi\}$, even though (obviously) the logical constant $\bot$ is neither a member of, nor a subset of, $\{\varphi\}$.

3.14 Appraising arguments

In putting forward, or endorsing, an argument $P_1, \ldots, P_n : Q$, one is claiming that one is entitled to make the transition *from* the premises $P_1, \ldots, P_n$ *to* the conclusion Q. There are various ways to express such a transitional claim. Here are three of them:

1. Suppose that $P_1, \ldots, P_n$. Then it follows that Q.

2. If $P_1, \ldots, P_n$ are all true, then Q is true also.

3. Suppose one has warrants for asserting $P_1, \ldots, P_n$. Then one can find a warrant for asserting Q.

We stress again that the statement of an argument is *bare*. In the foregoing three forms of expression, the inferential transition embodied by the argument $P_1, \ldots, P_n : Q$ is simply claimed in one step (indicated by the colon), rather than justified, or shown to be valid, by interpolating more obvious steps. The premises $P_1, \ldots, P_n$ and the conclusion Q can be very complicated sentences. The statement of the argument involving them need not give any indication at all as to how one might reason in a convincing way *from* those premises *to* that conclusion—that is, how one might *show* that any of (1), (2) or (3) holds. To put forward an argument is to claim something like (1), (2) or (3). But in general it is a claim made without (yet) offering a proof. Such a proof would be furnished by one's system of logic.

A system of logic codifies the kinds of inference that are permitted among sentences of a given language. In learning to do logic—that is, to reason logically—one learns how to make inferential steps of the permitted forms. By arranging inferential steps in an appropriate tree-like pattern, one can negotiate the passage from a set of *premises* to a *conclusion*. The pattern is a tree-like array of sentences, and is called a *proof*. A proof establishes that *whenever its premises are true, its conclusion is true also*. Put another way: the passage from the premises of a proof to its conclusion is *truth-preserving*. We also say that the proof shows that the argument in question (the alleged transition from premises to conclusion) is *valid*.

We have written $P_1, \ldots, P_n$ frequently in the foregoing, as if to commit ourselves to the view that arguments have at least one premise, and are therefore always *transitional* in nature. But this is not so; there are two very interesting and important extremal cases.

First extremal case

There are arguments *with no premises* (or, as one also says, with 'the empty set of premises'). In such a case the conclusion Q is really being asserted outright, and being maintained to hold *independently of any assumptions*. Synonymously: it is being maintained that Q is a *logical truth*. If the assertion is justified—if there is a proof of Q 'from no assumptions'—then Q is said to be a *theorem* of the logic concerned. If the beginner is wondering how that could come to pass, we shall just present them with the following very simple case, a proof of $A \rightarrow A$ from the empty set of premises. A is both the conclusion and the assumption-to-be-discharged of the subproof A:

$$\frac{\overline{\;A\;}^{(1)}}{A \rightarrow A}{(1)}$$

Note that a theorem of logic should therefore be distinguished from what is called a theorem in mathematics. Mathematical theorems are conclusions drawn from the *axioms* of the mathematical theory concerned. Thus mathematical theorems are *not* independent of any assumptions. Rather, they depend on the mathematical axioms featuring in the argument in question. Suppose that one has a mathematical argument

$$A_1, \ldots, A_n : B$$

where $A_1, \ldots, A_n$ are the axioms appealed to, and B is the *mathematical* theorem that is claimed to follow from those axioms. (Example: $A_1, \ldots, A_n$ could be the axioms of Euclidean plane geometry, and B could be Pythagoras's famous theorem about the areas of the squares on the sides of a right-angled triangle.) Then the way to obtain a *logical* theorem to match the mathematical one is to state the argument as a single conditional sentence, such as

$$(A_1 \wedge \ldots \wedge A_n) \rightarrow B$$

or

$$A_1 \rightarrow (A_2 \rightarrow \ldots (A_n \rightarrow B) \ldots).$$

Second extremal case

The other extremal case of an argument arises when the conclusion Q is the absurdity symbol $\perp$. And of course there is no interpretation in which $\perp$ is true. Hence, putting forward the argument

$$P_1, \ldots, P_n : \perp$$

is tantamount to claiming that there is no possible way that the sentences $P_1, \ldots, P_n$ can all be true together—that is to say, $P_1, \ldots, P_n$ form an *unsatisfiable set.* When there is a proof of the conclusion $\perp$ from $P_1, \ldots, P_n$ we say that the premises $P_1, \ldots, P_n$ are *jointly inconsistent*, or that the *set* $\{P_1, \ldots, P_n\}$ is inconsistent; and we call the proof in question a *dis*proof.

When n is small—that is, when there are relatively few premises—and each premise seems plausible when considered on its own, one calls (a proof of)

$$P_1, \ldots, P_n : \perp$$

an *aporia.* Philosophers delight in discovering aporias. A good aporia reveals a formerly hidden inconsistency among principles that had previously enjoyed widespread assent or that, upon first acquaintance, strike one as reasonable or plausible.[1].

Another term that philosophers use when discussing different kinds of arguments is *enthymeme.* An enthymeme is an argument that strikes one as valid but is not, strictly speaking, formally valid. To make it formally valid one needs to supply a missing premise (which the arguer is assuming to be known, and has therefore not bothered to state explicitly). An example of an enthymeme is the following argument:

All women are mortal
Xanthippe is Socrates's wife
———————————
Xanthippe is mortal

The missing premise is that the wife of anyone is a woman. It needs to

[1]See Tennant [2000b].

be supplied to make the argument *formally* valid:

The wife of anyone is a woman
All women are mortal
Xanthippe is Socrates's wife
———
Xanthippe is mortal

3.15 Validity of arguments

Validity does not require the truth of the premises, or the truth of the conclusion

The validity of an argument consists in the fact that truth is preserved from its premises to its conclusion, *under all possible interpretations of its non-logical expressions.* But that fact is compatible with various possible distributions of *truth values* (T for 'true', F for 'false') across the premises and the conclusion, under a particular interpretation. The only pattern ruled out by validity is the one in which all the premises are true, and the conclusion is false.

With a valid two-premise argument $P_1, P_2 : Q$, for example, only one of the eight possible distributions of truth values for P_1, P_2, and Q is not permitted:

	P_1	P_2	Q	
1.	T	T	T	permitted
2.	T	T	F	not permitted
3.	T	F	T	permitted
4.	T	F	F	permitted
5.	F	T	T	permitted
6.	F	T	F	permitted
7.	F	F	T	permitted
8.	F	F	F	permitted

In general, with a valid n-premise argument $P_1, \ldots P_n : Q$, only one of the 2^{n+1} possible distributions of truth values for $P_1, \ldots, P_n$ and Q is (in virtue of the argument's validity) not permitted: that in which the premises $P_1, \ldots, P_n$ are all true, but the conclusion Q is false.

Staying with $n = 2$ for the purposes of illustration, consider the valid argument-form

t is A
All As are B
t is B

Every instance of this argument-form is valid. For each of the seven permitted distributions of truth values across the premises and conclusion, we can find a corresponding instance of the argument-form in question. The truth values below are assigned under the obvious everyday interpretations of the English words involved:

	T : Socrates is a man	
1.	T : All men are mortal	
	T : Socrates is mortal	
	T : t is A	No possible instance with this
2.	T : All As are B	pattern of truth values, since
	F : t is B	the argument-form is valid
	T : Xanthippe is a woman	
3.	F : All women are Greek	
	T : Xanthippe is Greek	
	T : Xanthippe is a woman	
4.	F : All women are immortal	
	F : Xanthippe is immortal	
	F : Socrates is a Roman	
5.	T : All Romans are human	
	T : Socrates is human	
	F : Socrates was an ancient Roman	
6.	T : All ancient Romans spoke Latin	
	F : Socrates spoke Latin	
	F : Socrates was an ancient Roman	
7.	F : All ancient Romans drank hemlock	
	T : Socrates drank hemlock	
	F : Socrates was an ancient Roman	
8.	F : All ancient Romans were illiterate	
	F : Socrates was illiterate	

Not every two-premise argument-form, of course, actually admits all seven of the permitted distributions of T and F across its premises and conclusion. The foregoing argument-form was special in that regard. It is easy to see, however, that the simple two-premise argument-form

$$\frac{A \quad B}{A \wedge B}$$

cannot admit any of the patterns $(T, F/T)$, $(F, T/T)$ or $(F, F/T)$. This is because no conjunction with a false conjunct can be true.

The orthodox account of validity

On the orthodox account, validity of arguments is defined as follows:

> $P_1, \ldots, P_n : Q$ is *valid* just in case every interpretation (of non-logical vocabulary) that makes all of $P_1, \ldots, P_n$ true makes Q true.

In the case of multiple-conclusion sequents, this definition would read

> $P_1, \ldots, P_n : Q_1, \ldots, Q_m$ is *valid* just in case every interpretation (of non-logical vocabulary) that makes all of $P_1, \ldots, P_n$ true makes *at least one* of $Q_1, \ldots, Q_m$ true.

And this is (classically) equivalent to the construal of validity of multiple-conclusion sequents that was given in the text above:

> there is no interpretation that makes all of $P_1, \ldots, P_n$ true and makes all of $Q_1, \ldots, Q_m$ false.

Let us go back now to considering only single-conclusion sequents (or arguments). When $P_1, \ldots, P_n : Q$ is valid we write

$$P_1, \ldots, P_n \models Q.$$

(The metalinguistic symbol $\models$ is called the *double turnstile*.) We can also read this as

> The premises $P_1, \ldots, P_n$ *logically imply* the conclusion Q

or as

> The conclusion Q *is a logical consequence of* the premises $P_1, \ldots, P_n$.

Note that it is usually tacitly understood (and here we make it explicit) that a valid argument *need not require all of its premises* in order to be valid. Some of the premises available among $P_1, \ldots, P_n$ might turn out to be otiose, for the purpose of logically implying a particular conclusion Q (including $\bot$). We shall have more to say about this aspect of the orthodox notion of validity in §23.6.

The two extremal cases mentioned earlier would be read slightly differently. We can read '$\models Q$' as 'Q is logically true';[2] and we can read '$P_1, \ldots, P_n \models \bot$' as '$P_1, \ldots, P_n$ are not jointly satisfiable'.

The foregoing are *metalinguistic* claims, claims *about* whatever sentences $P_1, \ldots, P_n$ and Q of the language are in question. The language under consideration is called the *object language*; while the language in which the consideration takes place is called the *metalanguage*. If we use the label L for the object language, we should bear in mind that a claim of validity of argument, or of logical consequence, is actually relativized to L. Thus we should write

$$P_1, \ldots, P_n \models_L Q$$

in order to acknowledge this. Often, however, the subscript L is suppressed, especially when the language in question is known from the context.

Relativizing the relation $\models$ of logical consequence (or the validity of arguments) to a language L is more subtle and profound than might at first appear. The language L, for the purposes of such relativization, is not to be taken simply as a set of sentences conceived as finite strings of symbols. It is more than that. It is the set of sentences in question *along with a preferred way of understanding the logical operators.* Logicians with differing understandings of the logical operators can therefore disagree on what it is that makes a given sentence true, and on whether a given argument is valid. Such is the case, for example, with constructive and classical logicians, who disagree about the precise meanings (or at least: logical behavior) of certain crucial logical operators, such as

[2] In the multiple-conclusion case, we would read '$\models Q_1, \ldots, Q_m$' as 'the truth always lies among $Q_1, \ldots, Q_m$'.

negation, disjunction and existential quantification (which will be explained in due course). They therefore have differing conceptions of the truth-conditions of various sentences, and disagree on the validity of certain forms of argument. We have not yet said anything about negation, disjunction or existential quantification in this connection, but we mention them here in order to foreshadow important foundational and methodological problems. *Is there a uniquely correct logic? Is there a univocal conception of logical validity of argument?* These questions will be addressed in due course.

Validity can be an elusive property of arguments. It is often extremely difficult to tell whether a given argument is valid. Take, for example, the argument whose conclusion Q is Euclid's Parallels Postulate:

> for any line l, and any point p not on l, there is, in the plane containing l and p, exactly one line that contains p and does not intersect l,

and whose premises $P_1, \ldots, P_n$ are all the other first principles of Euclidean geometry (such as 'Any two distinct points lie on a unique straight line'). The question 'Is the argument $P_1, \ldots, P_n : Q$ valid?' vexed geometers for over two millenia, and was eventually answered negatively.[3]

For another example, let Q be the statement of Fermat's Last Theorem:

> for all natural numbers $n > 2$, and for all natural numbers a, b, and c, it is not the case that $a^n + b^n = c^n$,

and let $P_1, \ldots, P_n, \ldots$ be the first principles of number theory. The question 'Is the argument $P_1, \ldots, P_n, \ldots : Q$ valid?' vexed number-theorists

---

[3]The argument

Euclid's first four axioms : *Parallels postulate*

was infinitely counter-exemplified by Eugenio Beltrami's construction of a pseudo-spherical surface within Euclidean space (thereby giving a model of Lobachevskian plane geometry, in which there exists more than one line parallel to any given line and through a given point off the latter). See Beltrami [1868].

for over three centuries, and was eventually answered positively.[4]

3.16 Provability of arguments

A convincing *positive* answer to any question of the form 'Is the argument $P_1, \ldots, P_n : Q$ valid?' must take the form of a *proof* of the conclusion Q from the premises $P_1, \ldots, P_n$. A proof is finite. It is an appropriate arrangement of steps of inference, each in accordance with a basic truth- or validity-preserving *rule* of inference, taking one from one's chosen premises to the desired conclusion. A proof fills in the missing steps of reasoning, each of which can be effectively checked for correctness. This is why it can *show* convincingly that the overall transition from $P_1, \ldots, P_n$ to Q is indeed valid. So a proof establishes the *validity* of the bare argument $P_1, \ldots, P_n : Q$ that consists of just the premises and the conclusion. A proof 'perfects' the argument.

Proofs that are able to *convince* one of the validity of an argument have to be *finite* pieces of discourse. They cannot be infinite, since one has to be able to check them for correctness—they must be *surveyable.* It must be possible in principle, given any finite piece of discourse, to *tell by inspection* whether it is a proof. Moreover, one must be able to tell by inspection what premises the proof uses, and what conclusion the proof establishes on the basis of those premises.

When one can 'tell by inspection' that a symbolic construct has a certain property, the property in question is said to be *effectively decidable.* Another way of putting this is to say that it is *mechanically checkable.* The idea is that one could specify a finite set of rules that anyone could follow in order to determine whether a given symbolic construct has the property in question. The agent could follow the rules, and arrive at the correct decision—'yes' or 'no'—in a finite time, without needing to exercise any creativity or ingenuity. All such agents would arrive at the same decision; the rules themselves uniquely determine

[4] See Wiles [1995]. Note that we are allowing here for the first principles of number theory to be infinitely numerous. This is because the well-known Axiom Schema of Mathematical Induction:

$$(\Phi(0) \land \forall n(\Phi(n) \rightarrow \Phi(sn))) \rightarrow \forall m\Phi(m)$$

has infinitely many substitution instances (of arithmetical formulae for Φ).

the answer (for the given construct that is being checked for possession of the property). On the basis of *such* rules, the proofs constructed in accordance with them would be 'mechanically checkable' by a fully automated proof-checker.

The crucial thing about fully formal proofs is that the question whether a given finite piece of discourse qualifies as such a proof is effectively decidable. It might take ingenuity or creativity to find or construct a fully formal proof; but subsequently *checking it for proofhood*, or *verifying that it is indeed a proof*, is an effectively decidable matter. With this as the *goal* or *summum bonum* of a proof system, for the purposes of epistemology, it is easy to understand why logicians seek an 'ultimate' or 'fully detailed' regimentation or formalization of the proofs that typically occur in mathematical texts. The latter proofs meet the requirement of what is known as *informal* rigor, because they are addressed to specialists with expert knowledge of the fundamentals of the theory in which they are working. These experts are able to 'supply missing steps' that are obvious to the prover and their audience but which are frequently left out when all concerned can 'mentally supply' the missing bits. But even the most sophisticated automated proof-checker would often be at a loss confirming that such a proof 'works'. To *that* end, one requires *fully detailed formal* proofs, constructed in accordance with primitive rules of deduction whose correct application the proof-checker can mechanically verify.

When the argument $P_1, \ldots, P_n : Q$ admits of fully formal proof we write

$$P_1, \ldots, P_n \vdash Q.$$

The metalinguistic symbol $\vdash$ is called the *single turnstile*. With subsequent uses of the unadorned single turnstile, the context will always determine whether it stands for constructive deducibility or for classical deducibility. We can also read the last displayed deducibility statement as

1. There is a proof of the conclusion Q from the premises $P_1, \ldots, P_n$
2. $P_1, \ldots, P_n$ (suffice to) *prove* Q
3. The argument $P_1, \ldots, P_n : Q$ *is provable*
4. The premises $P_1, \ldots, P_n$ *provably imply* the conclusion Q

5. The conclusion Q *is deducible from* (or *derivable from*) the premises $P_1, \ldots, P_n$.

In the extremal case where the argument that is proved involves no premises, we would write '$\vdash Q$', and read this as 'Q is a logical theorem'. In the extremal case where the argument that is proved has $\perp$ for its conclusion, we would write '$P_1, \ldots, P_n \vdash \perp$', and read this as '$P_1, \ldots, P_n$ are jointly inconsistent'.

Now that we have defined the 'semantic' double turnstile $\models$ (for a language L) and the 'syntactic' single turnstile $\vdash$ (for a fully formal proof system $\mathcal{S}$ dealing with that language's logical operators), we can reformulate our earlier statements of the soundness and completeness of a logical system for a language. We can use L and $\mathcal{S}$ as subscripts on the double and single turnstiles respectively.

Recall that we said

> A system $\mathcal{S}$ of rules of inference that furnishes proofs *only* for valid arguments is said to be *sound.*

This can now be formalized (in the metalanguage) as follows:

SOUNDNESS of $\mathcal{S}$ for L. $\quad P_1, \ldots, P_n \vdash_{\mathcal{S}} Q \;\Rightarrow\; P_1, \ldots, P_n \models_L Q.$

Recall that conversely, we said

> A system $\mathcal{S}$ of rules of inference that furnishes proofs for *all* valid arguments is said to be *complete.*

This too can now be formalized (in the metalanguage), as follows:

COMPLETENESS of $\mathcal{S}$ for L. $\quad P_1, \ldots, P_n \models_L Q \;\Rightarrow\; P_1, \ldots, P_n \vdash_{\mathcal{S}} Q.$

Note that it is usually tacitly understood (and here we make it explicit) that the proof(s) in question *need not use all of the premises* $P_1, \ldots, P_n$. All that is required is that all the premises (or so called 'undischarged assumptions') on which the conclusion of the proof rests are drawn from among the available premises $P_1, \ldots, P_n$. So, some of the premises available among $P_1, \ldots, P_n$ might turn out to be otiose, for the purpose of deducing a given conclusion (including $\perp$). This comment echoes, for the relation of deducibility, what we said on p. 76 about the relation of logical consequence.

Note further that, on the more general reading $\Delta \vdash \varphi$, where Δ is a (possibly infinite) set of premises, all but finitely many of those premises will be otiose. This is because every proof is finite, hence uses only finitely many of the available premises. So, if Δ is infinite, *infinitely many* of its members will be otiose. This leads to the following important statement about deducibility:

COMPACTNESS OF DEDUCIBILITY.
If $\Delta \vdash \varphi$, then, for some *finite* subset Γ of Δ, we have $\Gamma \vdash \varphi$.

The foregoing, once again, are *metalinguistic* claims, claims *about* whatever sentences of the object language are in question.

The means of fully formal proof at our disposal make up a *proof system* for the object language. Typically these means include *logical axioms* (sentences that can be asserted outright, because of their logico-grammatical form) and *rules of inference* (formal patterns of permitted inferential transition). An example of a logical axiom (of *classical* logic) is the 'Law of Excluded Middle'. In logician's English it says

> Either it is the case that φ or it is not the case that φ.

Written in completely formal terms, it is

$$\frac{}{\varphi \vee \neg\varphi}\,.$$

By writing the horizontal inference stroke over $\varphi \vee \neg\varphi$, with nothing above it, one is in effect saying that any sentence of this form can be 'asserted outright'. It does not depend on any assumption 'above the line'—for there is no such assumption. This is what it is for a sentence of the form $\varphi \vee \neg\varphi$ to be treated as a logical axiom. (Note, however, that it is an axiom of *classical* logic. It is *not* an axiom of constructive logic.)

The reader should be advised to distinguish *logical* axioms from *mathematical* axioms. Logical axioms hold for all subject matters. Mathematical axioms do not purport to be *logical* truths. They divide, however, into two interestingly different kinds. First, certain mathematical axioms are the 'self-evident' deductive starting-points for the investigation of truths in particular branches of mathematics, such as number theory and geometry. In those branches the intention is to describe particular 'unique', or 'intended', structures, such as the natural numbers or Euclidean space. Second, other branches of mathematics have sets of axioms that can be true of many different structures. Such is the case with, say, the theory of groups, or the theory of Boolean algebras. There, the mathematical interest is in relations among different structures that make the axioms true. But let us return now to matters logical.

An example of a rule of inference *with* premises, we remind the reader, is that of $\wedge$-Introduction, which is usually stated as

$$\frac{\varphi \quad \psi}{\varphi \wedge \psi} .$$

Here, the rule allows one to infer the conjunctive conclusion $\varphi \wedge \psi$ from the pair of premises immediately above the line, namely φ and ψ. The assumptions on which $\varphi \wedge \psi$ ultimately rests will therefore be the combination of the assumptions on which φ and ψ respectively rest.

To repeat: a proof system consists of such axioms and rules as just illustrated. By taking inferential steps in accordance with such axioms and rules, one is able to negotiate a deductive passage from given premises $P_1, \ldots, P_n$ to such conclusions Q as follow from them.

Henceforth we shall suppress the adjectival phrase 'fully formal' in our discussions of proof systems. It is to be understood as always there, like the music of the spheres, even if unheard.

3.17 A collection of various formal translations of valid arguments in English, with proofs of them

We translate the following twenty five formal arguments into English, using the translation manual

Fx : x is friendly
Gx : x is generous
Lxy : x loves y.

Then we give formal proofs of the formal arguments, using only basic deductive rules. These are the introduction and elimination rules for the logical operators; and the classical rules of Double Negation Elimination, Dilemma, and Classical Reductio. We shall be using the serial forms of elimination for the operators $\wedge$, $\rightarrow$, and $\forall$ that were discussed in §3.7.

The reader should study carefully the steps in the proofs of the following formal arguments, to try to work out for themselves the forms of the inference rules that are being applied. Try to familiarize yourself with the 'rules in action', in order to be better able to appreciate their fully formal and precise statements later (in §7.4), *qua* rules of inference.

This is the first occasion on which the reader will be acquainted with the Introduction and Elimination rules for the existential quantifier $\exists$. So particular attention should be paid to those two 'rules in action' in the proofs below. By studying the proofs involving $\exists$, the reader should be able to form a shrewd idea of how the rules ($\exists$-I) and ($\exists$-E) should be formally stated. That will eventually happen on p. 262 in Chapter 13.

---

1. $\dfrac{\forall x Fx}{\exists x Fx}$ $\quad$ $\dfrac{\text{Everyone is friendly}}{\text{Someone is friendly}}$

$$\dfrac{\dfrac{\forall x Fx}{Fa}}{\exists x Fx}$$

---

2. $\dfrac{\forall x Fx}{\forall x \neg\neg Fx}$ $\quad$ $\dfrac{\text{Everyone is friendly}}{\text{Everyone is not unfriendly}}$

$$\dfrac{\dfrac{\dfrac{{}^{(1)}\overline{\neg Fa} \qquad \dfrac{\forall x Fx}{Fa}}{\bot}}{\neg\neg Fa}{}^{(1)}}{\forall x \neg\neg Fx}$$

---

3. $\dfrac{\exists x \neg Fx}{\neg\forall x Fx}$ $\quad$ $\dfrac{\text{Someone is unfriendly}}{\text{Not everyone is friendly}}$

$$\dfrac{\dfrac{\exists x \neg Fx \qquad \dfrac{{}^{(2)}\overline{\neg Fa} \qquad \dfrac{\overline{\forall x Fx}^{(1)}}{Fa}}{\bot}}{\bot}{}^{(2)}}{\neg\forall x Fx}{}^{(1)}$$

---

4. $\dfrac{\neg\exists xFx}{\forall x\neg Fx}$ $\quad$ $\dfrac{\text{No-one is friendly}}{\text{Everyone is unfriendly}}$

$$\dfrac{\dfrac{\dfrac{\neg\exists xFx \qquad \dfrac{\dfrac{}{Fa}\,(1)}{\exists xFx}}{\bot}}{\neg Fa}\,(1)}{\forall x\neg Fx}$$

---

5. $\dfrac{\forall x\neg Fx}{\neg\exists xFx}$ $\quad$ $\dfrac{\text{Everyone is unfriendly}}{\text{No-one is friendly}}$

$$\dfrac{\dfrac{(1)\dfrac{}{\exists xFx} \qquad \dfrac{\dfrac{\forall x\neg Fx}{\neg Fa} \qquad \dfrac{}{Fa}\,(2)}{\bot}}{\bot}\,(2)}{\neg\exists xFx}\,(1)$$

---

6. $\dfrac{\neg\forall xFx}{\exists x\neg Fx}$ $\quad$ $\dfrac{\text{Not everyone is friendly}}{\text{Someone is unfriendly}}$

$$\dfrac{\dfrac{\neg\forall xFx \qquad \dfrac{\dfrac{\dfrac{(1)\dfrac{}{\neg\exists x\neg Fx} \qquad \dfrac{\dfrac{}{\neg Fa}\,(2)}{\exists x\neg Fx}}{\bot}}{Fa}\,(2)}{\forall xFx}}{\bot}}{\exists x\neg Fx}\,(1)$$

Note that the step labeled (2) is an application of the rule of Classical Reductio, discharging the assumption $\neg Fa$.

---

7. $\dfrac{\forall x\neg\neg Fx}{\forall xFx}$ $\quad$ $\dfrac{\text{Everyone is not unfriendly}}{\text{Everyone is friendly}}$

$$\dfrac{\dfrac{\dfrac{\forall x\neg\neg Fx}{\neg\neg Fa}}{Fa}}{\forall xFx}$$

Note that the second step in this proof is an application of the classical rule of *Double Negation Elimination* (DNE)—to be covered in detail, with other strictly classical rules, in Chapter 8.

---

8. $$\frac{\exists xFx}{\exists x\neg\neg Fx} \qquad \frac{\text{Someone is friendly}}{\text{Someone is not unfriendly}}$$

$$\dfrac{\exists xFx \qquad \dfrac{\dfrac{\dfrac{{}^{(2)}\overline{\neg Fa} \quad \overline{Fa}^{(1)}}{\bot}{}^{(2)}}{\neg\neg Fa}}{\exists x\neg\neg Fx}}{\exists x\neg\neg Fx}{}^{(1)}$$

---

9. $$\frac{\exists x\neg\neg Fx}{\exists xFx} \qquad \frac{\text{Someone is not unfriendly}}{\text{Someone is friendly}}$$

$$\dfrac{\exists x\neg\neg Fx \qquad \dfrac{\dfrac{\overline{\neg\neg Fa}^{(1)}}{Fa}}{\exists xFx}}{\exists xFx}{}^{(1)}$$

Note the use, once again, of DNE.

---

10. $$\frac{\exists x(Fx\wedge Gx)}{\exists xFx} \qquad \frac{\text{Someone is both friendly and generous}}{\text{Someone is friendly}}$$

$$\dfrac{\exists x(Fx\wedge Gx) \qquad \dfrac{\dfrac{\overline{Fa\wedge Ga}^{(1)}}{Fa}}{\exists xFx}}{\exists xFx}{}^{(1)}$$

---

11. $$\frac{\forall x(Fx\wedge Gx)}{\forall xFx} \qquad \frac{\text{Everyone is both friendly and generous}}{\text{Everyone is friendly}} \qquad \dfrac{\dfrac{\dfrac{\forall x(Fx\wedge Gx)}{Fa\wedge Ga}}{Fa}}{\forall xFx}$$

---

12. $\dfrac{\exists x(Fx \vee Gx)}{\exists xFx \vee \exists xGx}$ $\dfrac{\text{Someone is either friendly or generous}}{\text{Either someone is friendly or someone is generous}}$

$$\dfrac{\exists x(Fx \vee Gx) \qquad \dfrac{\overline{Fa \vee Ga}^{(1)} \quad \dfrac{\dfrac{\overline{Fa}^{(2)}}{\exists xFx}}{\exists xFx \vee \exists xGx} \quad \dfrac{\dfrac{\overline{Ga}^{(2)}}{\exists xGx}}{\exists xFx \vee \exists xGx}}{\exists xFx \vee \exists xGx}{\scriptstyle (2)}}{\exists xFx \vee \exists xGx}{\scriptstyle (1)}$$

---

13. $\dfrac{\exists xFx \vee \exists xGx}{\exists x(Fx \vee Gx)}$ $\dfrac{\text{Either someone is friendly or someone is generous}}{\text{Someone is either friendly or generous}}$

$$\dfrac{\exists xFx \vee \exists xGx \qquad \dfrac{\overline{\exists xFx}^{(1)} \quad \dfrac{\dfrac{\overline{Fa}^{(2)}}{Fa \vee Ga}}{\exists x(Fx \vee Gx)}}{\exists x(Fx \vee Gx)}{\scriptstyle (2)} \qquad \dfrac{\overline{\exists xGx}^{(1)} \quad \dfrac{\dfrac{\overline{Ga}^{(3)}}{Fa \vee Ga}}{\exists x(Fx \vee Gx)}}{\exists x(Fx \vee Gx)}{\scriptstyle (3)}}{\exists x(Fx \vee Gx)}{\scriptstyle (1)}$$

---

14. $\dfrac{\forall xFx}{\forall x(Fx \vee Gx)}$ $\dfrac{\text{Everyone is friendly}}{\text{Everyone is either friendly or generous}}$

$$\dfrac{\dfrac{\dfrac{\forall xFx}{Fa}}{Fa \vee Ga}}{\forall x(Fx \vee Gx)}$$

---

15. $\dfrac{\forall xFx \vee \forall xGx}{\forall x(Fx \vee Gx)}$ $\dfrac{\text{Either everyone is friendly or everyone is generous}}{\text{Everyone is either friendly or generous}}$

$$\dfrac{\forall xFx \lor \forall xGx \qquad \dfrac{\dfrac{\dfrac{}{\forall xFx}{\scriptstyle(1)}}{Fa}}{Fa \lor Ga} \qquad \dfrac{\dfrac{\dfrac{}{\forall xGx}{\scriptstyle(1)}}{Ga}}{Fa \lor Ga}}{\dfrac{Fa \lor Ga}{\forall x(Fx \lor Gx)}}{\scriptstyle(1)}$$

---

16. $\dfrac{\forall x(Fx \to Gx)}{\forall x(\neg Gx \to \neg Fx)}$ $\quad$ $\dfrac{\text{Everyone who is friendly is generous}}{\text{Everyone who isn't generous is unfriendly}}$

$$\dfrac{\dfrac{\dfrac{\dfrac{{\scriptstyle(1)}\dfrac{}{\neg Ga} \qquad \dfrac{\dfrac{\forall x(Fx \to Gx)}{Fa \to Ga} \qquad \dfrac{}{Fa}{\scriptstyle(2)}}{Ga}}{\bot}}{\neg Fa}{\scriptstyle(2)}}{\neg Ga \to \neg Fa}{\scriptstyle(1)}}{\forall x(\neg Gx \to \neg Fx)}$$

---

17. $\dfrac{\exists x\forall yLxy}{\forall y\exists xLxy}$ $\quad$ $\dfrac{\text{Someone loves everyone}}{\text{Everyone is loved by someone}}$

$$\dfrac{\exists x\forall yLxy \qquad \dfrac{\dfrac{\dfrac{}{\forall yLay}{\scriptstyle(1)}}{Lab}}{\exists xLxb}}{\dfrac{\exists xLxb}{\forall y\exists xLxy}}{\scriptstyle(1)}$$

---

18. $\dfrac{\forall x\forall yLxy}{\forall xLxx}$ $\quad$ $\dfrac{\text{Everyone loves everyone}}{\text{Everyone loves themselves}}$

$$\dfrac{\dfrac{\dfrac{\forall x\forall yLxy}{\forall yLay}}{Laa}}{\forall xLxx}$$

---

19. $\dfrac{\exists xLxx}{\exists x\exists yLxy}$ $\dfrac{\text{Someone loves themselves}}{\text{Someone loves someone}}$

$$\dfrac{\exists xLxx \qquad \dfrac{\dfrac{\dfrac{}{Laa}{\scriptstyle(1)}}{\exists yLay}}{\exists x\exists yLxy}}{\exists x\exists yLxy}{\scriptstyle(1)}$$

---

20. $\dfrac{\forall x(Fx\rightarrow Gx)}{\forall xFx\rightarrow\forall xGx}$ $\dfrac{\text{Everyone who is friendly is generous}}{\text{Everyone is friendly only if everyone is generous}}$

$$\dfrac{\dfrac{\dfrac{\dfrac{\forall x(Fx\rightarrow Gx)}{Fa\rightarrow Ga} \qquad \dfrac{\dfrac{}{\forall xFx}{\scriptstyle(1)}}{Fa}}{Ga}}{\forall xGx}}{\forall xFx\rightarrow\forall xGx}{\scriptstyle(1)}$$

---

21. $\dfrac{\exists xFx\rightarrow\exists xGx}{\exists x(Fx\rightarrow Gx)}$ $\dfrac{\text{If someone is friendly, then someone is generous}}{\text{Someone is such that s/he is friendly only if s/he is generous}}$

$$\dfrac{\dfrac{\dfrac{\exists xFx\rightarrow\exists xGx \qquad \dfrac{\dfrac{}{Fb}{\scriptstyle(3)}}{\exists xFx}}{\exists xGx} \qquad \dfrac{\dfrac{\dfrac{}{Ga}{\scriptstyle(1)}}{Fa\rightarrow Ga}}{\exists x(Fx\rightarrow Gx)}}{\exists x(Fx\rightarrow Gx)}{\scriptstyle(1)} \qquad \dfrac{\dfrac{\dfrac{{\scriptstyle(3)}\dfrac{}{\neg Fb} \quad \dfrac{}{Fb}{\scriptstyle(2)}}{\bot}}{Fb\rightarrow Gb}{\scriptstyle(2)}}{\exists x(Fx\rightarrow Gx)}}{\exists x(Fx\rightarrow Gx)}{\scriptstyle(3)}$$

Note that in this proof the terminal step, labeled (3), is an application of yet another 'strictly classical' rule, this one called *Dilemma.* In the left-hand immediate subproof the 'positive horn' assumption Fb was used to derive the sought conclusion $\exists x(Fx\rightarrow Gx)$; and in the right-hand immediate subproof the 'negative horn' assumption $\neg Fb$ was used to derive that same conclusion. $\exists x(Fx\rightarrow Gx)$ was then brought down as the main conclusion of the whole proof; and the 'horn assumptions' Fb and $\neg Fb$ were *discharged* at their respective occurrences, as indicated by each of them being overlined with its own 'discharge stroke' labeled (3).

---

22. $\exists x(Fx \wedge \neg Gx)$ / $\neg\forall x(Fx \rightarrow Gx)$ — Someone is friendly but not generous / Not everyone who is friendly is generous

$$\frac{\exists x(Fx \wedge \neg Gx)}{\neg\forall x(Fx \rightarrow Gx)} \qquad \frac{\text{Someone is friendly but not generous}}{\text{Not everyone who is friendly is generous}}$$

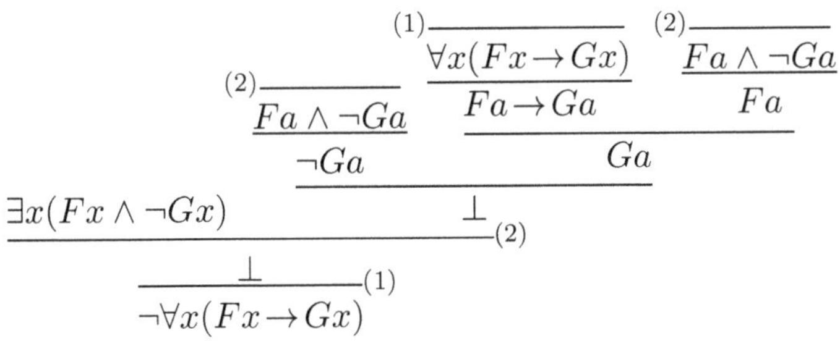

23.

$$\frac{\begin{array}{l}\exists x \neg Fx \\ \forall x(Fx \vee Gx)\end{array}}{\exists x Gx} \qquad \frac{\begin{array}{l}\text{Someone is unfriendly} \\ \text{Everyone is either friendly or generous}\end{array}}{\text{Someone is generous}}$$

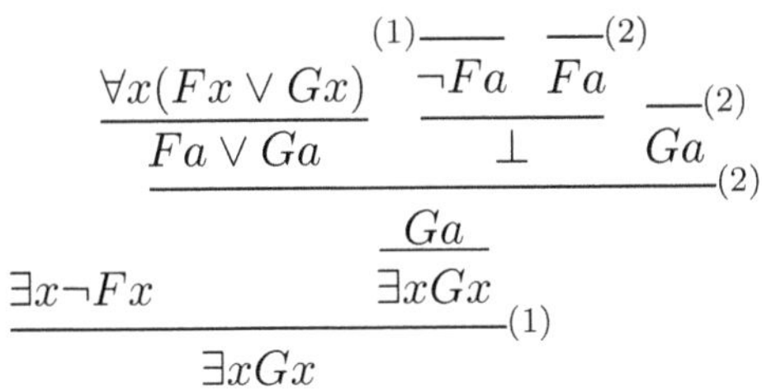

24.

$$\frac{\begin{array}{l}\exists x \neg Gx \\ \forall x(Fx \rightarrow Gx)\end{array}}{\neg\forall x Fx} \qquad \frac{\begin{array}{l}\text{Someone isn't generous} \\ \text{Everyone who is friendly is generous}\end{array}}{\text{Not everyone is friendly}}$$

$$\dfrac{\dfrac{\exists x\neg Gx \qquad \dfrac{{\scriptstyle(2)}\dfrac{}{\neg Ga} \qquad \dfrac{\dfrac{\forall x(Fx\to Gx)}{Fa\to Ga} \qquad \dfrac{\dfrac{}{\forall xFx}{\scriptstyle(1)}}{Fa}}{Ga}}{\bot}}{\bot}{\scriptstyle(2)}}{\neg\forall xFx}{\scriptstyle(1)}$$

25.

$\neg\exists xGx$	No-one is generous
$\forall x(Fx\to Gx)$	Everyone who is friendly is generous
$\forall x\neg Fx$	Everyone is unfriendly

$$\dfrac{\dfrac{\dfrac{\neg\exists xGx \qquad \dfrac{\dfrac{\dfrac{\forall x(Fx\to Gx)}{Fa\to Ga} \qquad \dfrac{}{Fa}{\scriptstyle(1)}}{Ga}}{\exists xGx}}{\bot}}{\neg Fa}{\scriptstyle(1)}}{\forall x\neg Fx}$$

Learning to do logic involves learning how to construct completely formal proofs, like the twenty five foregoing ones. For that to be possible, the argument in question has to be valid. Naturally, the beginning student would not be expected to be able to deal with arguments whose proofs are as long, subtle and complex as those that mathematicians took centuries to discover. Rather, the student is expected only to reach an appreciation of what sorts of things proofs *are*, in principle, and to be able to find proofs of arguments whose validity is reasonably apparent.

At this stage we expect the student to be able to tell by inspection, of each of the foregoing twenty five proofs, what rule of inference is being applied at each step therein. The student is not yet being expected to find such proofs for themselves upon being presented with a valid argument to be proved. The first taste of that will come with Exercise 40 on p. 266.

The Maxim of Shallow Analysis

When trying to prove an argument, it is useful to see whether not all of the logical structure that one *could* expose within its sentences *needs* to be exposed for the purpose of constructing a proof. To take a trivial example, suppose one is asked to translate into logical notation, in order to *prove*, the argument

> Allspice is appropriate despite the fact that it's banana bread we're baking, but if it's banana bread we're baking then we'll need clotted cream.
> *So*, allspice is appropriate.

If one is asked to find a proof of this argument by formalizing it, then there is no need to furnish the premise with a logical form any more detailed than the one used in this proof:

$$\cfrac{\cfrac{(A \wedge B) \wedge D}{A \wedge B}}{A}$$

To assign the premise the logical form $(A \wedge B) \wedge (B \rightarrow C)$ would be overkill—it would be *too deep* an analysis, in the context of proving this particular argument. One can make do with a shallower analysis. Hence the name of the maxim.

Be mindful, though, that if one is required to find a *counterexample* to an *in*valid argument, then one *does* need to expose every possible scrap of logical form in the sentences involved. Some valid arguments can lose their validity if their logical structure is 'coarsened' in any way.

A common misconception

Beginners often confuse a deducibility claim '$P_1, \ldots, P_n \vdash Q$' with a proof of Q from $P_1, \ldots, P_n$. But remember that '$P_1, \ldots, P_n \vdash Q$' says only that

> *there is a proof* of Q from $P_1, \ldots, P_n$.

It does not say what any such proof is, or how to find one. Thus writing down '$P_1, \ldots, P_n \vdash Q$' is not to carry out the instruction

> *Give* a proof of Q from $P_1, \ldots, P_n$

or the instruction

> *Find* a proof of Q from $P_1, \ldots, P_n$.

The only way to comply with such an instruction is to write down a proof. All that one knows from the assurance that $P_1, \ldots, P_n \vdash Q$ is that one can comply with the instruction. But knowing that one can comply is not yet to comply; being told that something is there to be found is not the same as finding it.

3.18 Invalidity and counterexamples

If an argument is *invalid*, then there will be *some* situation (or, in the mathematical context, some *structure*) that would make its premises, but not its conclusion, true. Such a situation is called a *counterexample* to the argument. A counterexample can be an abstract mathematical structure containing infinitely many entities; or it can be a finite arrangement (real or imagined) of concrete objects possessed of various properties and standing in various relations to each other. Counterexamples are thought up with a definite purpose: to expose the invalidity of a given argument.

An argument's invalidity *consists in* the existence of a counterexample to it. Two famous and rather different counterexamples arose in the context of the argument discussed above, whose conclusion Q was Euclid's Parallels Postulate:

> for any line l, and any point p not on l, there is, in the plane containing l and p, exactly one line that contains p and does not intersect l,

and whose premises $P_1, \ldots, P_n$ were the other first principles of Euclidean geometry. Geometers were eventually able to describe structures in which $P_1, \ldots, P_n$ were true, but in which Q was false *for want of any line through the point p and not intersecting the line l*. They were also able to describe structures in which $P_1, \ldots, P_n$ were true, but in which Q was false *because there was more than one line through the point p and not intersecting the line l.*[5] These discoveries gave rise to two different kinds of non-Euclidean geometry.

We said earlier that learning to do logic involves learning how to construct proofs. But that is only half the job. Learning to do logic also involves learning how to find or construct counterexamples that will reveal the *invalidity* of a given argument. What counterexample one provides will depend, of course, on the invalid argument in question. The beginning student would not be expected to be able to find counterexamples of the same subtlety and complexity as those that mathematicians took centuries to discover. Rather, the student is expected only to reach an appreciation of what sorts of things counterexamples *are*, in principle, and to be able to find counterexamples to arguments whose invalidity is reasonably apparent.

What follows is a collection of invalid arguments, for each of which we furnish a very simple counterexample. The reader should check carefully how it is that the premise[s] of each argument is [are] true, while its conclusion is false. Later on we shall advance to slightly more complicated invalid arguments for each of which we shall provide a counterexample—call it M—along with M-relative *verifications* of its premises and an M-relative *falsification* of its conclusion. These M-relative verifications and falsifications—which are two different types of M-relative *evaluations*—are interesting tree-like combinatorial objects that bring out the ways in which a sentence receives (in M) the truth value that it does. The M-relative evaluations spell out reasons why, in the model M in question, the premises are true but the conclusion is false.

We begin with some invalid single-premise arguments in *propositional* logic, for which we provide counterexamples. Recall that propositional logic is the logic of the *connectives*—negation $\neg$, conjunction $\wedge$, disjunction $\vee$, and the conditional $\rightarrow$. Propositional logic does not deal with the quantifiers $\forall$ and $\exists$, and does not treat identity ($=$) either.

First, a propositional argument is stated. Next, a truth-value assignment is specified. Then an inverted-tree calculation is given for the truth of the premise under that assignment, and another one is given for the falsity of the conclusion. Note how in every case the colorations reveal

[5] See footnote 3.

that not every node needs to be evaluated as true or as false on the way to evaluating the whole sentence. Those nodes whose truth value is not needed for the overall calculation are in black. Atoms whose truth values are not used in the calculation are in black; so too are the dominant operators of the subsentences whose truth values need not be calculated.

$$\frac{(A\rightarrow C)\vee(B\rightarrow C)}{(A\vee B)\rightarrow C} \quad \left\{\begin{matrix} A & B & C \\ \downarrow & \downarrow & \downarrow \\ T & F & F \end{matrix}\right\}$$

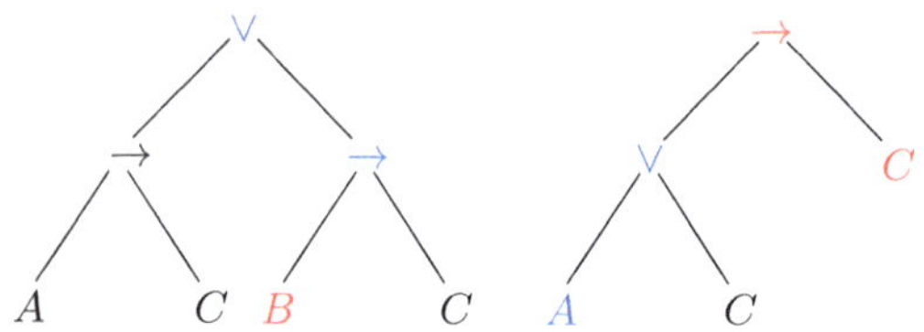

$$\frac{(A\rightarrow B)\vee(A\rightarrow C)}{A\rightarrow(B\wedge C)} \quad \left\{\begin{matrix} A & B & C \\ \downarrow & \downarrow & \downarrow \\ T & F & T \end{matrix}\right\}$$

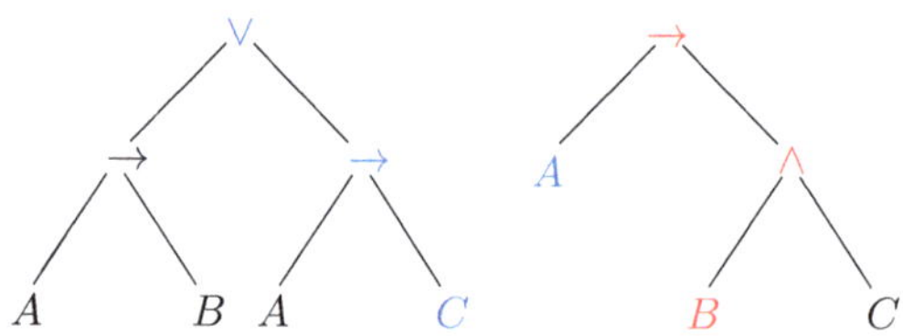

$$\frac{(A\wedge B)\rightarrow C}{(A\rightarrow C)\wedge(B\rightarrow C)} \quad \left\{\begin{matrix} A & B & C \\ \downarrow & \downarrow & \downarrow \\ F & T & F \end{matrix}\right\}$$

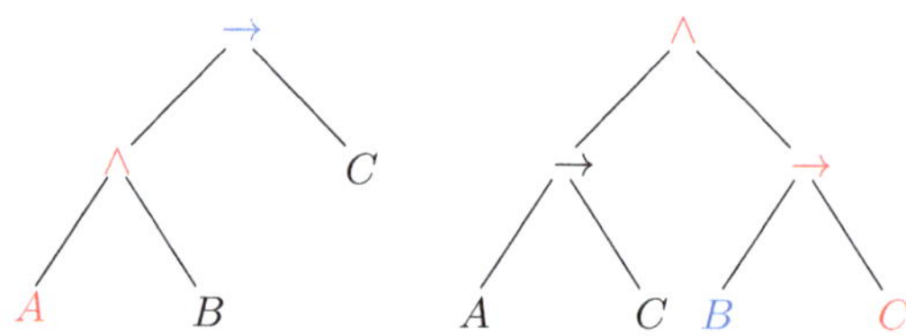

$$\frac{A\rightarrow(B\vee C)}{(A\rightarrow B)\wedge(A\rightarrow C)} \quad \left\{\begin{matrix} A & B & C \\ \downarrow & \downarrow & \downarrow \\ T & F & T \end{matrix}\right\}$$

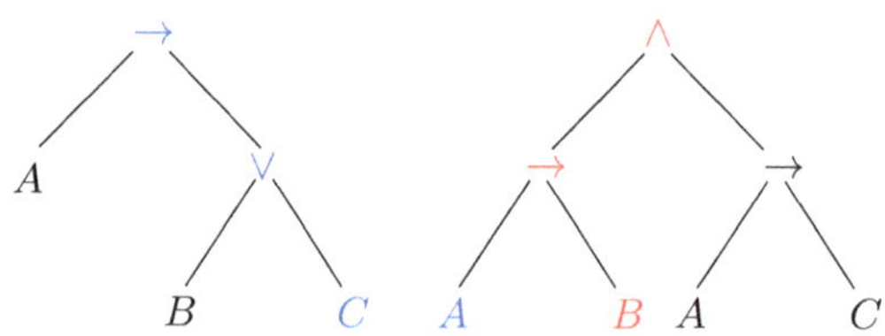

In due course we shall be employing a slightly different formatting of evaluations-as-true and evaluations-as false (in *propositional* logic) under a given assignment of truth values to atoms. But for the time being the foregoing 'inverted tree' calculations accomplish what is required: the *counterexemplification* of the invalid argument in question.

We return now to first-order logic, where interpretations of the language are not mere assignments of truth values to atomic sentences. Rather, first-order interpretations need to assign extensions to predicates (in some domain, no matter how simple or 'underpopulated' it might be). Remember that *just one* counterexample, no matter how small or silly-looking, is enough to show that an argument is invalid.

More diagrammatic counterexamples

For the immediately following collection of two dozen simple (and *invalid*) single-premise arguments in *first-order* logic, model-relative evaluations will not be provided, because the task of working out in each of the models pictorially supplied that the premise is true, but the conclusion is false, is so easy. One can simply 'read off' from the diagram supplied that the premise is true but the conclusion is false. Remember that the dots (or bullets) represent the *individuals* in the domain.

The keen reader may, however, wish to give labels to the dots in the diagrams, and then furnish for each argument a verification of its premise along with a falsification of its conclusion, in the manner illustrated in §3.10. In doing so the reader will have to devise for themselves the rules of verification and of falsification for negations and conditionals. (Or—*spoiler alert*!—they could take a quick look ahead to §6.2.)

$$\frac{\forall x Fx}{\forall x \neg Fx}$$

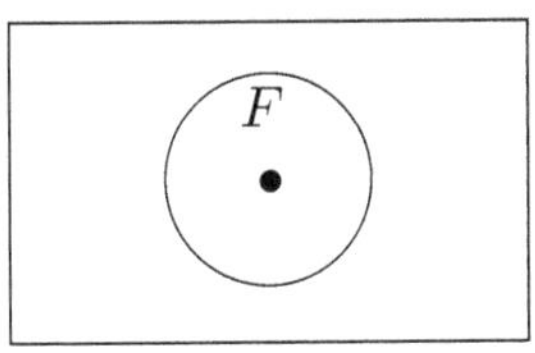

$$\frac{\forall x \neg Fx}{\exists x Fx}$$

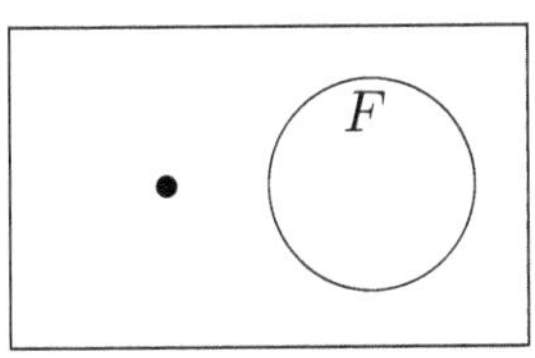

$$\frac{\forall x Fx}{\exists x \neg Fx}$$

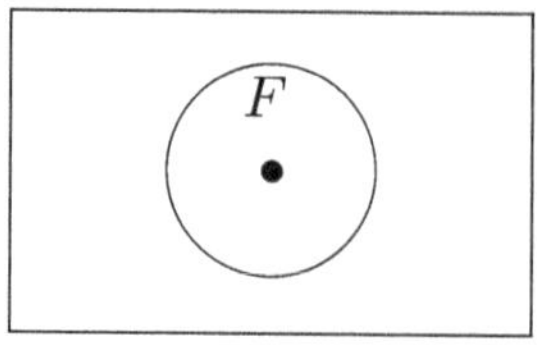

$$\frac{\exists x \neg\neg Fx}{\forall x Fx}$$

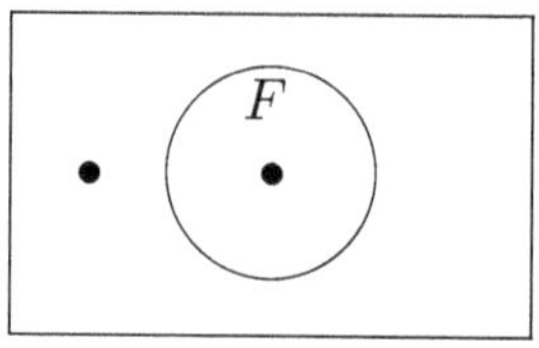

$$\frac{\neg\forall x Fx}{\neg\exists x Fx}$$

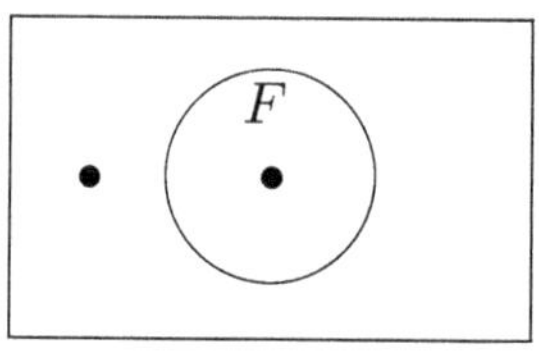

$$\frac{\exists x Fx}{\forall x Fx}$$

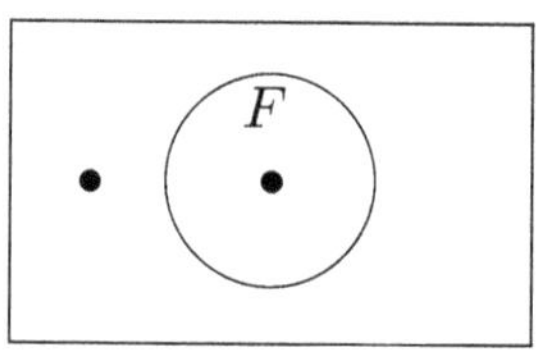

$$\frac{\exists x (Fx \wedge Gx)}{\forall x (Fx \vee Gx)}$$

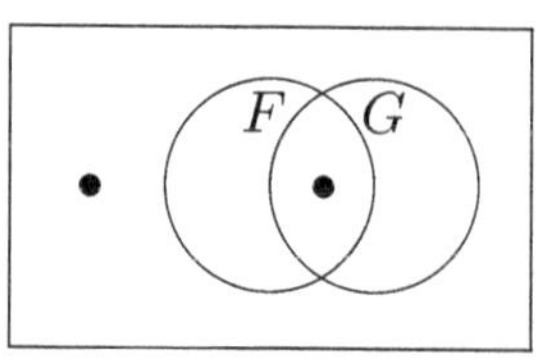

$$\frac{\exists x(Fx \lor Gx)}{\exists x Fx \land \exists x Gx}$$

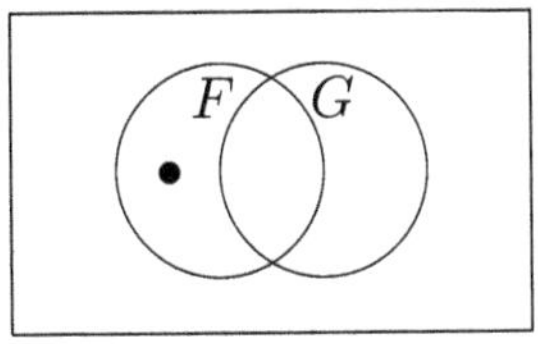

$$\frac{\exists x Fx \land \exists x Gx}{\exists x(Fx \land Gx)}$$

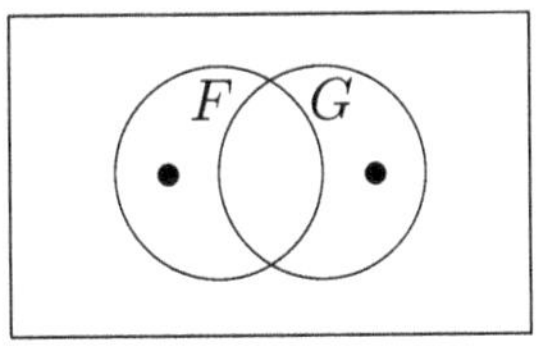

$$\frac{\forall x Fx}{\forall x(Fx \land Gx)}$$

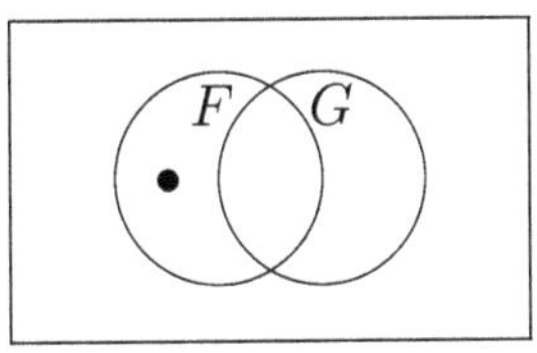

$$\frac{\forall x Fx \to \forall x Gx}{\forall x(Fx \to Gx)}$$

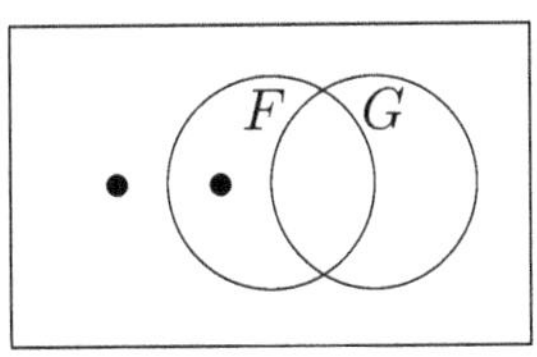

$$\frac{\forall x(Fx \to Gx)}{\forall x(Gx \to Fx)}$$

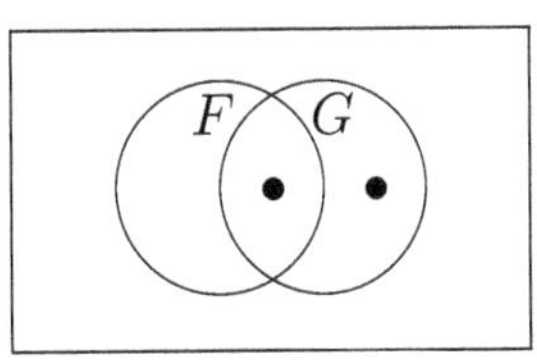

$$\frac{\forall y \exists x Lxy}{\exists x \forall y Lxy}$$

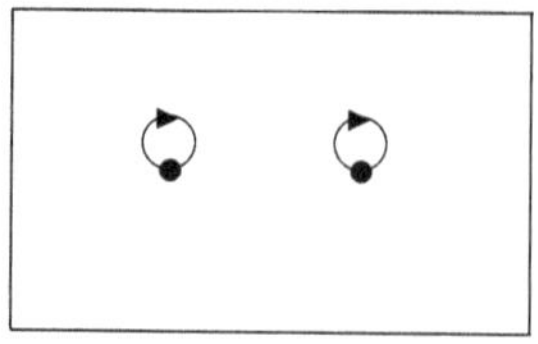

$$\frac{\forall x Lxx}{\forall x \forall y Lxy}$$

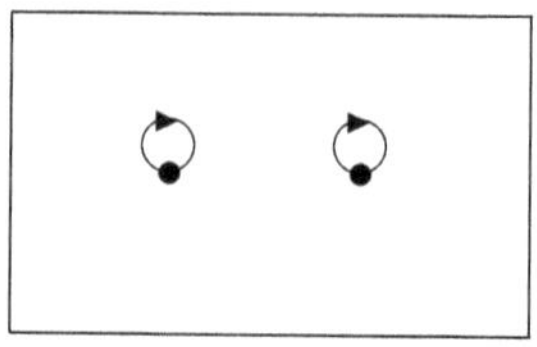

$$\frac{\exists x \exists y Lxy}{\exists x Lxx}$$

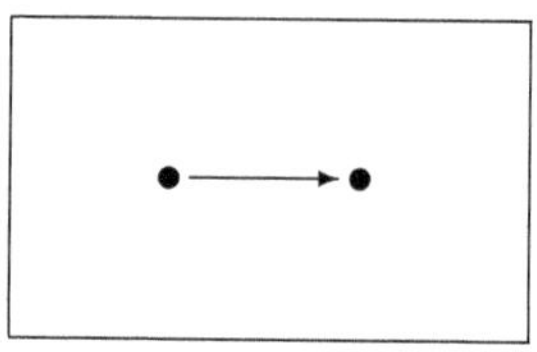

$$\frac{\exists x \neg Fx}{\forall x \neg Fx}$$

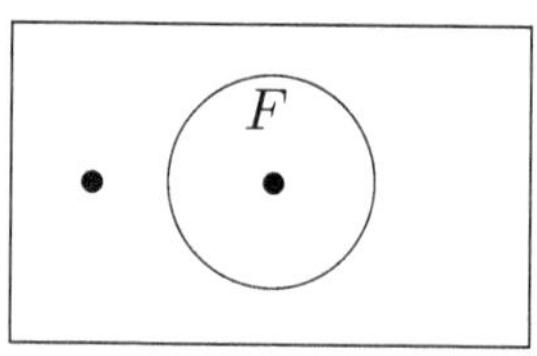

$$\frac{\neg \forall x Fx}{\forall x \neg Fx}$$

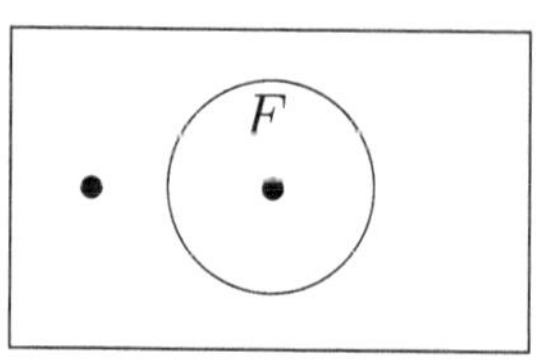

$$\begin{array}{l} \exists x \neg Fx \\ \hline \neg \exists x Fx \end{array}$$

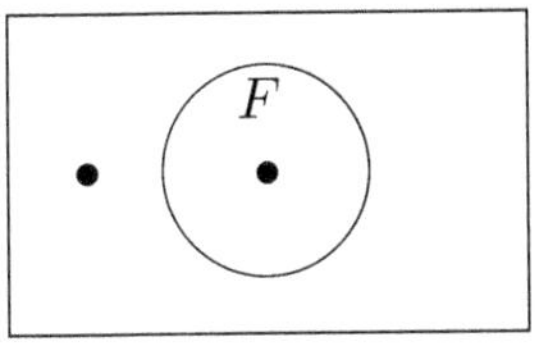

$$\begin{array}{l} \exists x \neg Gx \\ \forall x (Fx \to Gx) \\ \hline \neg \exists x Fx \end{array}$$

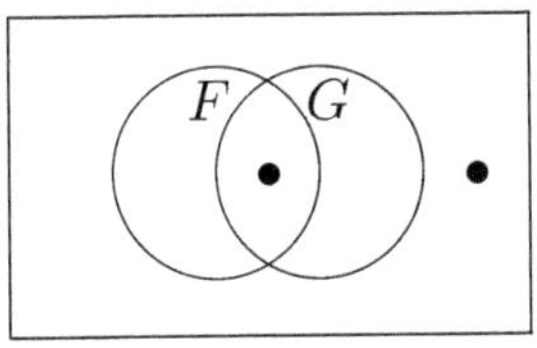

$$\begin{array}{l} \exists x \neg Fx \\ \exists x (Fx \vee Gx) \\ \hline \exists x \neg Gx \end{array}$$

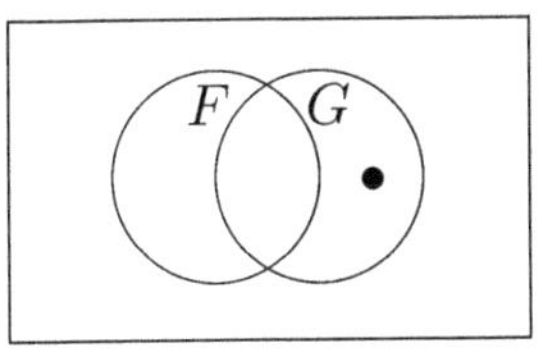

$$\begin{array}{l} \forall x (Fx \vee Gx) \\ \hline \forall x Fx \end{array}$$

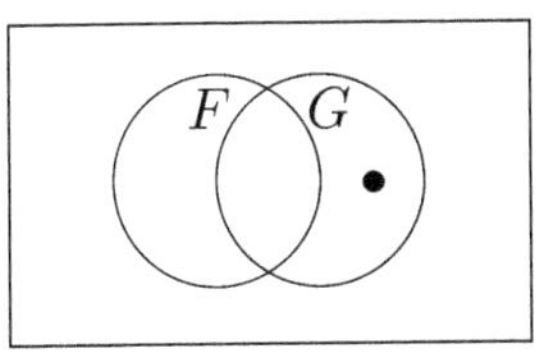

$$\frac{\forall x(Fx \to Gx)}{\neg\forall xFx \to \neg\forall xGx}$$

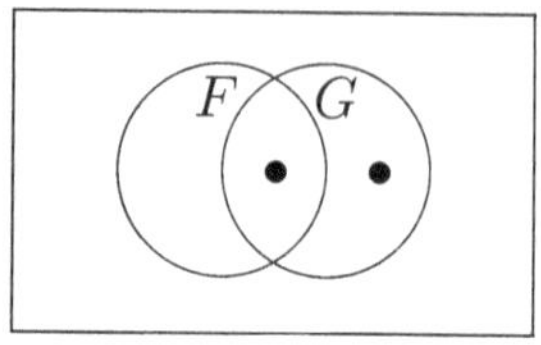

$$\frac{\exists xFx \to \exists xGx}{\forall xFx \to \forall xGx}$$

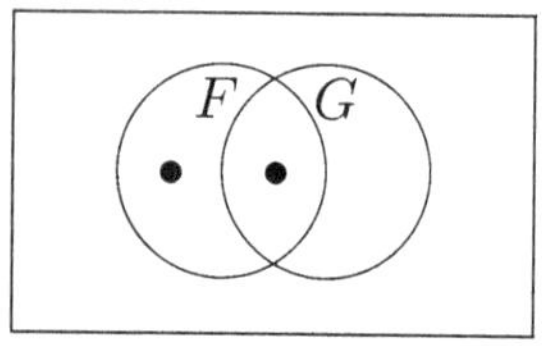

$$\frac{\exists x(Fx \wedge \neg Gx)}{\forall x(Fx \to \neg Gx)}$$

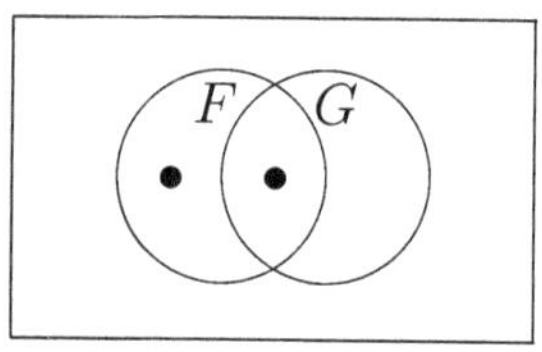

A fully worked-out solution to a counterexample problem, with important pedagogical asides along the way

(i) Translate the following argument into formal first-order notation.

Anyone who loves everyone loves themselves ;
Anyone loved by someone is loved by everyone ;
ergo, Anyone who loves someone loves themselves.

(ii) The argument is invalid. Describe a counterexample to it.

First, we translate the argument into logical notation:

$$\text{(i)}\qquad \frac{\begin{array}{l}\forall x(\forall y Lxy \to Lxx)\\ \forall x(\exists y Lyx \to \forall z Lzx)\end{array}}{\forall x(\exists y Lxy \to Lxx)}$$

Next, we specify a two-element counterexample

(ii) Let the model M contain α and β as its only individuals. Suppose α loves β, and β loves β, and that is all the loving there is. We can picture the situation as follows.

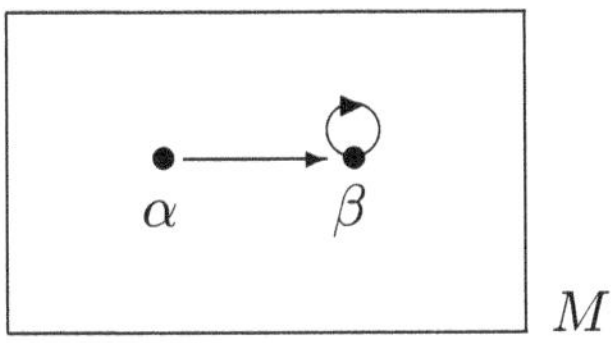

The premises of the argument are true in M, but the conclusion is false in M, as shown by the M-relative evaluations that will follow in due course, by way of a 'fully worked out' solution to this counterexample problem. For each *premise* we shall provide an M-relative *verification*; then we shall provide an M-relative *falsification* for the *conclusion*. The reader is invited to study these slightly more complex M-relative evaluations carefully, to try at this stage to improve further their understanding of 'how model-relative evaluations work'. This will be explained in Chapter 11.

We saw in §3.10 how the label M is used as a subscript for positive and negative 'M-factoids'. These take the respective positive or negative forms (for, say, a monadic predicate P)

$$\frac{}{P(\alpha)}{}^{M} \quad \text{or} \quad \frac{P(\alpha)}{\bot}{}_{M}\ .$$

In the positive case, $P(\alpha)$ 'comes from nowhere' within M. In the negative case, absurdity ($\bot$) *follows immediately* from $P(\alpha)$ within M. For any model M and for any monadic predicate P interpreted by M, and for any individual α in the domain of M, *exactly one* of the foregoing atomic factoids will be part of M. If α *is* in the extension of P in M, then the 'positive' atomic fact (the one on the left) is part of M. But if α is *not* in the extension of P in M, then the 'negative' atomic fact (the one on the right) is part of M. M cannot contain *both* of them. (We do not tolerate truth-value 'gluts'!) This fundamental insistence goes all the way back to Aristotle. It is called the Law of Non-Contradiction. It appears within any sensible logic as the theorem $\neg(\varphi \wedge \neg\varphi)$.

Let us return now to our exposition of the promised M-relative evaluations, which will be appearing below.

The reader has already seen how a label of the form $(@\mathcal{V})$ marks a step that *verifies* its *conclusion* (below the line) in which the operator @ is dominant; while a label of the form $(@\mathcal{F})$ marks a step that *falsifies* the *major premise* (above the line) in which the operator @ is dominant. We emphasize once again that it is *very* important to realize that these 'steps' in evaluations are applications of rules of *verification* or *falsification*; they are *not* to be thought of as applications of *deductive* rules of inference, no matter how beguilingly similar they might appear to be. Verifications and falsifications are *model-relative*; deductions (proofs) are *not*. The latter are *model-invariant*. This difference will come to the fore in due course, when we explain how the evaluation rules (that is, rules of verification or falsification) *give rise to* rules of deduction (rules of introduction or elimination). That will be the major topic of Chapter 7.

Note that our particular counterexemplifing model M here (pictured back on p. 101) has only two individuals—α and β—in its domain. Therefore any step of 'universal verification' $(\forall\mathcal{V})$ for a conclusion of the form $\forall x\psi(x)$ has to involve, 'above the line', a verification of the α-instance $\psi(\alpha)$ *and* a verification of the β-instance $\psi(\beta)$.[6] The 'atomic

[6]Likewise, any step of 'existential falsification' $(\exists\mathcal{F})$ of a major premise of the form $\exists x\psi(x)$ would have to involve, 'above the line', a falsification of the α-instance $\psi(\alpha)$ *and* a falsification of the β-instance $\psi(\beta)$. (The example at hand, however, does not happen to call for any steps of existential falsification.) We underscore the M-relativity of these two evaluation rules for the quantifiers by adding the label 'M' to the label of the rule being applied. In the examples at hand, this will happen at the final step of the verification of each of the two premises, which are applications of the rule $\forall\mathcal{V}$.

facts' (or *factoids*) of this model M are the following:

$$\dfrac{L\alpha\alpha}{\bot}{\scriptstyle M} \qquad \dfrac{}{L\alpha\beta}{\scriptstyle M} \qquad \dfrac{L\beta\alpha}{\bot}{\scriptstyle M} \qquad \dfrac{}{L\beta\beta}{\scriptstyle M}$$

The first premise of the argument, $\forall x(\forall yLxy \to Lxx)$, is true in M. Here is an M-relative verification of that premise:

$$\dfrac{\dfrac{\dfrac{\overset{(1)}{\overline{\forall yL\alpha y}} \qquad \dfrac{\dfrac{}{L\alpha\alpha}{\scriptstyle (2)}}{\bot}{\scriptstyle M}}{\bot}{\scriptstyle (2)\,(\forall\mathcal{F})}}{\forall yL\alpha y \to L\alpha\alpha}{\scriptstyle (1)\,(\to\mathcal{V})} \qquad \dfrac{\dfrac{}{L\beta\beta}{\scriptstyle M}}{\forall yL\beta y \to L\beta\beta}{\scriptstyle (\to\mathcal{V})}}{\forall x(\forall yLxy \to Lxx)}{\scriptstyle (\forall\mathcal{V})\,M}$$

Likewise, the second premise of the argument, $\forall x(\exists yLyx \to \forall zLzx)$, is true in M. Here is an M-relative evaluation of *it*:

$$\dfrac{\dfrac{\dfrac{\overset{(1)}{\overline{\exists yLy\alpha}} \qquad \dfrac{\dfrac{}{L\alpha\alpha}{\scriptstyle (2)}}{\bot}{\scriptstyle M} \qquad \dfrac{\dfrac{}{L\beta\alpha}{\scriptstyle (2)}}{\bot}{\scriptstyle M}}{\bot}{\scriptstyle (2)\,(\exists\mathcal{F})}}{\exists yLy\alpha \to \forall zLz\alpha}{\scriptstyle (1)\,(\to\mathcal{V})} \qquad \dfrac{\dfrac{\dfrac{}{L\alpha\beta}{\scriptstyle M} \qquad \dfrac{}{L\beta\beta}{\scriptstyle M}}{\forall zLz\beta}{\scriptstyle (\forall\mathcal{V})}}{\exists yLy\beta \to \forall zLz\beta}{\scriptstyle (3)\,(\to\mathcal{V})}}{\forall x(\exists yLyx \to \forall zLzx)}{\scriptstyle (\forall\mathcal{V})\,M}$$

But the conclusion of the argument, $\forall x(\exists yLxy \to Lxx)$, is false in M:

$$\dfrac{\forall x(\exists yLxy \to Lxx) \qquad \dfrac{\overset{(1)}{\overline{\exists yL\alpha y \to L\alpha\alpha}} \qquad \dfrac{\dfrac{}{L\alpha\beta}{\scriptstyle M}}{\exists yL\alpha y} \qquad \dfrac{\dfrac{}{L\alpha\alpha}{\scriptstyle (2)}}{\bot}{\scriptstyle M}}{\bot}{\scriptstyle (2)\,(\to\mathcal{F})}}{\bot}{\scriptstyle (1)\,(\forall\mathcal{F})}$$

Note how the sentence being falsified stands as the premise of the final step of falsification, dealing with its dominant operator. Such a premise always 'stands proud', with no evaluation-work above it. This is because of the inherent design of falsifications (as opposed to verifications). By contrast, sentences that are *verified* always stand as conclusions of a final step of verification, and therefore (when they are complex) always have evaluation-work above them.

Exercise 9. The reader may at this point find some benefit in revisiting each of the two dozen examples of invalid arguments given earlier (pp. 95-100), to see whether they can fashion for their premises and conclusions dispositive M-relative verifications and falsifications, respectively, with respect to the model M that was depicted beside each of them.

Exercise 10. Here is a 1-element counterexample to this invalid argument.

$$\begin{array}{l} \neg\exists x\forall y Lxy \\ \neg\exists x\forall y Lyx \\ \hline \forall x\exists y(Lxy \wedge Lyx) \end{array}$$

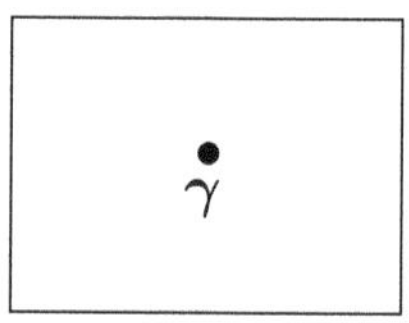

M

Construct the dispositive M-relative verifications of premises and falsification of the conclusion, with respect to the model M depicted.

Exercise 11. Each of the next two invalid arguments has a 2-element counterexample, as displayed.

$$\begin{array}{l} \neg\exists x\forall y Lxy \\ \neg\exists x\forall y Lyx \\ \exists x\exists y Lxy \\ \hline \forall x\exists y(Lxy \wedge Lyx) \end{array}$$

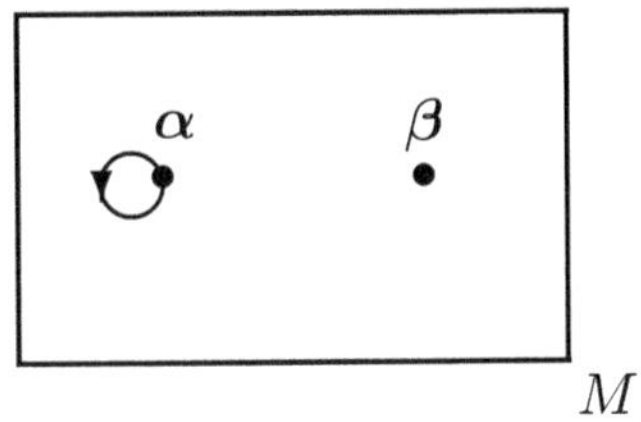

$$\begin{array}{l} \neg\exists x\forall y Lxy \\ \neg\exists x\forall y Lyx \\ \exists x\exists y Lxy \\ \forall x(\exists y Lxy \vee \exists y Lyx) \\ \forall x(Lxx \to \exists y(\neg y{=}x \wedge Lxy)) \\ \forall x(Lxx \to \exists y(\neg y{=}x \wedge Lyx)) \\ \hline \forall x\exists y(Lxy \wedge Lyx) \end{array}$$

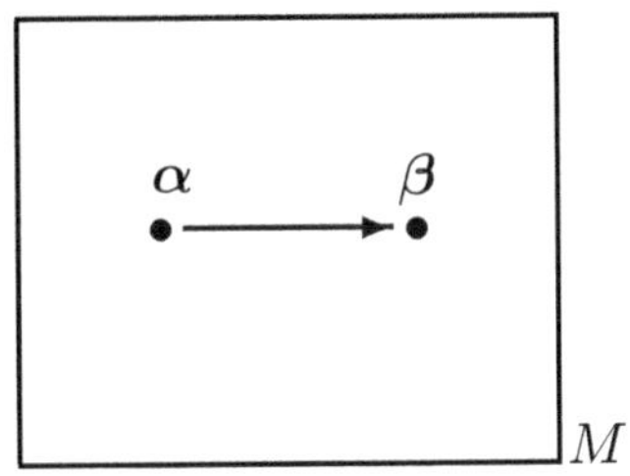

In each case, construct the dispositive M-relative verifications of premises and falsification of the conclusion, with respect to the model M depicted.

Exercise 12. *Show by means of appropriate M-relative evaluations that the 4-element model M depicted below is a countermodel to the nine-premise argument beside it.*

$$\begin{array}{l} \neg\exists x\forall y Lxy \\ \neg\exists x\forall y Lyx \\ \exists x\exists y Lxy \\ \forall x(\exists y Lxy \vee \exists y Lyx) \\ \forall x(Lxx \to \exists y(\neg y{=}x \wedge Lxy)) \\ \forall x(Lxx \to \exists y(\neg y{=}x \wedge Lyx)) \\ \forall x(\exists y Lxy \to Lxx) \\ \forall x Lxx \to \forall x\exists y(\neg y{=}x \wedge Lxy) \\ \forall x Lxx \to \forall x\exists y(\neg y{=}x \wedge Lyx) \\ \hline \forall x\exists y(Lxy \wedge Lyx) \end{array}$$

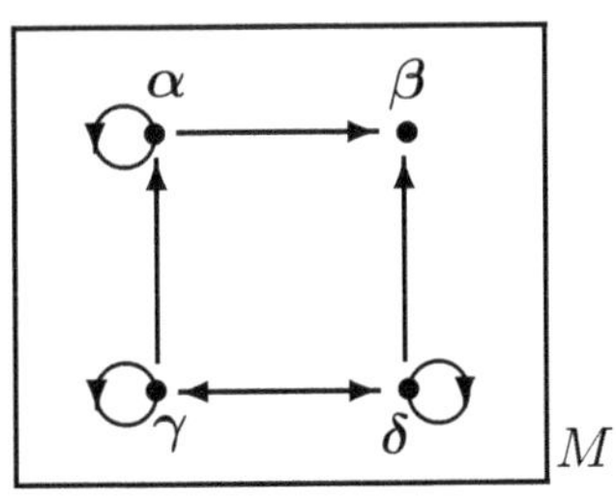

Chapter 4

The Formal Language of First-Order Logic

Now that we have explained the origin and motivation for the variable-binding operators $\exists$ and $\forall$—which form sentences (or formulae) from formulae containing free variables—it is time to give a precise, rigorous definition of *well-formedness*, or grammaticality, of all the complex expressions of a formal language for logic. Our language of choice is that of *first-order logic with identity.* At first order, one can generalize about, or have one's variables 'ranging over', only *individuals.* We do *not* have any 'higher-order' variables ranging over properties of, or relations among, individuals, or ranging over functions that map individuals to individuals. To put it starkly, once again: at first order we can say 'John loves everyone but Mary'; but we cannot say 'John is everything but a lover to Mary'.

In this book we *do*, however, provide for variable-binding, *term-forming* operators (*vbto*s). These are different from the quantifiers $\exists$ and $\forall$, in that they bind variables in formulae to produce *singular terms*, not formulae. They are often neglected in introductory texts in logic. And this is a great pity—because the most fundamental aim in Logic, we shall emphasize once again, is to provide proper means of fully formal regimentation of the language of *mathematics.* In mathematics, there are important variable-binding term-forming operators that are treated as grammatically primitive. Among them (in logician's English) are the following three: the definite-description operator 'the x such that $F(x)$'; the number-abstraction operator 'the number of x such that $F(x)$'; and the set-abstraction operator 'the set of all and only those x such that $F(x)$'. In our completely formal language singular terms that are formed by using these operators will be rendered, respectively, as $\iota x F(x)$; $\# x F(x)$; and $\{x|F(x)\}$.

Here now is a list of all the primitive expressions from which we shall be building up more complex expressions (both *terms* and *formulae*). The subsequent definitions of the notions *term* and *formula* provide the rules of grammatical combination of expressions in general, to form more complex expressions.

Logical symbol	**Logical operation/relation**	**English correlate**
$\neg$	Negation	it is not the case that ...
$\vee$	Disjunction	either ... or ... (or both)
$\wedge$	Conjunction	... and ...
$\rightarrow$	Implication	... only if .../ if ... then ...
$\exists x$	Existential quantification	there exists x such that ... x...
$\forall x$	Universal quantification	for every x ... x...
$=$	Identity	... is identical to ...
$\# x(\ldots x \ldots)$	Number abstraction	the number of x such that ... x...
$\iota x(\ldots x \ldots)$	Definite description	the x such that ... x...
$\{x \mid \ldots x \ldots\}$	Set abstraction	the set of all x such that ... x...
$\in$	Membership	... is a member of ...

The *extra-logical* vocabulary can contain any names, and any function-symbols and predicates of any finite adicities or arities. Adicities (or arities) are the number of singular terms that are 'taken' by a predicate to form a sentence, or by a function sign to form a functional term. Thus 1 is the arity of the one-place predicate $F(\ )$, and of the one-place function sign $f(\ )$; 2 is the arity of the two-place predicate $L(\ ,\ )$, and of the two-place function sign $+$ for addition; and so on. We shall write f for a generic n-place function-symbol, and write P for a generic n-place predicate.

At the outset we need to stress that the following sets of vocabulary items are pairwise disjoint. This is in order to ensure that there can be no structural ambiguities in complex formal expressions.

- The set of names
- The set of parameters
- The set of individual variables
- The set of primitive function signs
- The set of primitive predicates
- The set of connectives: $\{\neg, \wedge, \vee, \rightarrow\}$
- The set of quantifiers: $\{\exists, \forall\}$
- The set of variable-binding term-forming operators: $\{\iota, \#, \{\ |\ \}\}$
- The set of parentheses $\{(,)\}$

4.1 Terms and formulae

We shall use the abbreviation 'f.o.v.' for 'free occurrence of a variable'. The plural form 'f.o.v.s' will abbreviate 'free occurrences of variables'.

We are now going to define the notions of 'term' and 'formula' in one long comprehensive statement, consisting of basis clauses, inductive (or formation) clauses, and closure clauses. In the following co-inductive definitions of these notions, the so-called *basis clauses* tell one how to obtain the simplest objects of the kinds in question. The so-called *closure clauses* are intended to ensure that the kinds in question contain only such objects as can be constructed in accordance with the other clauses. Clauses other than basis and closure clauses are called *formation clauses*, or *inductive clauses*. These describe how expression-forming operators can be applied to components assumed given ('already formed'), in order to form a new compound.

In the definitions of term and of formula we emphasize in boldface the appeal made by some clauses to formulae and terms, respectively. Definitions like these (here, of *term* and of *formula*) are said to be *co-inductive*. The two definitions *intercalate*, since the application of either one of the two definitions to license a construct as properly formed can involve appeal to provisions in the other definition. Thus, the closure clause in the definition of term can make indirect appeal to the clauses in the definition of formula; and *vice versa.* The mutual appeals eventually 'wind down' to basic cases, so that no conceptual (or ontological) circularity is involved.

Definition 1 (Terms: basic, and functional).

1. (Basis clause) *Any name is a term, with no f.o.v.s*

2. (Basis clause) *Any parameter is a term, with no f.o.v.s*

3. (Basis clause) *Any variable is a term, with just itself as a f.o.v.*

4. *If f is an n-place function-symbol, and $t_1, \ldots t_n$ are terms, then $f(t_1, \ldots, t_n)$ is a term, whose f.o.v.s are those of $t_1, \ldots, t_n$.*

(Terms: clauses for variable-binding term-forming operators)

5. *If φ is a* **formula** *with x free, then $\#x\varphi$ is a term, whose f.o.v.s are those in φ, save of x, whose free occurrences in φ are now bound.*

6. *If φ is a* **formula** *with x free, then $\iota x\varphi$ is a term, whose f.o.v.s are those in φ, save of x, whose free occurrences in φ are now bound.*

7. *If φ is a* **formula** *with x free, then $\{x|\varphi\}$ is a term, whose f.o.v.s are those in φ, save of x, whose free occurrences in φ are now bound.*

8. (Closure clause) *Every term can be shown to be a term (with a specification of its f.o.v.s) by appeal to the foregoing (and following) clauses.*

Definition 2 (Formulae: atomic, and connected).

1. (Basis clause) *If P is an n-place predicate, and $t_1, \ldots t_n$ are* **terms**, *then $P(t_1, \ldots, t_n)$ is a(n atomic) formula, whose f.o.v.s are those of $t_1, \ldots, t_n$.*

2. *If φ is a formula, then so is $\neg\varphi$, with the same f.o.v.s as φ.*

3. *If φ and ψ are formulae, then so is $(\varphi \vee \psi)$, whose f.o.v.s are those in φ and those in ψ.*

4. *If φ and ψ are formulae, then so is $(\varphi \wedge \psi)$, whose f.o.v.s are those in φ and those in ψ.*

5. *If φ and ψ are formulae, then so is $(\varphi \rightarrow \psi)$, whose f.o.v.s are those in φ and those in ψ.*

(Formulae: quantified)

6. *If φ is a formula with x free, then $\exists x\varphi$ is a formula whose f.o.v.s are those in φ, save of x.*

7. *If φ is a formula with x free, then $\forall x\varphi$ is a formula whose f.o.v.s are those in φ, save of x.*

8. (Closure clause) *Every formula can be shown to be a formula (with a specification of its f.o.v.s) by appeal to the foregoing clauses.*

Definition 3 (Closed Terms, and Sentences).

1. *A* closed term *is a term with no free occurrences of variables.*

2. *A* sentence *is a formula with no free occurrences of variables.*

Definition 4 (Substitutions of terms).

1. *Let φ be a formula with x free, and let t be a term whose only free variable—if it has one—is x. Then φ^x_t is the formula that results from φ by replacing each free occurrence of x in φ with an occurrence of t. The term t is called the* substituend *used in the substitution.*

2. *Let φ be a formula in which the parameter a occurs. Let x be a variable such that no occurrence of a in φ lies within the scope of an occurrence of an operator binding the variable x ($\exists x$, $\forall x$, ιx, $\#x$, or $\{x| \ldots x \ldots\}$). Then φ^a_x is the formula that results from φ by replacing each occurrence of the parameter a in φ with an occurrence of the variable x. Here, the variable x is the substituend.*

3. *Let* φ *be a formula in which the closed term* t *occurs. Let* u *be a closed term. Then* φ^t_u *is the formula that results from* φ *by replacing each occurrence of the term* t *by an occurrence of the term* u*. Here, the term* u *is the substituend.*

We now define the interdependent notions of a positive v. a negative occurrence of a subformula within a formula.

Definition 5 (Positive and negative occurrences). *Any formula occurs positively within itself.* ψ *occurs negatively both in* $\neg\psi$ *and in* $\psi \rightarrow \theta$*.* ψ *and* θ *occur positively in* $\psi \wedge \theta$ *and in* $\psi \vee \theta$*.* θ *occurs positively in* $\psi \rightarrow \theta$*.* ψ *occurs positively both in* $\forall x\psi$ *and in* $\exists x\psi$*. If* χ *is a non-immediate subformula of* θ*, then it is easy to determine, of any of* χ*'s such occurrences in* θ *whether it is a positive, or negative, occurrence. Trace upwards in the analysis tree of* θ *from the occurrence of* χ *in question. Count how many times the branch passes through an occurrence of* $\neg$ *or (from the left) an occurrence of* $\rightarrow$*. If that number is even, the occurrence of* χ *is positive within* θ*; but if that number is odd, the occurrence is negative.*

Positive and negative occurrences of formulae will play an important role in our discussion of the *relevance* of premises of proofs to their conclusions. See §24.2. They will also find mention in our recipe for constructing proofs, in §13.4.

Exercise 13. *Consider all the formal sentences in Exercise 5 above and Exercise 19 below. Determine, of each atomic formula occurrence within each of those sentences, whether it is a positive or negative occurrence therein.*

Certain atomic formulae are of special interest: those of the form

$$P(x_1, \ldots x_n)$$

or of the form

$$x = f(x_1, \ldots x_n),$$

where $x, x_1, \ldots x_n$ are individual variables (and of course are free in the formulae in question). We shall call these *primitive* formulae.[1] They

[1] Another term is 'flat' formulae. This term is sometimes used in the literature on computational logic, but will not be used here.

are of interest because they can be used to state the 'simplest possible' predicative and functional facts within an interpretation.

Free occurrences of x in φ are said to be *bound* in $\forall x\varphi$ and in $\exists x\varphi$ by the respective initial quantifier occurrences. Likewise, free occurrences of x in φ are said to be *bound* in $\iota x\varphi$, in $\#x\varphi$, and in $\{x|\varphi\}$ by the respective occurrences of those *vbto*s.

Any student with a background in mathematics will notice that there is a much-used connective in mathematical discourse that is conspicuously absent from our official list of primitive connectives. It is the biconditional, the 'if and only if' that is so often abbreviated in mathematicians' parlance as 'iff'. Logicians use the two-way arrow $\leftrightarrow$ for this connective. They call it the *biconditional.* But they usually omit it from their official list of primitives because it is so easily and obviously definable in terms of $\wedge$ and $\rightarrow$, thus:

$$\varphi \leftrightarrow \psi \quad \equiv_{df} \quad (\varphi \rightarrow \psi) \wedge (\psi \rightarrow \varphi).$$

4.2 Binding variables

The formal sentence $\exists x \forall y\ x{=}y$ is formed as follows.

> First, form the atomic formula $x = y$, in which both the occurrence of x and the occurrence of y are free.
>
> Next, form the universal quantification $\forall y\ x{=}y$, in which y is now bound by the universal quantifier, but the occurrence of x is still free.
>
> Finally, form the existential quantification $\exists x \forall y\ x = y$, in which x is now bound by the existential quantifier, so that no occurrences of variables remain free. This is now a sentence.

Alphabetic choice of bound variable is irrelevant

Consider the two formal sentences

$$\exists x \exists y \neg x{=}y$$

and

$$\exists z \exists w \neg z{=}w.$$

They are distinct, since they involve different alphabetic choices of variables. Yet in an obvious sense they are 'really the same' claim—to the effect that there are at least two things (something is distinct from something else). When we view these two formal sentences as (not just logically equivalent, but) the 'very same' sentence in some sense, we are abstracting to a level at which the actual alphabetic choice of bound variables is irrelevant, and all that matters is the *pattern of binding* that the repeated variable-occurrences establish. The common pattern in this case is

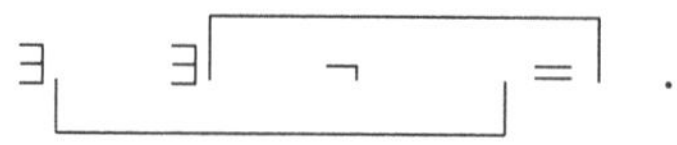

The identity predicate is a two-place predicate, bringing with it a *left argument-place* and a *right argument-place* into which one can substitute a term or one can quantify. Here, both argument-places are 'quantified into'. The first (i.e. leftmost) existential quantifier quantifies into the left argument-place (as indicated by the lower link), and the second existential quantifier quantifies into the right argument-place (as indicated by the upper link). *That* is really what the underlying logical form involves: a pattern of *linkages* between quantifier-occurrences and the argument-places that they quantify into.

The sole function of the bound-variable notation, we stress once more, is to *indicate the pattern of linkages established by quantifications.* In order to do this, it suffices to use, for each linkage, occurrences of some same easily recognizable emblem in order to mark its ends. From those emblem-occurrences the pattern of linkages can be uniquely recovered.

Thus, for example, one could place copies of convenient emblems at the ends of the links:

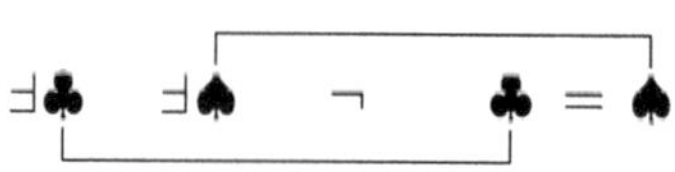

and then drop the links themselves:

$$\exists\clubsuit \quad \exists\spadesuit \quad \neg \quad \clubsuit = \spadesuit$$

in the knowledge that the links can be restored, whenever needed, *because we know where they begin and end.* But now of course we have restored linearity to our notation. The formal sentence, squishing symbols closer together, is

$$\exists\clubsuit\,\exists\spadesuit\,\neg\clubsuit = \spadesuit.$$

Logicians being prosaic, and printers and publishers penny-pinching, and LaTeX not yet having been invented, the eye-catching emblems were dutifully replaced by occurrences of letters straggling at the end of the alphabet (the ones we have been calling *variables*). Thus we learned to make do with formal sentences like

$$\exists x \exists y \neg x = y$$

—and with alphabetical variants such as

$$\exists z \exists w \neg z = w,$$

according to the formula-constructor's momentary whims.

4.3 Analysis trees

Every formula (hence, every sentence) has a unique *pedigree of construction* out of primitive expressions, both extra-logical and logical. The pedigree reflects the (partial) order in which one builds up the formula according to the formation rules just stated. It can be made very clear, in the case of any formula, by providing a two-dimensional *analysis tree.* Let us illustrate by means of some simple exercises.

Three versions of linear logical notation, illustrated with Ekman's problem

Ekman's problem was thought up in connection with proof-finding, which is not our topic of concern at present. Rather, we wish to use the sentence involved as an example of a formal sentence which (like all formal sentences) can be re-written in different ways, using particular formal grammars, while still representing a unique logical form.

In parentheses-plus-infix notation the sentence involved in Ekman's problem is

$$((p \to (p \to q)) \land ((p \to q) \to p)) \to q.$$

In Polish notation (look ahead to §4.7) the sentence becomes:

$$\to \land \to p \to pq \to \to pqpq.$$

And as a Prolog term it would be rendered thus:

if(and(if(p,if(p,q)),if(if(p,q),p)),q).

Note how the Prolog term has a surfeit of parentheses, even though it is 'really' in Polish notation!

The two-dimensional 'tree' form of logical notation, illustrated with Ekman's problem

Exercise 14. Give a tree-analysis of the logical form of the sentence involved in Ekman's problem.

Solution:

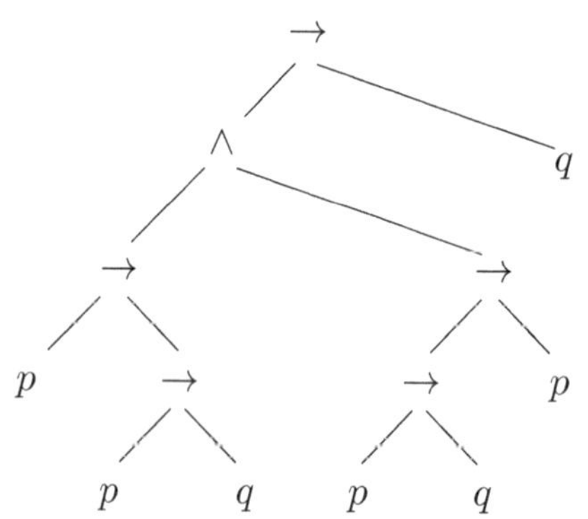

Note that the leaf nodes are labeled by descriptive primitive expressions (the ps and qs), while the branching nodes are labeled by primitive logical expressions. This is true of analysis trees for grammatically complex expressions of all descriptive languages, including languages with quantifiers. The only exception is when a leaf node is labeled with a variable (free or bound).

Exercise 15. Consider the formula $\exists x \exists y(\neg x = y \wedge \forall z(z = x \vee z = y))$. What does it say? How is it constructed?

Solution: $\exists x \exists y(\neg x = y \wedge \forall z(z = x \vee z = y))$ says that there are exactly two things. (There are two distinct things such that everything is identical to one or the other of them.) The formal sentence has the *analysis tree*

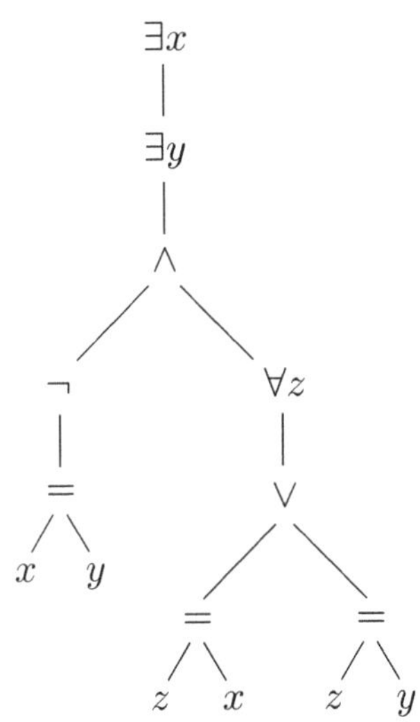

Note how the leaf nodes are labeled with variables, which are bound by quantifiers at nodes above them. All the other nodes—both monofurcating and bifurcating—are labeled with logical expressions.

If instead of using variables we were to use links from quantifier occurrences to argument-places of predicates (and in this sentences the only predicate is the binary predicate of identity), then the analysis tree would become

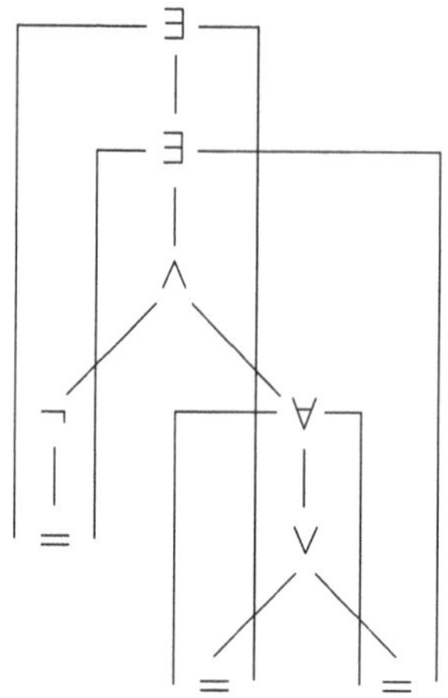

This formal sentence is in what is known as *the pure first-order language of identity*. This is an important fragment of the formal language of first-order logic. Let us now investigate it further, to see what it can be used to *say*..

4.4 The pure first-order language of identity

The pure first-order language of identity is a very simple language, and provides an ideal setting in which to introduce and illustrate syntactic concepts that have wider application to other, more powerfully expressive first-order languages. In this language, *the only predicate is the (logical) identity predicate; there are no names; and there are no function signs.* So the language is devoid of any extra-logical, or 'descriptive', vocabulary.

Numerosity claims

Note that the pure first-order language of identity does not allow one to *refer to* any particular thing. The language contains no singular terms (such as '0', '∅', 'John', 'Mary's father', '$\sqrt{2}$', 'the number of Jupiter's moons', 'the tallest spy', etc.). The language permits one only to generalize about things (existentially and universally). Yet despite that limitation it succeeds in formulating some important claims. In the pure first-order language of identity, one can express thoughts about the (finite) *numerosity* of things. For example:

$\forall x_1 \forall x_2\, x_1 = x_2$	There is at most one thing
$\forall x_1 \forall x_2 \forall x_3 (x_1 = x_2 \vee x_1 = x_3 \vee x_2 = x_3)$	There are at most two things
$\forall x_1 \forall x_2 \forall x_3 \forall x_4 (x_1 = x_2 \vee x_1 = x_3 \vee x_1 = x_4 \vee x_2 = x_3 \vee x_2 = x_4 \vee x_3 = x_4)$	There are at most three things
$\vdots$	$\vdots$
$\exists x_1\, x_1 = x_1$	There is at least one thing
$\exists x_1 \exists x_2\, \neg x_1 = x_2$	There are at least two things
$\exists x_1 \exists x_2 \exists x_3 (\neg x_1 = x_2 \wedge \neg x_1 = x_3 \wedge \neg x_2 = x_3)$	There are at least three things
$\vdots$	$\vdots$
$\neg \exists x_1\, x_1 = x_1$	Nothing exists
$\exists x_1 \forall x_2\, x_1 = x_2$	There is exactly one thing
$\exists x_1 \exists x_2 (\neg x_1 = x_2 \wedge \forall x_3 (x_3 = x_1 \vee x_3 = x_2))$	There are exactly two things
$\exists x_1 \exists x_2 \exists x_3 (\neg x_1 = x_2 \wedge \neg x_1 = x_3 \wedge \neg x_2 = x_3 \wedge \forall x_4 (x_4 = x_1 \vee x_4 = x_2 \vee x_4 = x_3))$	There are exactly three things
$\vdots$	$\vdots$

One can easily see that if one were to continue these lists indefinitely, one would incur the need for an ever-increasing number of distinct variables for use within a single sentence. That is why we assume at the outset that we have available infinitely many distinct variables. (Naturally, we need only *countably* infinitely many—that is, exactly as many as there are counting numbers 0, 1, 2, ...)

Exercise 16. *Which of the following sentences in the first-order language of arithmetic is true just in case there are exactly two things?*

1. $\forall x (x = 2 \vee x = 1 + 1)$

2. $\exists x \forall y\, y = 2 \times x$

3. $\exists x \exists y (\neg x = y \wedge \forall z (z = x \vee z = y))$

4. $\forall x \exists y \neg x = y \wedge \exists x \exists y\, x = y$

5. $\exists x \forall y\, x = y \vee \exists y \forall x\, x = y$

Logical relationships

One can see that even with such simple-minded lists as those just given, multifarious *logical relationships* arise among their members. These are the relationships of *logical implication, logical inconsistency*, and *logical equivalence*.

Suppose, for example, that $0 < m < n$. Then the claim 'There are at least n things' *logically implies* the claim 'There are at least m things'. (For, whatever the situation, if the first claim is true, then the second claim is true also.) Likewise, the claim 'There are at most m things' logically implies the claim 'There are at most n things'.

We are still assuming that $0 < m < n$. The claim 'There are exactly m things' is *logically inconsistent with* the claim 'There are exactly n things'. (That is, they cannot be true together, no matter what the situation.) Likewise, the claim 'There are at most m things' is logically inconsistent with the claim 'There are at least n things'.

Finally, for any n, the claim 'There are exactly n things' is *logically equivalent to* the conjunctive claim 'There are at least n things *and* there are at most n things'. (That is, these claims are true in exactly the same situations, or under exactly the same conditions.) It is also logically equivalent to the conjunctive claim 'There are at least n things and it is not the case that there are at least $(n + 1)$ things'.

We shall take this opportunity to foreshadow some notation that will be rigorously defined later. Logicians (as we have frequently seen since p. 45) use the symbol $\perp$ to represent absurdity (or constant falsehood). And, as already mentioned at the very beginning of this work, they use the symbol $\models$ to represent 'logically implies'. So we have, for $m < n$:

There are at least n things $\models$ There are at least m things.

There are at most m things $\models$ There are at most n things.

There are exactly m things, There are exactly n things $\models \perp$.

There are at most m things, There are at least n things $\models \perp$.

There are exactly n things $\models$ There are at least n things.

There are exactly n things $\models$ There are at most n things.

There are at least n things, There are at most n things $\models$ There are exactly n things.

4.5 The notions of scope and dominant occurrence

Analysis trees make vivid an important relation among occurrences of logical operators and of other expressions (such as variables) within formulae. This is the relation of *one occurrence containing another occurrence within its scope.* Consider once again this analysis tree:

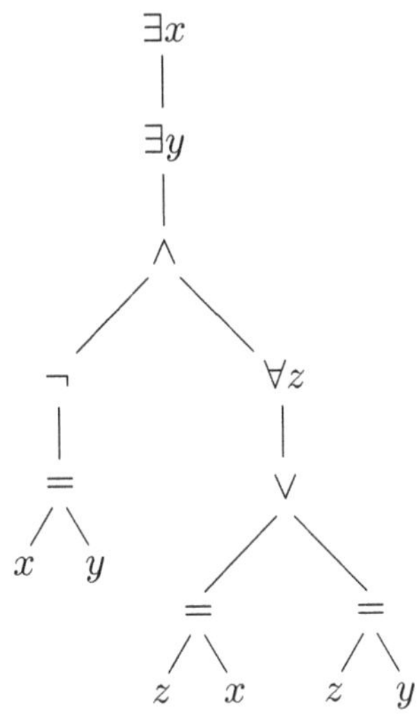

The occurrence of ∨ is contained within the scope of the occurrence of ∀, but it is not contained within the scope of the occurrence of ¬.

In general, an occurrence α of some expression is within the scope of an occurrence β of that same or some other expression just in case there is a branch running upwards within the analysis tree from (the node labeled) α to (the node labeled) β.

In applying the definition of scope, we need to remember that analysis trees are inverted—they have their root-nodes at the top, and their leaf-nodes at the bottom.

The topmost operator-occurrence—in this example, the topmost occurrence of ∃—is said to be the *dominant* operator-occurrence of the formula. It is the operator-occurrence that enjoys the *widest possible scope* within the formula, by virtue of being applied last in the formula's construction. It has within its scope every other expression-occurrence within the formula. We shall see in due course just how crucial a role is played by dominant operators in natural deductions. Introduction rules for an operator govern how one can infer *to* a sentence with that operator dominant. Elimination rules govern how one can infer *from* a sentence with that operator dominant. Being able to identify the dominant operator in a sentence—especially when it is written in infix notation—is vital for one's ability to check whether a purported proof is correctly formed. In this regard, Polish notation (see §4.7) is a boon—one just looks for the initial (leftmost) logical operator occurrence.

4.6 Substitution instances of sentences

Consider the following sentences involving only connectives. The letters A, B, and C are placeholders for atomic sentences.

$$\neg(A \to (B \vee C))$$
$$\neg((B \vee C) \to A)$$
$$\neg(A \to \neg A)$$
$$\neg(\neg\neg B \to A)$$

One can see that these sentences have in common the 'gross form'

$$\neg(\varphi \to \psi).$$

and that each of them is obtained from this gross form by replacing φ and ψ with appropriately chosen sentences. That is, each of the four examples is a *substitution instance* of the sentence $\neg(\varphi \to \psi)$.

Exercise 17. Determine the four substitutions for φ and ψ in $\neg(\varphi \to \psi)$ that respectively yield the four substitution instances

$$\neg(A \to (B \vee C))$$
$$\neg((B \vee C) \to A)$$
$$\neg(A \to \neg A)$$
$$\neg(\neg\neg B \to A)$$

4.7 Polish notation

We remarked on p. 115 that in Frege's day logicians were prosaic, printers and publishers were penny-pinching, and LaTeX had not yet been invented. That led to as much linearization as possible of logical notation. The result nowadays is that it is only because we are so accustomed to writing sentences in a linear notation that we do not use the analysis trees themselves *as* the formulae of the formal language. For, in a vivid sense brought out by the tree diagrams (with or without linkages), the logical structure of a formula is really 'two-dimensional'. We are able to persist with a linear notation, writing the sentence above as

$$\exists x \exists y(\neg x = y \wedge \forall z(z = x \vee z = y))$$

only because we can work out in what unique (partial) order its constituents were assembled—a partial order made explicit by either kind of analysis tree.

By introducing linkages in a two-dimensional notation for formulae, we were able to do away with variables. There is another device that is best explained by appeal to analysis trees, only this time it is a device that enables one to do away with parentheses, while maintaining linearity of notation. The first analysis tree given above 'is' the last formal sentence displayed, as can be seen by a spatial transformation that 'flattens the tree' by applying, from the bottom upwards, local transformations τ of the form

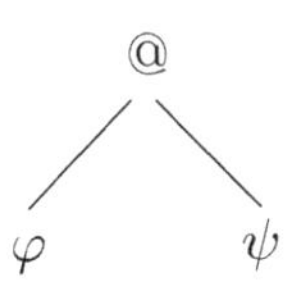

$$\stackrel{\tau}{\longmapsto} \quad (\varphi^\tau @ \psi^\tau) \qquad (@ = \vee, \wedge, \rightarrow)$$

and of the form

$$\stackrel{\tau}{\longmapsto} \quad @\varphi^\tau \qquad (@ = \neg, \exists x, \forall x).$$

Note that it is only the first kind of transformation—the one for the bifurcating case—that introduces parentheses. Parentheses would be rendered unnecessary if we were to employ instead the following kind of transformation, in the bifurcating case, from tree-notation to linear notation:

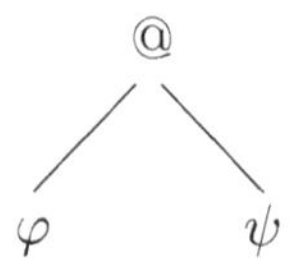

$$\stackrel{\mathcal{T}}{\longmapsto} \qquad @\varphi^{\mathcal{T}}\psi^{\mathcal{T}} \qquad\qquad (@ = \vee, \wedge, \rightarrow)$$

along with an initial transformation on atomic formulae that would place their predicates leftmost:

$$t = u \qquad \longmapsto \qquad = tu$$

Applying the parenthesis-eschewing transformations to the first analysis tree above, we would obtain the linear transform

$$\exists x \exists y \wedge \neg = xy \forall z \vee = zx = zy.$$

This is a sentence of the first-order language of pure identity in what is known as (left) *Polish notation.* The notational method obviously extends to any first-order language with descriptive vocabulary. Our example here just happens to be 'pure'.[2] Expression-forming operators are placed leftmost, followed by the component expressions from which they form a new compound. Provided only that one knows *how many* components, and of what kind, each expression-forming operator calls for, one can recover the (unique) analysis tree for any complex formula in Polish notation. So we see that parentheses can, in principle, be dispensed with.

The weight of tradition, however, favors the notation that uses parentheses and 'infixes' the two-place connectives between the component

[2]Originally, Polish notation used different symbols for the quantifiers and connectives; but this difference is inessential. The essential feature of Polish notation is its method of arranging the component expressions when forming compound expressions—placing the expression-forming operators leftmost at every stage of formation, so as to avoid the need for parentheses.

formulae when forming compounds. In this work we adopt the customary notation, except when there are definite advantages that accrue by departing from it.

The reader should be on the lookout for the occasional suppression of parentheses where their explicit occurrence would be superfluous. Thus, for example, stand-alone conjunctions that are not proper subformulae of other formulae are usually displayed as $\varphi \wedge \psi$ rather than as $(\varphi \wedge \psi)$; and similarly for disjunctions and conditionals. Repeated conjunctions, such as $(\varphi \wedge (\theta \wedge \psi))$ or $((\varphi \wedge \theta) \wedge \psi)$, can be rendered without the inner parentheses: $(\varphi \wedge \theta \wedge \psi)$. So can repeated disjunctions. This is for the obvious reason that the different groupings resulting from inserting parentheses make no difference to the logical force of the conjunction or disjunction in question. (In technical parlance, conjunction and disjunction are *associative* operations. Note, however, that the conditional is *not* associative.)

There is also a convention to the effect that the conditional has wider scope than conjunction and disjunction. Thus the parenthesis-free 'formula' $\varphi \wedge \theta \rightarrow \psi \vee \xi$ should be read as $(\varphi \wedge \theta) \rightarrow (\psi \vee \xi)$. Another convention provides for negation and the quantifiers to have the narrowest scope possible. Thus $\varphi \wedge \theta \rightarrow \neg\psi \vee \xi$ should be read as $(\varphi\wedge\theta) \rightarrow (\neg\psi\vee\xi)$ rather than as $(\varphi\wedge\theta) \rightarrow \neg(\psi\vee\xi)$. And $\forall x F(x) \rightarrow G(x)$ should be read as a conditional, with the last occurrence of x free—and not as the quantification $\forall x(F(x) \rightarrow G(x))$, in which all occurrences of x are bound. By following such conventions one may judiciously omit parentheses without incurring the risk of syntactic ambiguities.

4.8 Formulae of descriptive first-order languages

In order to make claims in a formal language about states of affairs other than the mere numerosity of things, one needs at least some *extra-logical* vocabulary items, which can stand for, or hold of, various things. Among these are *names* such as 'John', '0' etc. Names are both primitive and simple. They do not serve to form more complex grammatical expressions. But other descriptive primitives do. To these we now turn.

Descriptive function-symbols and predicates

Descriptive function-symbols and predicates can be of any finite *adicity.* The function-symbols '$\sqrt{\ldots}$', 'the father of ...', and the predicate '... is a person', for example, are one-place, or *monadic*; the function-symbol '...+...' and the predicate '... loves ...' are two-place, or *dyadic*; the predicates '... gave ... to ...' and '... is between ... and ...' are three-place, or *triadic*; and so on. Since one of the main purposes of formalization is to achieve succinctness of expression, such function-symbols and predicates will always be written as single letters, rather than spelled out in full. Thus: '$f(x)$' for 'the father of x'; '$N(x)$' for 'x is a number'; '$P(x)$' for 'x is a person; '$L(x, y)$' for 'x loves y'; etc. Moreover, when no confusion can result we shall often suppress the parentheses. We shall also follow the conventional practice of writing familiar dyadic mathematical predicates *between* their argument-terms, rather than leftmost. Thus:

$$x=y\,,\quad x\in y\,,\ \text{and}\ x<y$$

rather than the 'Polish' atomic formulae

$$=xy\,,\quad \in xy\,,\quad \text{and}\ <xy\,.$$

The first-order language of arithmetic

The most extensively studied first-order theories are those that are expressed in the *language of arithmetic.* Among these theories the most famous one is so-called Peano Arithmetic (PA). It contains many subtheories that have names of their own in the research literature, which will not detain us here. They all involve the same extra-logical primitives: the name 0; the one-place function sign s for 'successor' (typically prefixed); and the familiar two-place function signs $+$ and $\times$ for addition and multiplication (typically infixed between their two terms). PA is based on the following axioms:

$$\begin{aligned}&\forall x \neg 0 = sx\\&\forall x \forall y(sx = sy \rightarrow x = y)\\&\forall x\ x + 0 = x\\&\forall x\ x \times 0 = 0\\&\forall x \forall y\ x + sy = s(x + y)\\&\forall x \forall y\ x \times sy = (x \times y) + x\end{aligned}$$

and the infinitely many instances of the following *axiom scheme* of Mathematical Induction:

$$(\Phi_0^x \wedge \forall x(\Phi \rightarrow \Phi_{sx}^x)) \rightarrow \forall x \Phi,$$

where Φ is a formula of the language of arithmetic with just x free. (The reader should recall clause (1) of Definition 4 on page 112, concerning substitutions of terms for free variables. The substituend term may contain x free.)

The relation between formal and natural languages

Note that we have given here examples of expressions both from ordinary discourse about people and things, which are concrete; and from mathematical discourse, which is about abstract entities, such as numbers. It would be a mistake to think that formal languages can be used only to speak about mathematical subject matters, or about matters only of the utmost generality (such as the numerosity of things). Rather, appropriate descriptive formal languages can be devised for use in connection with any subject-matter at all.

Formal languages were consciously designed. They afford a degree of logical detail, precision and rigor that is harder to achieve within any evolved natural language such as English or Chinese. Formal languages distill the most essential features of a medium of thought in general. They enable one to make reference to individual things; to classify and correlate things; to discriminate among them, and to identify some of them; and to generalize about them. Even if formal languages do not (yet) cater for every kind of expressive subtlety in natural language, they do at least cater for the features just listed, which are most central to rational thought. In particular, formal languages cater for *all* the expressive resources needed for the formulation and deductive development of mathematics.

One distinguishing feature of a formal language provides sufficient reason to use it for more mundane applications. Each sentence of a formal language has *exactly one grammatical analysis*, revealing how it is formed from the primitive expressions within it. A formal language contains no syntactic ambiguities. Nor does it contain any lexical ambiguities (such as 'bank' or 'duck'). Natural languages, by contrast, are rife with syntactic ambiguities, making them less suitable for the expression of rigorous scientific thought. Users of natural languages have recourse to the context of utterance, to common background knowledge, and to the speaker's gestures and intonation contours, in order to glean the valuable extra information that is often needed in order to work out which reading of an ambiguous utterance the speaker might have intended. But scientific languages are typically written and read, not spoken and heard; and scientific theories appear in books and journals that are independent of any context of utterance. Thus there is a premium on syntactic exactitude in the case of scientific languages; one can at best resort to the textual context to resolve any ambiguities. Great value therefore attaches to syntactic univocity; and formal languages satisfy this need.

Sentences of formal languages can be thought of as representing the 'deep structures' or 'logical forms' of sentences of natural language (upon suitable disambiguation, if necessary). So formal sentences can serve as surrogates for sentences of natural-language argumentation, and formal methods of proof can be used in order to justify passages of reasoning expressed in natural language. One can 'regiment' sentences of a natural language like English by translating them into a formal language. One can then appraise arguments in which these sentences feature, by bringing methods of formal logic to bear. To this end, one needs to be assured that crucial expressive resources in the English sentences, on which the validity of an argument might turn, are catered for with formal analogues in the vocabulary (logical or extra-logical) of the formal language.

Sentences of formal languages can also be thought of as autonomous and non-parasitic, as sentences *sui generis*, whose role is not to represent or mimic or do duty for sentences of a natural language. On this conception, sentences of formal languages are to be *used* on their own, to express thoughts and to advance trains of reasoning about their subject matters, in accordance with their own rules for interpretation and manipulation. This is perhaps the best way to conceive of modern mathematics, when it is fully formalized and made independent of its early roots in natural language.

Formal first-order languages can also be used as high-level computer programming languages. This is the case with Prolog, which is based on a significant and powerful fragment of first-order language, consisting of so-called Horn clauses. These are clauses of the form

$$\forall x_1 \ldots \forall x_n((F_1(x_1, \ldots, x_n) \wedge \ldots \wedge F_m(x_1, \ldots, x_n)) \rightarrow G(x_1, \ldots, x_n)).$$

In Prolog these are used the other way round, with lowercase and uppercase swapped, and without the quantifiers explicit, to state what subgoals $\mathrm{f}_1, \ldots \mathrm{f}_m$ one can try to achieve in order to achieve the head goal g:

$$\begin{aligned} \mathrm{g}(\mathrm{X}_1, \ldots, \mathrm{X}_n) \text{ :- } & \mathrm{f}_1(\mathrm{X}_1, \ldots, \mathrm{X}_n), \\ & \vdots \\ & \mathrm{f}_m(\mathrm{X}_1, \ldots, \mathrm{X}_n). \end{aligned}$$

We shall not discuss Prolog and Horn clauses further at this stage, since the aim here is only to indicate to the reader the various ways in which one can conceive of the purpose and the role of formal first-order languages (and of their expressively weaker fragments or expressively stronger extensions).

Sortal predicates

Philosophers call certain one-place predicates—such as 'number', 'person', 'animal' etc.—*sortal* predicates, because they supply a criterion of identity for the things to which they apply: 'x is the same number as y', 'x is the same person as y', 'x is the same animal as y', etc. The relevant criterion of identity then enables one to *count* how many things satisfy the sortal predicate in question. By contrast, some examples of descriptive predicates that are *not* sortal predicates are: 'x is solid', 'x

is wooden', 'x is red', and 'x is hot'. One cannot count the number of solid things, or the number of wooden, red or hot things, since we have no criterion of identity to tell us where one such thing ends and another begins.

Let F be a sortal predicate. Our earlier lists of numerosity claims can now be adapted so as to talk about Fs rather than about things in general. We have

$\forall x_1 \forall x_2((Fx_1 \wedge Fx_2) \rightarrow x_1 = x_2)$	There is at most one F
$\forall x_1 \forall x_2 \forall x_3((Fx_1 \wedge Fx_2 \wedge Fx_3) \rightarrow (x_1 = x_2 \vee x_1 = x_3 \vee x_2 = x_3))$	There are at most two Fs
$\vdots$	$\vdots$

$\exists x_1 Fx_1$	There is at least one F
$\exists x_1 \exists x_2(\neg x_1 = x_2 \wedge Fx_1 \wedge Fx_2)$	There are at least two Fs
$\vdots$	$\vdots$

$\neg \exists x_1 Fx_1$	No F exists
$\exists x_1(Fx_1 \wedge \forall x_2(Fx_2 \rightarrow x_1 = x_2))$	There is exactly one F
$\exists x_1 \exists x_2(\neg x_1 = x_2 \wedge Fx_1 \wedge Fx_2 \wedge \forall x_3(Fx_3 \rightarrow (x_3 = x_1 \vee x_3 = x_2)))$	There are exactly two Fs
$\vdots$	$\vdots$

4.9 Variable-binding term-forming operators

In this section we consider ways in which one can form singular terms from formulae, by binding a free variable. The operators that do this are the ones described earlier—*variable-binding term-forming operators*, or *vbto*s for short. We shall concentrate on the three already introduced: the definite description operator; the number-abstraction operator; and the set-abstraction operator.

The definite description operator

This important variable-binding term-forming operator is the one that forms definite descriptions, and corresponds to the definite article 'the' in English. Let $\varphi(x)$ be a formula with the variable x free. Then $\iota x\varphi(x)$ is a term, whose free occurrences of variables are those in $\varphi(x)$, save those of x, which are now bound by ιx. The term $\iota x\varphi(x)$ can be read as 'the φ', or (more wordily) as 'the sole φ'; or 'the unique φ'; or 'the one and only thing that is φ'.

The iota is a *logical* operator. As with all the other logical operators (and the identity predicate), its interpretation is fixed. The truth-conditions for any atomic claim of the form $\psi(\iota x\varphi(x))$ ('The φ has property ψ') are that *there is exactly one* φ, *and it has property* ψ.[3] Using other logical symbols already in use, we can write this as

$$\exists x(\forall y(x = y \leftrightarrow \varphi(y)) \wedge \psi(x)).$$

Note that the variable y here needs to be chosen so that there are no inadvertent bindings of y by quantifier- or other operator-occurrences within $\varphi(x)$. By the notation $\varphi(y)$ we mean the result of replacing all free occurrences of x in $\varphi(x)$ by occurrences of this appropriately chosen variable y.

Another notation that one can use for this purpose is φ^x_y, with the substituting expression as the subscript and the free variable substituted for as the superscript. The last formal sentence could then be written as

$$\exists x(\forall y(x = y \leftrightarrow \varphi^x_y) \wedge \psi(x)).$$

As was stipulated in Definition 4, if x is free in the formula φ, and t is a (possibly complex) term, we write φ^x_t for the result of substituting occurrences of t for all free occurrences of x in φ.

---

[3]This is the standard reading of sentences involving definite descriptive phrases, following a tradition set by Bertrand Russell's famous ordinary-language analysis in his landmark 1905 paper 'On denoting'. In that paper, however, he did not introduce the iota as a formal symbol. He did so only later, in his 1908 paper 'Mathematical Logic as Based on the Theory of Types' (at p. 253)—and then only by means of contextual definition, not by adoption as a primitive variable-binding term-forming operator.

The number-abstraction operator

Note that claims of the form 'There are at least n Fs', 'There are at most n Fs' and 'There are exactly n Fs' do not contain any expression that can be construed as a (complex) name of the natural number n in question. In these fluent English renderings of the formal sentences, the apparent number word 'n' is in *adjectival*, not *substantival*, position. Its logical analysis turns out to involve a suitable constellation of quantifiers, coupled with the use of the identity predicate and connectives. For example, 'There are exactly two Fs' is analyzed logically as

$$\exists x \exists y(\neg x = y \wedge Fx \wedge Fy \wedge \forall z(Fz \rightarrow (z = x \vee z = y))),$$

in which there is no singular term corresponding to '2', since the occurrence of 'two' in the English sentence is adjectival, not substantival.

Matters would be different if, instead of 'There are exactly two Fs' one were to express the same thought by saying 'The number of Fs is (identical to) 2'. In the latter sentence, the symbol '2' is substantival. It functions as a singular term, standing for the individual number 2. So the logical form in question should be something like

$$\#xFx \;=\; 2$$

('the number of x such that Fx is identical to 2'). This formal sentence contains the *variable-binding term-forming operator* $\#$. In general, this operator binds a variable x free in a formula φ to produce the *singular term* $\#x\varphi$, rather than a formula. (One reads $\#x\varphi$ as 'the number of φs'.) In the application made of it here, $\#$ binds the (free) variable x in the formula Fx to produce the closed term $\#xFx$ ('the number of Fs'). This term, and the term '2', can then flank the identity predicate so as to produce the identity statement '$\#xFx = 2$'.

There is no substantival occurrence of the number-word 'n' in the English sentence 'There are exactly n Fs'. But its truth-conditions are the same as those of the sentence 'The number of Fs is (identical to) n', in which the number-word (or *numeral*) 'n' *does* occur substantivally.

There is not much point in being able to speak of the number of Fs without also being able to say what that number might be: zero, or one, or two, ... To this end we introduce the name 0, and the one-place successor function-sign s. The *numeral* for the number n will be a functional term of the form $s \ldots s0$, with n occurrences of the function-sign s. The numeral for the number n will be called $\underline{n}$. Thus, for example, $\underline{3}$ is the numeral $sss0$.

In general, the formal sentence regimenting 'There are exactly n Fs' should be provably equivalent—assuming the language contains the operator #—to the formal identity-claim '$\#xFx = \underline{n}$'. For example, the formal sentence

$$\exists x \exists y(\neg x = y \wedge Fx \wedge Fy \wedge \forall z(Fz \rightarrow (z = x \vee z = y)))$$

(regimenting 'There are exactly two Fs') should be provably equivalent to the formal identity-claim

$$\#xFx = ss0$$

(regimenting 'The number of Fs is 2').

When all such equivalences hold, one is able to move back and forth between the substantival and adjectival uses of number-words in the English sentences that exhibit these logical forms. We take the view here that the application of numbers to count finite collections of things is a *logical* operation, subject to suitable logical constraints. Exactly what these constraints are have been investigated in the author's books *Anti-Realism and Logic* and *The Logic of Number*.

There is an important difference, however, between the definite description operator and the number-abstraction operator. The apple in the basket (if there is exactly one) is an apple in the basket. But the number of apples in the basket is not an apple. Likewise, the set of apples in the basket is not an apple. This is why we speak of *abstraction* in the last two cases—number-abstraction, and set-abstraction. To the latter we now turn.

The set-abstraction operator

Our final example of a variable-binding term-forming operator is that of set-abstraction. Given any complex predicate $\varphi(x)$, one might wish to refer to *the set of all and only those things that have the property* φ. By strict analogy with the terms $\#x\varphi(x)$ and $\iota x\varphi(x)$ one could adopt a formal expression such as $\sigma x\varphi(x)$ for this set-term, with a set-term forming operator σ placed leftmost.[4] But there is an established convention nowadays among mathematicians to write, instead,

$$\{x \mid \varphi(x)\},$$

with the strictly superfluous curly parentheses. We too will persist in this habit, and even pass it on to others.

There is little point in having a notation for the set of φs unless one is able also to talk about what is *in*, or *is a member of*, such a set. So we adopt also the all-important two-place predicate $\in$ of *set-membership.* The formula '$x \in y$' is to be read as 'x is a member of y'. One obvious principle that ought to hold is

$$\forall y(y \in \{x \mid \varphi(x)\} \rightarrow \varphi(y)).$$

An equivalent *inferential* expression of this principle would be

$$\frac{t \in \{x \mid \varphi(x)\}}{\varphi(t)}$$

where 't' is a placeholder for any closed term. This inference rule ensures that every member of the set of φs has the property φ. And provided that there is indeed such a thing as the set of φs, everything with the property φ is in it:

$$\exists z\ z = \{x \mid \varphi(x)\} \quad \rightarrow \quad \forall y(\varphi(y) \rightarrow y \in \{x \mid \varphi(x)\}).$$

The expression $\exists z\ z = \{x \mid \varphi(x)\}$ may be read as '$\{x \mid \varphi(x)\}$ exists', and abbreviated as $\exists!\{x \mid \varphi(x)\}$. Note that the leftmost occurrence of the conditional arrow is the dominant operator-occurrence in this formal sentence. The principle may be expressed in inferential form as follows:

$$\frac{\exists!\{x \mid \varphi(x)\} \quad \exists!t \quad \varphi(t)}{t \in \{x \mid \varphi(x)\}}$$

[4]Frege had an extension-forming operator vaguely similar to this in his *Grundgesetze der Arithmetik.* Frege's class-abstraction term $\dot{\varepsilon}\Phi(\varepsilon)$ would be written today as $\{\varepsilon \mid \Phi(\varepsilon)\}$. It is the little 'breathing comma' superscript over the variable ε that is Frege's *vbto*. See Tennant [2024].

where 't' is again a placeholder for any closed term. We shall learn in due course the deep reason why one has to be cautious in formulating this last principle, by conditioning the truth of

$$\forall y(\varphi(y) \rightarrow y \in \{x \,|\, \varphi(x)\})$$

—or the validity of the inference

$$\frac{\exists! t \qquad \varphi(t)}{t \in \{x \,|\, \varphi(x)\}}$$

—on there being such a thing as the set of φs.

4.10 Free logic

The presence, in a formal language, of the last two of these variable-binding term-forming operators (ι and $\{\ |\ \}$) enforces a subtle but important change in the underlying logic. This is because terms formed by means of them can fail to denote. Indeed, some such terms do not just *happen* not to denote; they are *necessarily* non-denoting. Examples are the descriptive term $\iota x \neg x = x$ (the thing that is not identical to itself'); any descriptive term of the form $\iota x(Fx \wedge \neg Fx)$ ('the thing that is both F and not-F); and the set-abstraction term $\{x|\neg x \in x\}$ ('the set of all things that are not members of themselves'); and yet another set-abstraction term in the language of set theory: $\{x|x = x\}$. (This latter purports to refer to 'the whole universe'—which turns out to be impossible. The universe itself does not exist; only objects *in* it do.)

These are all singular terms which are formed by means of *vbtos*, and which fail to denote for logical or conceptual reasons. But there are even simpler kinds of term which fail (contingently) to denote. Consider names of fictional or mythical individuals, such as 'Pegasus' or 'Santa Claus'. There just happen to be no such individuals as Pegasus or Santa Claus. What Russell applauded as a 'robust sense of reality' enjoins one to ensure that the logic one uses, especially for mathematics, is a *free* logic, freed of the catastrophic background assumption that every grammatically well-formed singular term must be guaranteed a denotation.

Any language containing non-denoting terms *must* be furnished with a *free logic*, which allows one to make the right inferential moves among sentences containing such terms. Standard logic is unfree: it is based on the background assumption that all singular terms of the language denote. This ideal of a 'logically perfect' language in unattainable if we wish to have certain variable-binding term-forming operators in our language as primitive. The ideal should also be abandoned if one wishes to accommodate function signs representing functions that are only *partially* defined, as is often the case in mathematics; or if we wish to accommodate 'empty names', such as 'Pegasus' or 'Santa Claus', without indulging little children's superstitious fantasies.

In the present Chapter, we are concerned only to understand the *grammatical* complications that arise when we have *vbto*s as primitives in our formal language. In Chapter 16 we shall be treating free logic more thoroughly, by modifying the inference rules for identity, and those for the two quantifiers, so as to be able to accommodate non-denoting singular terms.

4.11 Fragments of formal languages

§4.1 provided a co-inductive definition of *term* and of *formula* for the most inclusive possible 'standard' descriptive language of first order. That definition afforded us *all* the possibly complex terms and formulae that can be built up using *only* the following *primitive* expressions: names, function signs, and predicates; identity; all the usual connectives and quantifiers; and the three most important variable-binding term-forming operators.

We ignored non-standard first-order quantifiers such as 'there exist infinitely many x such that ...'; or 'there exist uncountably many x such that ...'; or 'there exist domain-many x such that ...'. These *could* be added straightforwardly (to the formal grammar) with formation clauses like those for the existential and universal quantifiers. The reader should be warned, however, that there could be offsetting metalogical costs in doing so. For example: the logical truths of the language with the quantifier 'there exist infinitely many x such that ...' cannot be axiomatized.

From our co-inductive definition in §4.1 one can obtain definitions of *term* and/or *formula* for *fragments* of our standard inclusive language. One simply suppresses or deletes any inductive clause pertaining to any expression, or kind of expression, that is not available in the fragment in question. Thus, for example, one could consider the fragment based on $\neg$ and $\wedge$ as its only connectives, and $\exists$ as its only quantifier, by deleting the formation clauses pertaining to $\vee$, $\rightarrow$ and $\forall$. Or one could consider the fragment without any variable-binding term-forming operators, by suppressing the formation clauses pertaining to them. The result in this second case would then be that the definitions of *term* and of *formula* would be separate inductive definitions, and no longer co-inductive. This indeed is the standard situation in most textbooks of formal logic, very few of which consider *vbto*s as primitives. So much the worse, then, for their handling of the language of mathematics.

If one wished to confine one's attention to a quantifier-free fragment, one might still wish to be able to distinguish predicates and terms within atomic sentences. In that case one would delete only the inductive clauses pertaining to the quantifiers. But if one were to go further and dispense with any discernment of structure within atomic sentences, one would delete the clause describing how atomic formulae are formed from predicates and terms. One would help oneself instead to a countably infinite supply of *atoms* $A, B, C, \ldots$;[5] simplify the clause for atomic formulae by stipulating that any atom is a formula; and confine oneself to the formation clauses pertaining to the connectives. That choice yields the language of *propositional logic.*

Many textbooks begin with a thorough treatment of propositional logic before proceeding to, or even mentioning as forthcoming, the further complexities of first-order logic. We shall embark on the same two-step approach presently (in Chapter 5), with an initial meta-theoretical focus on propositional logic. But at least we have already delivered on our promise to let connectives and quantifiers be co-equals in translational matters. Here, with both connectives and quantifiers in play, the beginner has been brought to appreciate the concept of *logical form* of sentences. And we have seen also how easily the beginner can be brought to understand—by being given the 'numerosity sentences' that crucially involve the quantifiers along with identity—the important logical concepts of consistency, inconsistency, invalidity (fallacy), and validity (logical consequence).

[5] Atoms are also sometimes called *atomic sentences*, or *propositional variables*, or *placeholders for sentences* (as they were on page 122).

With propositional logic, we have the immediate simplification that there is no need for any definition of *term*, since terms are not regarded as logically salient. There is also no need for the distinction between *formula* and *sentence*, since no use is made of free variables. In propositional logic, every formula is a sentence, and is called a *Boolean combination* of its atoms.

Note that the closure clause of any inductive definition of a language-fragment is context-sensitive. It refers to the basis- and inductive clauses that pertain to the fragment in question.

4.12 The Sheffer language

Particular interest attaches to a formal language not yet provided for. This is the *Sheffer language*, which has as its only logical operators *what has come to be called* the *Sheffer connective stroke*,[6] and the *Sheffer quantifier stroke*. Their formation rules are as follows:

> *If φ and ψ are formulae, then so is $(\varphi|\psi)$, whose f.o.v.s are those in φ and those in ψ.*
>
> *If φ and ψ are formulae and at least one of them has x free, then $(\varphi \mid_x \psi)$ is a formula whose f.o.v.s are those in φ and those in ψ, save of x.*

$(\varphi|\psi)$ is to be read as 'It is not the case that both φ and ψ', and $(\varphi \mid_x \psi)$ is to be read as 'Nothing is both φ and ψ'. In the case of *classical* (but not: *constructive*) logic, the Sheffer operators can be used to define all the usual connectives and quantifiers. The definitions are as follows:

$$
\begin{array}{lll}
\neg\varphi & \equiv_{df} & \varphi|\varphi \\
\varphi \wedge \psi & \equiv_{df} & (\varphi|\psi)|(\varphi|\psi) \\
\varphi \vee \psi & \equiv_{df} & (\varphi|\varphi)|(\psi|\psi) \\
\varphi \rightarrow \psi & \equiv_{df} & \varphi|(\psi|\psi) \\
\forall x\varphi & \equiv_{df} & (\varphi|\varphi) \mid_x (\varphi|\varphi) \\
\exists x\varphi & \equiv_{df} & (\varphi \mid_x \varphi)|(\varphi \mid_x \psi)
\end{array}
$$

[6]Interestingly, in Sheffer's original article he defined '$p\,|\,q$' as 'neither p nor q', *not* as 'not both p and q'. See Sheffer [1913] at p. 487. Since then, however, '$p\,|\,q$' has come to mean the latter, with Sheffer's name attached to the stroke.

One advantage of the Sheffer language and its associated proof system is that it affords shorter proofs of metatheorems, by virtue of the reduced case-load in the inductive step of any meta-proof by induction (on either the complexity of formulae or the complexity of proofs).[7]

This theoretical gain, however, is outweighed by two considerations. First, the gain is to be had only in the case of classical logic. Second, the theoretical brevity is offset by the difficulties the student will have in following various passages of reasoning *about* or *with* the Sheffer operators.

Simple though the Sheffer natural deduction system might be, and despite the fact that its rules of inference are beautifully elegant and symmetric[8] (and homologous as between the connective and the quantifier strokes), it is nevertheless intuitively unwieldy. This is a pity, because, strangely enough, the electrical engineer will assure one that the *NAND* 'logic gate' (designed to transmit electical pulses in the manner of the truth-table for the Sheffer stroke)[9] is the simplest of all logic gates to build. That is why, in the computer-science community, there was for a time great interest in the question how to find the shortest way to express any Boolean combination in terms of the Sheffer stroke alone.

Exercise 18.

(i) Translate the following argument into formal first-order notation.

Anyone who loves everyone loves themselves;
Anyone loved by someone is loved by everyone;
ergo, Anyone who loves someone loves themselves.

(ii) The argument is invalid. Describe a counterexample to it.

Exercise 19.
Give fluent English translations for the following formal sentences. Also, provide for each formal sentence its analysis tree.

---

[7]The beginner cannot be expected to understand this point. For we have not yet encountered any meta-proofs by induction. They are, however, a staple in metalogic, and will be covered in the intended sequel to this work.

[8]This symmetry is most apparent when one uses infix notation rather than Polish notation. See Tennant [1979b].

[9]Again, the beginner cannot be expected to understand this point. Truth tables will be discussed in §5.2.

1. $\exists x \exists y Lxy$
2. $\forall x \forall y Lxy$
3. $\neg \exists x \forall y Lxy$
4. $\neg \exists x \forall y Lyx$
5. $\forall x(\exists y Lxy \vee \exists y Lyx)$
6. $\forall x(\exists y Lxy \wedge \exists y Lyx)$
7. $\forall x(\exists y Lxy \rightarrow \exists y Lyx)$
8. $\forall x(Lxx \rightarrow \exists y(\neg y = x \wedge Lxy))$
9. $\forall x(Lxx \rightarrow \exists y(\neg y = x \wedge Lyx))$
10. $\forall x(\exists y Lxy \rightarrow Lxx)$
11. $\forall x Lxx \rightarrow \forall x \exists y(\neg y = x \wedge Lxy)$
12. $\forall x Lxx \rightarrow \forall x \exists y(\neg y = x \wedge Lyx)$
13. $\forall x(\exists y Lyx \rightarrow Lxx)$
14. $\forall x(\exists y Lxy \rightarrow Lxx)$
15. $\forall x \exists y(Lxy \wedge Lyx)$
16. $\forall x \exists y(Lxy \vee Lyx)$
17. $\neg(\exists x Fx \vee \exists y Gy)$
18. $\neg \exists x Fx \vee \exists y Gy$
19. $\neg \exists x(Fx \wedge \forall y Lxy)$
20. $\neg \exists x(Fx \wedge \forall y Lyx)$
21. $\neg \exists x Fx \wedge \exists x \forall y Lxy$
22. $\forall x(Fx \rightarrow \exists y Lxy)$
23. $\forall x Fx \rightarrow \exists y \forall x Lxy$
24. $\forall x \forall y(Lxy \rightarrow Fx)$
25. $\forall x(\exists y Lxy \rightarrow Fx)$
26. $\forall x((Fx \wedge Gx) \rightarrow Lxx)$
27. $\forall x((Fx \vee Gx) \rightarrow Lxx)$
28. $\forall x((Fx \rightarrow Gx) \rightarrow Lxx)$
29. $\exists x((\neg Fx \wedge \neg Gx) \wedge \neg Lxx)$
30. $\exists x((\neg Fx \vee \neg Gx) \wedge \neg Lxx)$

Chapter 5

Propositional Logic

In this chapter we shall confine our attention to propositional logic. It affords a very accessible approach to an important aspect of our overall treatment of logic. This is to reveal how the primitive rules of natural deduction do justice to our understanding of what each logical operator means. The case can be made most effectively at the propositional level, to fix ideas. The treatment can then be extended to first-order logic, with its more complicated grammar, more powerful modes of expression, and more intricate modes of interpretation.

Propositional logic is often called 'baby logic'. But this is too disparaging. At its level of simplification, propositional logic allows one to devise manageable introductions to the main features of certain concepts and methods that attain their full fruition at the level of first-order logic. Propositional logic also has some nice features that are *lost* in the transition to first-order logic. An example is the effective decidability of the validity of arguments with finitely many premises.

Once again, we justify a certain degree of repetition in the interests of the reader who might choose this later chapter as their point of entry for formal logic in general.

5.1 Formal sentences of propositional logic

The following sets of vocabulary items are pairwise disjoint:

- The set of propositional atoms $A, B, C, \ldots$ (with numerical subscripts if necessary)

- The set of connectives: $\{\neg, \wedge, \vee, \rightarrow\}$
- The set of parentheses $\{(,)\}$

Parentheses are used in order to ensure that there is no *structural ambiguity* in any formal sentence, when using the *infix* notation for connected sentences.

Infix notation

This definition uses the *infix* notation for connected sentences. This is the notation standardly used by logicians nowadays.

Definition 6 (Formal sentences using infix notation).
Basis clause: *Every propositional atom is a sentence.*
Inductive clauses:

- *If φ is a sentence, then so is $\neg\varphi$.*
- *If φ and ψ are formal sentences, then so is $(\varphi \vee \psi)$.*
- *If φ and ψ are formal sentences, then so is $(\varphi \wedge \psi)$.*
- *If φ and ψ are formal sentences, then so is $(\varphi \rightarrow \psi)$.*

Closure clause: *Every formal sentence of infix notation can be shown to be so by appeal to the foregoing Basis and Inductive clauses.*

Prefix notation

This definition uses the *prefix* notation for connected sentences that is known as *left Polish.* It has no need for parentheses.

Definition 7 (Formal sentences using prefix notation).
Basis clause: *Every propositional atom is a sentence.*
Inductive clauses:

- *If φ is a sentence, then so is $\neg\varphi$.*
- *If φ and ψ are formal sentences, then so is $\wedge\varphi\psi$.*

- *If φ and ψ are formal sentences, then so is $\vee\varphi\psi$.*
- *If φ and ψ are formal sentences, then so is $\rightarrow\varphi\psi$.*

Closure clause: *Every formal sentence in left Polish notation can be shown to be so by appeal to the foregoing Basis and Inductive clauses.*

Conventions for suppressing parentheses

The reader who has chosen to enter this later chapter early, because of the simpler setting of propositional logic, might not have read the earlier discussion of suppression of parentheses. At the risk of being repetitious once again, we enter the following remarks for such a reader.

Logicians using infix notation employ some conventions to reduce the number of parentheses that they use when writing down formal sentences. The most important one is that *outermost parentheses* can be suppressed for non-embedded sentences whose dominant operator is $\wedge$, $\vee$, or $\rightarrow$. Thus, logicians will write

$$A \wedge B,\ A \vee B,\ A \rightarrow B$$

instead of

$$(A \wedge B), (A \vee B), (A \rightarrow B).$$

Another context in which logicians sometimes omit parentheses is when their order of nesting would not affect the sentences' truth conditions. Thus, for example, they will write

$$A \wedge B \wedge C$$

instead of either

$$A \wedge (B \wedge C)$$

or

$$(A \wedge B) \wedge C,$$

because the conjunction connective $\wedge$ is associative. That is, the arguments

$$\frac{A \wedge (B \wedge C)}{(A \wedge B) \wedge C} \quad \frac{(A \wedge B) \wedge C}{A \wedge (B \wedge C)}$$

are valid. Similarly with $\vee$ in place of $\wedge$ throughout.

Since the conditional connective $\rightarrow$ is *not* associative, however—witness the invalidity of *both* the arguments

$$\frac{A \rightarrow (B \rightarrow C)}{(A \rightarrow B) \rightarrow C} \quad \text{and} \quad \frac{(A \rightarrow B) \rightarrow C}{A \rightarrow (B \rightarrow C)}$$

—one cannot suppress the parentheses involved without specifying a convention for re-inserting them. One could, say, opt always to read $A \rightarrow B \rightarrow C$ as the 'right embedded' $A \rightarrow (B \rightarrow C)$ rather than the 'left embedded' $(A \rightarrow B) \rightarrow C$.

When a connected sentence whose dominant operator is $\wedge$, $\vee$, or $\rightarrow$ occurs as a *subsentence*, it is wise to keep its parentheses in place. Consider, for example, the difference between $\neg A \wedge B$ and $\neg(A \wedge B)$.

5.2 Truth tables for the connectives

There are only two truth values: T (for Truth) and F (for Falsity).

The traditional way of interpreting the connectives of propositional logic is by stating the so-called *truth tables* for them. These tables tell us how the truth values of a compound (connected) sentence depends on (is determined by) the truth values of its immediate constituents. In general, of course, these immediate constituents need not be atoms.

A *truth-tabular computation* proceeds from a truth-value assignment (to the atoms occurring in a given sentence) to the correct truth value to be assigned to that sentence. We shall use τ as a (metalinguistic) sortal variable for truth-value assignments. Any such τ serves as an *interpretation* of the propositional language. It tells one, of each atom, which truth value—T or F—it has. *Relative to such an assignment,* one can then determine the truth value of any more complex sentence involving the atoms in question. That process of determination involves just 'applying the truth tables' in such a way as to transmit values from the embedded atoms to the whole sentence containing them.

With each connective, the construction of its truth table involves considering all possible assignments of truth values to immediate constituents. Each such assignment determines a *row* in the truth table. Obviously, for a one-place connective (here, negation) there will be only two rows; and for a two-place connective (such as conjunction, disjunction, and the conditional) there will be four rows.

We now introduce an important symbol for the notion of 'making true'. When a truth-value assignment τ makes a sentence φ true according to the truth tables, we write

$$\tau \Vdash \varphi.$$

In stating the truth tables, we shall take the connectives one by one.

Negation ('it is not the case that ...'; Logical connective: $\neg$)

A	$\neg A$
T	F
F	T

The truth table for negation 'flips' the two truth values. If A is true, then its negation $\neg A$ is false. And if A is false, then $\neg A$ is true. One can call these the 'left-to-right' readings of the rows in the truth table.

Conjunction ('and'; Logical connective: $\wedge$)

A	B	$A \wedge B$
T	T	T
T	F	F
F	T	F
F	F	F

With conjunction, the left-to-right readings of the rows can be summarized as follows.

> If A and B are both true, then their conjunction $A \wedge B$ is true. But if either of them is false, then $A \wedge B$ is false.

This means that the *fourth* row is otiose; its result (F for the conjunction) is over-determined, since that result would be secured by the falsity of either one of A or B. The 'real action' of the truth table lies in its first, second and third rows.

Disjunction ('or'; Logical connective: $\vee$)

A	B	$A \vee B$
T	T	T
T	F	T
F	T	T
F	F	F

With disjunction, the left-to-right readings of the rows can be summarized as follows.

> If the disjuncts A and B are both false, then their disjunction $A \vee B$ is false. But if either of them is true, then $A \vee B$ is true.

This means that the *first* row is otiose; its result (T for the disjunction) is over-determined, since that result would be secured by the truth of either one of A or B. The 'real action' of the truth table lies in its second, third, and fourth rows. The operator $\vee$ is often clarified as 'inclusive "or"'—$A \vee B$ means 'either A or B *or both*'. This contrasts with 'exclusive "or"', which is 'either A or B *but not both*'. Exclusive disjunction is not taken as a primitive connective in propositional logic, although of course it would be *definable* in terms of our chosen primitives, as $(A \vee B) \wedge \neg(A \wedge B)$. The exclusive disjunction is often symbolized as $A \otimes B$.

Conditional ('only if'; Logical connective: $\rightarrow$)

A	B	$A \rightarrow B$
T	T	T
T	F	F
F	T	T
F	F	T

With the conditional, the left-to-right readings of the rows can be summarized as follows.

> If the antecedent A is true and the consequent B is false, then the conditional $A \rightarrow B$ is false. But if either A is false or B is true, then $A \rightarrow B$ is true.

This means that the *third* row is otiose; its result (T for the conditional) is over-determined, since that result would be secured by either one of the falsity of the antecedent A or the truth of the consequent B. The 'real action' of the truth table lies in its first, second, and fourth rows.

If there are n atoms in the sentence for which one has to construct a truth table, then the n leftmost columns will deal with all the values assigned to those atoms, row by row. So, for example, the truth table for any sentence $\varphi(A, B, C)$ involving just those three atoms will begin, to the left, as follows:

A	B	C	
T	T	T	...
T	T	F	...
T	F	T	...
T	F	F	...
F	T	T	...
F	T	F	...
F	F	T	...
F	F	F	...

To the right, as indicated by the ellipses, will be the results of computing the intermediate truth values of all the proper subsentences of $\varphi(A, B, C)$. To flesh out the 3-atom example further, suppose φ is the sentence

$$\neg A \rightarrow \neg(B \wedge C).$$

The non-atomic subsentences that need to be dealt with are

$$\neg A,\ B \wedge C,\ \text{and}\ \neg(B \wedge C).$$

The foregoing incomplete truth table could then be 'completed' as follows. Note that '@' signifies that one does not need to know the truth value for that spot, since it is not needed for the final determination of a truth value in the rightmost column. The values that are dispositive are highlighted in red (for falsity) or blue (for truth).

A	B	C	$\neg A$	$B \wedge C$	$\neg(B \wedge C)$	$\neg A \rightarrow \neg(B \wedge C)$
T	T	T	F	T	F	T
T	T	F	F	F	T	T
T	F	T	F	F	T	T

T	F	F	F	F	T	T
F	T	T	T	T	F	F
F	T	F	T	F	T	T
F	F	T	T	F	T	T
F	F	F	T	F	T	T

The rightmost column in a fully completed truth table tells one what truth value the connected sentence enjoys under the respective combinations of truth values for its atomic constituents as represented respectively by its rows.

5.3 Truth tables reconceived as truth functions

Here, each operator @ has its asterisked companion @* as its *truth function.*

φ	$\neg\varphi$	$\neg^*$
T	F	$\neg^*(T) = F$
F	T	$\neg^*(F) = T$

φ	ψ	$\varphi \wedge \psi$	$\wedge^*$
T	T	T	$\wedge^*(T,T) = T$
T	F	F	$\wedge^*(T,F) = F$
F	T	F	$\wedge^*(F,T) = F$
F	F	F	$\wedge^*(F,F) = F$

φ	ψ	$\varphi \vee \psi$	$\vee^*$
T	T	T	$\vee^*(T,T) = T$
T	F	T	$\vee^*(T,F) = T$
F	T	T	$\vee^*(F,T) = T$
F	F	F	$\vee^*(F,F) = F$

φ	ψ	$\varphi \rightarrow \psi$	$\rightarrow^*$
T	T	T	$\rightarrow^*(T,T) = T$
T	F	F	$\rightarrow^*(T,F) = F$
F	T	T	$\rightarrow^*(F,T) = T$
F	F	T	$\rightarrow^*(F,F) = T$

How, now, does one compute the truth value of a compound sentence under an assignment of truth values to its atoms? Just a single example will reveal all. Suppose the compound sentence φ is

$$A \rightarrow (B \wedge (C \vee \neg D)).$$

Suppose the assignment of truth values to the atoms is τ (whose details we do not yet specify). Thus for any atom A, the 'computed value'

$\tau^*(A)$ is simply $\tau(A)$. The overall value $\tau^*(\varphi)$ can then be computed as follows, 'from inside out'.

$$\rightarrow^*(\tau^*(A), \wedge^*(\tau^*(B), \vee^*(\tau^*(C), \neg^*(\tau^*(D))))).$$

In general, the computational steps on compounds are distributive:

$$\begin{aligned} \tau^*(\neg\varphi) &= \neg^*(\tau^*(\varphi)); \\ \tau^*(\varphi \wedge \psi) &= \wedge^*(\tau^*(\varphi), \tau^*(\psi)); \\ \tau^*(\varphi \vee \psi) &= \vee^*(\tau^*(\varphi), \tau^*(\psi)); \\ \tau^*(\varphi \rightarrow \psi) &= \rightarrow^* (\tau^*(\varphi), \tau^*(\psi)) \end{aligned}$$

Matters are more perspicuous, though, if one uses a tree analysis to compute the truth value of a compound sentence.

Since every truth-value calculation begins with at least some of the values assigned to the atoms involved, a word is in order about how we depict their enjoying the values (according to the assignment τ) that they do.

Consider now the following way of representing the various 'basic moves' of truth-tabular evaluation when dealing with *compound* sentences. Each inverted tree below illustrates how the value or values 'lower down' (for the components) determine the value 'at the next level up' for the compound containing them. The symbol $\ominus$ indicates that *it does not matter* what truth value (T or F) is put in its place. The upper truth value for the compound is already determined by the truth values displayed 'lower down'. The convention here for coloration is that 'truth-determining' arrows are blue, while 'falsity-determining' arrows are red. Accordingly also the truth value T is colored blue and the truth value F is colored red.

Note how with each of the binary connectives ($\wedge$, $\vee$, and $\rightarrow$) there are only *three* box diagrams, not four. This is because we can ignore, with this form of representation of truth-value transmission, the otiose rows of these connectives' truth tables that were identified above—the fourth row in the truth table for $\wedge$; the first row in the truth table for $\vee$; and the third row in the truth table for $\rightarrow$.

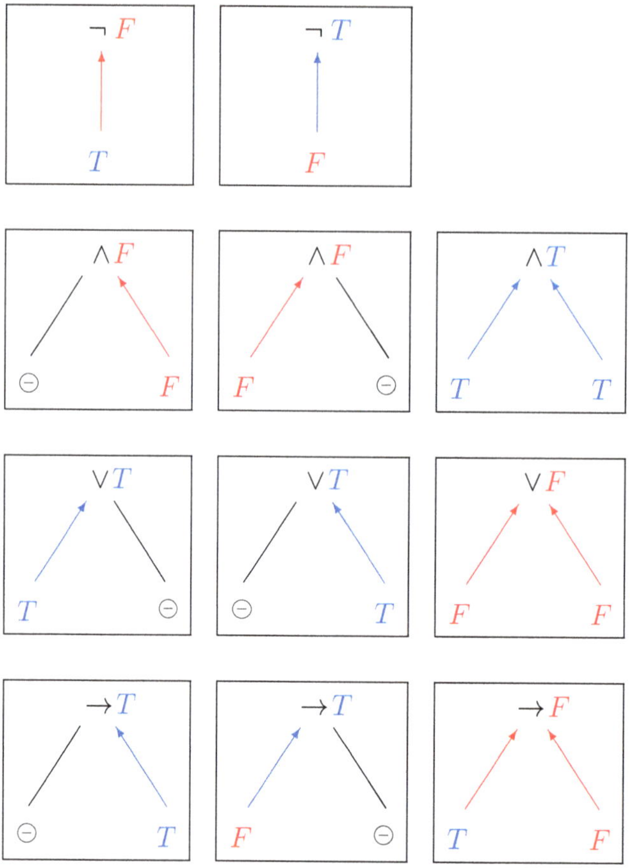

With *atoms* A, which of course are not compound, we shall write

$$\frac{}{A}\,\tau$$

to register the fact that the truth-value assignment τ assigns the value T to the atom A. And we shall write

$$\frac{A}{\bot}\,\tau$$

to register the fact that the truth-value assignment τ assigns the value F to the atom A. We call these 'positive factoids' and 'negative factoids' respectively.

We now introduce the reader to a slightly different but clearly equivalent way of depicting the same 'local actions' of the connectives, as depicted in the box diagrams above, in determining the truth value of a compound from the truth values of its immediate constituents. We preserve the same pattern of boxes for easy comparison.

In the boxes below, if the compound is determined to be *true*, then it appears *below* the bottom horizontal line. This shows that it is revealed as *verified* by the truth-tabular computations that have been carried out higher up. But if the compound is determined to be *false*, then it appears leftmost and *immediately above* the bottom line, while the symbol $\bot$ is placed below that line. This shows that it is revealed as *falsified* by the truth-tabular computations that have been carried out higher up. (This mode of representation generalizes to compounds the kind of displays we have just explained for positive and negative factoids.)

To falsify $\neg\varphi$, verify φ; but to verify $\neg\varphi$, falsify φ:

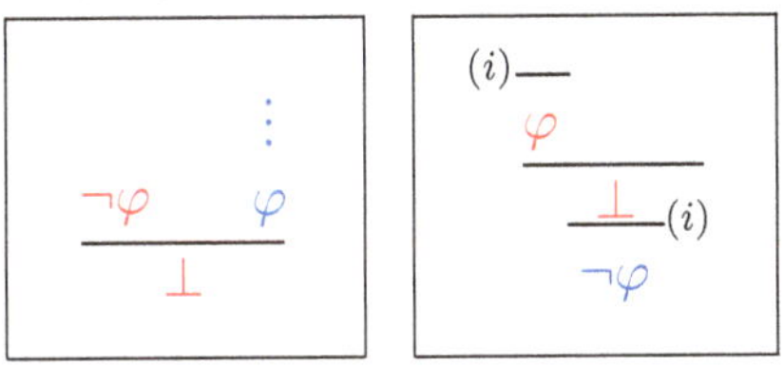

To falsify $\varphi\wedge\psi$, falsify φ or falsify ψ; but to verify $\varphi\wedge\psi$, verify φ and verify ψ:

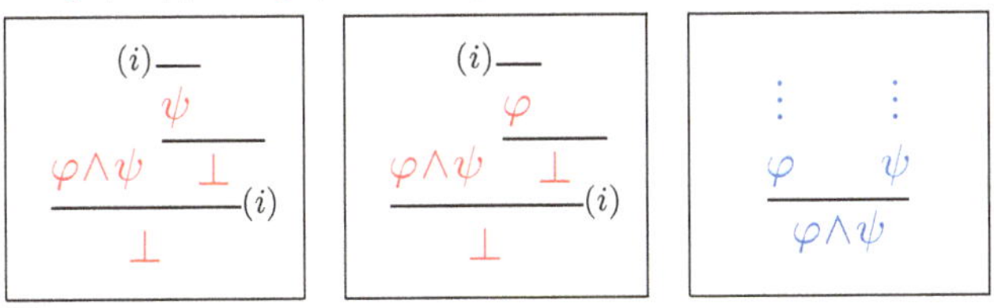

To verify $\varphi\vee\psi$, verify φ or verify ψ; but to falsify $\varphi\vee\psi$, falsify φ and falsify ψ:

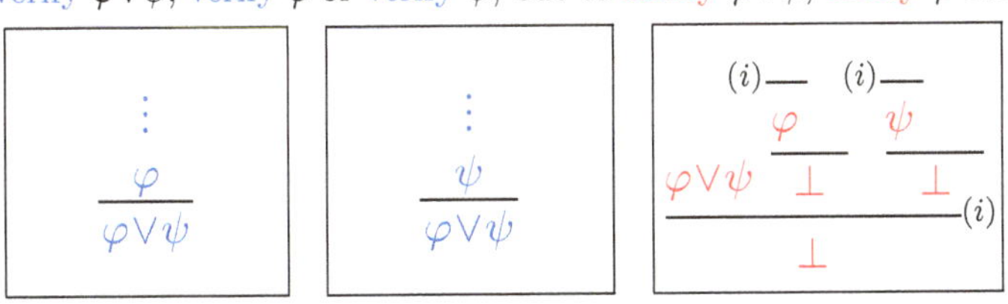

To verify $\varphi \rightarrow \psi$, verify ψ or falsify φ; but to falsify $\varphi \rightarrow \psi$, verify φ and falsify ψ:

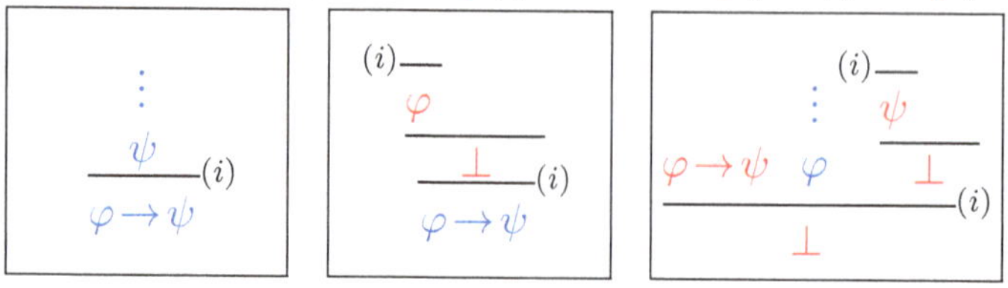

Let us now revisit our earlier example sentence

$$A \rightarrow (B \wedge (C \vee \neg D)).$$

Its tree analysis is on the left below; and before even considering the details of a particular truth-value assignment, the pattern of computation that would have to be undertaken is shown on the right:

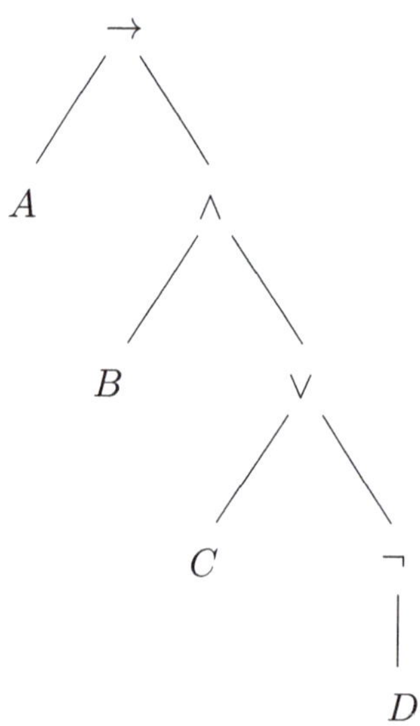

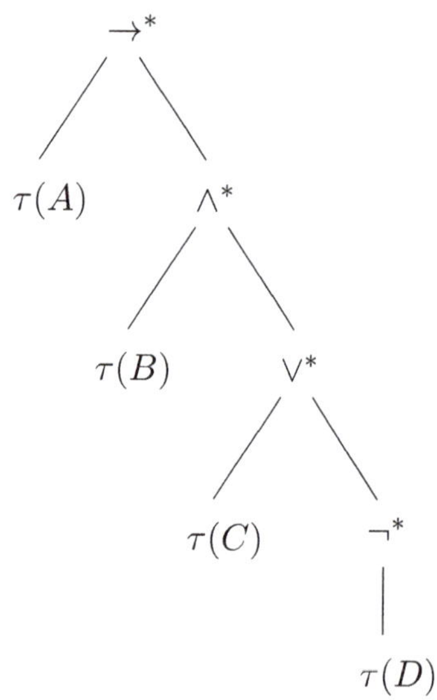

Thanks to the inverted-tree analysis, the previous 'inside out' process can now become a 'bottom up' process. Let us see how this works upon the adoption of a particular truth-value assignment to atoms.

Suppose the assignment τ is $\begin{array}{cccc} A & B & C & D \\ \downarrow & \downarrow & \downarrow & \downarrow \\ T & T & F & T \end{array}$. The computation-tree then becomes (at the start) the labeled tree

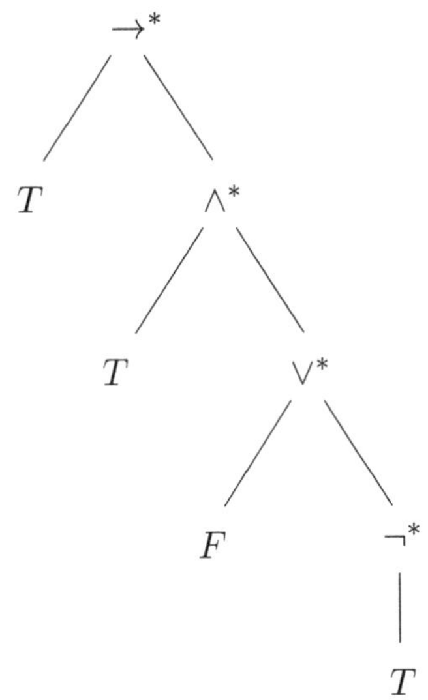

The computation then proceeds as follows, from left to right, by invoking the appropriate rows in the truth-tables of the respective connectives. Note how the downward branchings are snipped away as soon as a value of T or F replaces an asterisked connective occurrence.

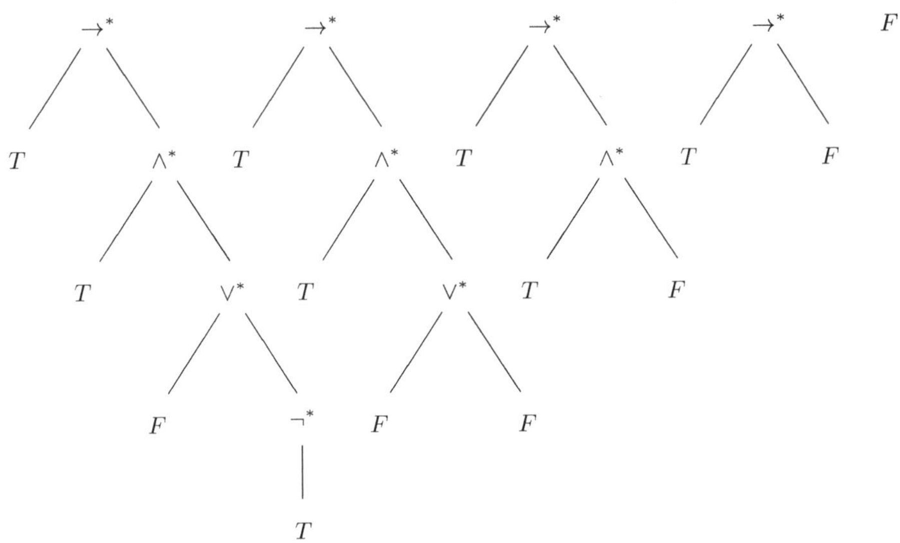

As it happens, in this example all the 'computational action' proceeds upwards along the rightmost branch. And there does not happen to be any other way to determine that the overall truth value is F on this particular assignment τ.

In a new kind of tree-notation to be introduced in due course, the very same computation will be represented by what we shall call a τ-*relative falsification* of the sentence $A \rightarrow (B \wedge (C \vee \neg D))$. It will represent this particular way in which the sentence

$$A \rightarrow (B \wedge (C \vee \neg D))$$

can be seen to have the value F under the assignment

$$\tau: \quad \begin{array}{cccc} A & B & C & D \\ \downarrow & \downarrow & \downarrow & \downarrow \\ T & T & F & T \end{array}$$

that we are working with. In the promised falsification (see below), we apply the 'local steps' explicated in the last pattern of boxes above. As already explained, an overlining with subscript τ indicates that τ assigns the value T to the atom below the overlining. But if the atom in question is underlined with subscript τ, and has $\perp$ immediately below it, this indicates that τ assigns that atom the value F.

Here now is the τ-relative falsification in question. Constructs like this will receive a full theoretical treatment in Chapter 6. This 'premature' example serves to give the reader a general idea of what is involved, before explicit formal details are presented in due course:

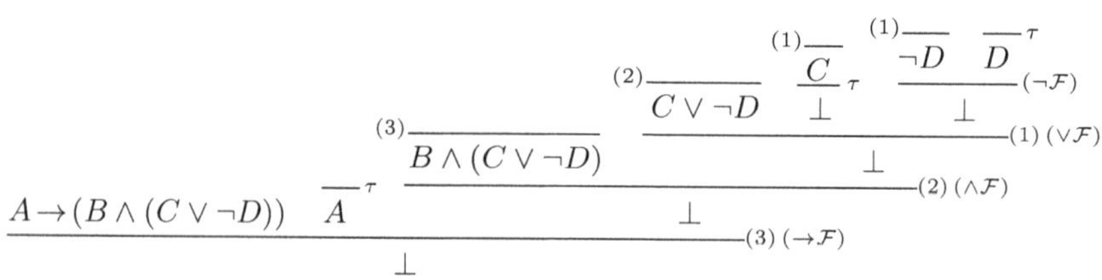

We have inserted labels on the right next to the individual steps. As it happens, they are all steps of falsification, as indicated by the calligraphic $\mathcal{F}$ in the labels, alongside whatever connective is dominant in the major premise that is being falsified.

Note that no use has been made, in this evaluation-as-false, of the information that τ assigns to B the value that it does. Only the values that τ assigns to A, C, and D were used in determining the result that the compound sentence $A \rightarrow (B \wedge (C \vee \neg D))$ is false under τ.

5.4 Fell swoops

A truth-value computation like this, in which one can get by without appealing to *all* the atoms' truth values, is what Quine called a 'fell swoop'.

A falsification like the foregoing one uses just the 'positive and negative' atomic information (and, as we have just seen, not necessarily all of it) furnished by the truth-value assignment in question. And it is built up by applying rules for verifying, or falsifying, compound sentences. Those rules are tailored to the operators enjoying dominant occurrences. Each logical operator @ has its own specific rule of verification @-$\mathcal{V}$, and its own specific rule of falsification @-$\mathcal{F}$. These will be expounded in due course; they are already virtually explicit in the last set of boxes above. We shall see that they amount, in the case of propositional logic, to *no more than an inferential representation of the left-to-right readings of the rows of the truth tables.* So the promised rules of verification and falsification succeed in providing an *inferential semantics* for the connectives.

More on fell swoops, using upright trees

In the interests of developing some intellectual agility, we shall make use here of a tree analysis of a sentence with the tree upright, not inverted. The leaf nodes will be at the top, and the root node at the bottom. Nothing essential changes. All this does is make it metaphorically more faithful to speak of truth values 'trickling down' from the atomic sentences involved to the whole sentence in which they are embedded.

Consider the sentence $\neg(A \vee B) \rightarrow (\neg C \wedge \neg D)$ of propositional logic. The 'upright' tree-analysis of this sentence is as follows.

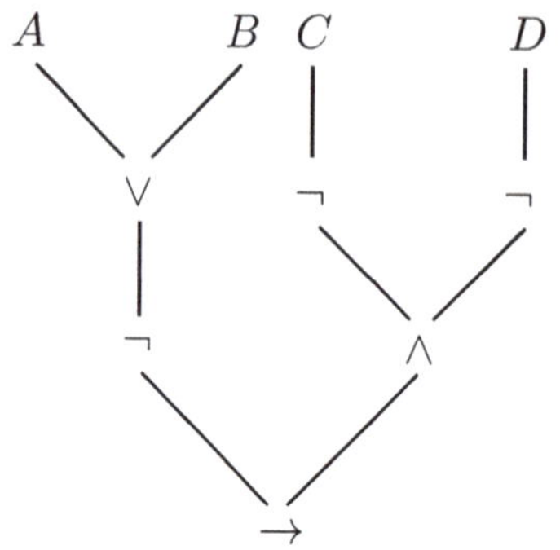

We shall work with the following assignment of truth values to atoms:

$$\tau = \left\{ \begin{array}{cccc} A & B & C & D \\ \downarrow & \downarrow & \downarrow & \downarrow \\ F & F & T & T \end{array} \right. ,$$

The leaf nodes of the tree are therefore primed as follows:

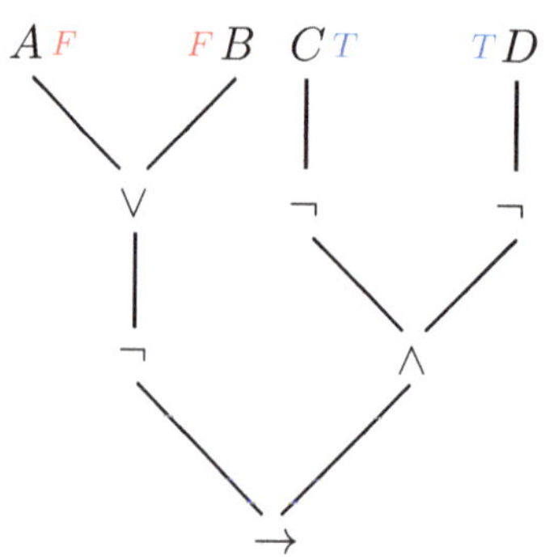

Now let us see how the truth values assigned to atoms by τ 'trickle down'. Particular interest would attach to any 'fell swoop' that does not need to evaluate every subsentence on the way down to the root of the tree. Here is one fell swoop that does the job. It makes no use of the truth value assigned to the atom D:

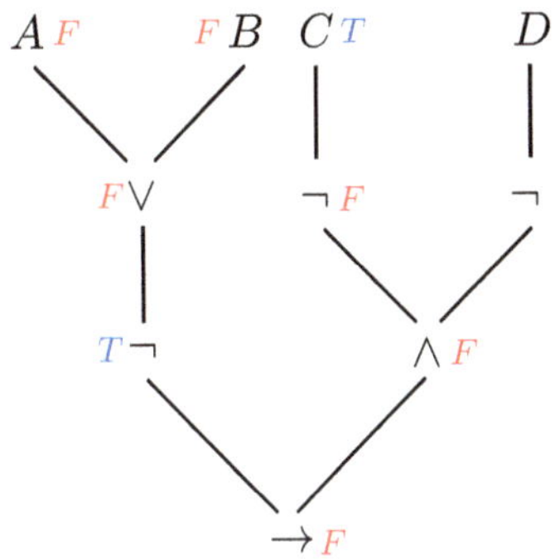

And here is a different fell swoop, one which makes no use of the truth value assigned to the atom C:

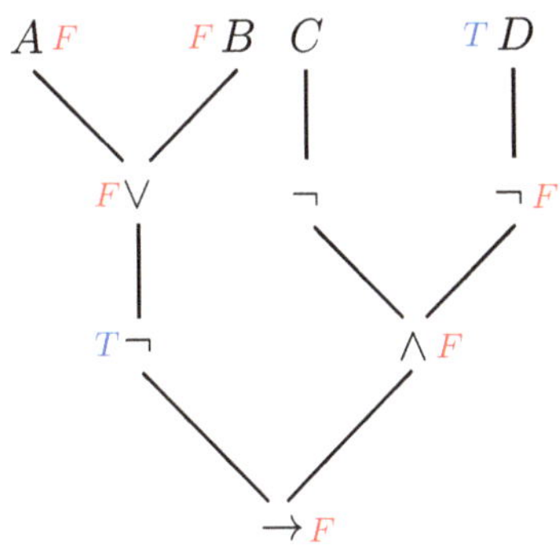

Corresponding to the first fell swoop given above, which made no use of the truth value assigned to D, is the τ-relative falsification

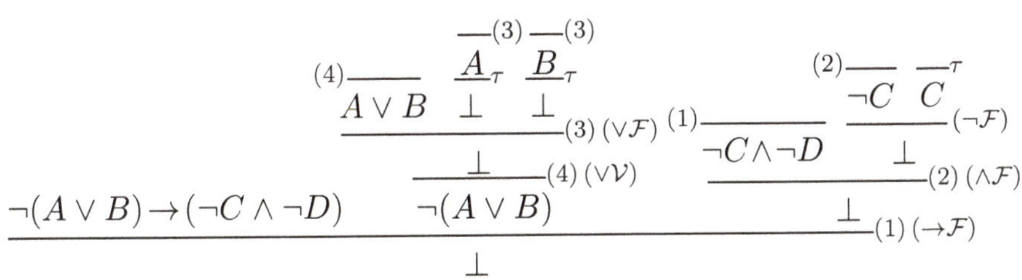

Corresponding to the second fell swoop above, which made no use of the

truth value assigned to C, is the the τ-relative falsification

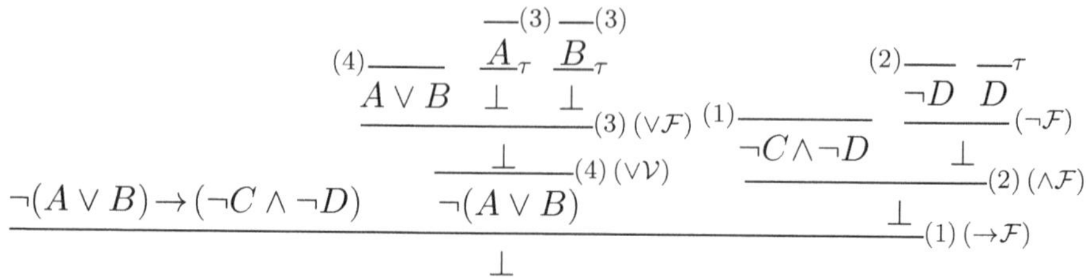

This simple example shows that one and the same sentence φ, relative to one and the same truth-value assignment τ, can enjoy the overall truth value that it does *in more than one way*. Corresponding to each distinct way in which a given sentence φ might be *false* (say, as with the example here) relative to a given τ, there is a distinct *τ-relative falsification* of φ representing that way. And corresponding to each distinct way in which a given sentence φ might be *true* (say) relative to a given τ, there is a distinct *τ-relative verification* of φ representing that way.

We shall be defining these evaluation constructs (τ-relative verifications and falsifications) in due course. The reader can think of them as formal representations of *truthmakers* and *falsitymakers*. The truth of a sentence φ under an assignment τ will then *consist in* the existence of a τ-relative truthmaker for φ; while the falsity of φ under an assignment τ will likewise *consist in* the existence of a τ-relative falsitymaker for φ. For no sentence φ and for no truth-value assignment τ will φ have both a τ-relative truthmaker and a τ-relative falsitymaker. This defers to Aristotle's Law of Non-Contradiction. Moreover, the *classical* logician will seek to ensure in addition that for every sentence φ and for every truth-value assignment τ dealing with the atoms therein, φ will have *either* a τ-relative truthmaker *or* a τ-relative falsitymaker. This is the classicist's Principle of Bivalence. The constructivist, however, does not defer to it.

Exercise 20. *Go back to the quartet of truth-value calculations that were made in §3.18, on p. 94. Re-express each of them as a τ-relative evaluation (either a verification or a falsification), for the truth-value assignment τ in question.*

5.5 Logical truths and falsehoods, and contingency

Sentences of propositional logic can be partitioned into the following three classes.

> **Logical truths** are those sentences that are true under every assignment of truth values to their atoms.

> **Logical falsehoods** are those sentences that are false under every assignment of truth values to their atoms.

> **Contingent sentences** are those that are true under at least one assignment of truth values to their atoms, and false under at least one assignment of truth values to their atoms.

We proceed now to give the beginner their first 'general result' about a logical system (here, the system of propositional logic). It involves some mathematical reasoning at the meta-level. It will give the reader their first taste of the kind of mathematical reasoning typically employed when doing 'metalogic'. The reasoning here is 'by induction on the complexity of sentences'. There are many significant metalogical results that are proved by induction on the complexity of abstract objects—sentences, say, or proofs—that are built up in methodical ways from specified simplest parts.

Lemma 1. *Any sentence with no repetitions of atoms is contingent. That is, it can be made true and it can be made false.*

Proof. By induction on complexity of sentences.
Basis. Every atom can be made true (simply assign it T) and can be made false (simply assign it F).
Inductive Hypothesis. Suppose the result holds for all sentences less complex than φ.
Inductive step. Consider φ of the forms $\neg\psi$, $\psi \wedge \theta$, $\psi \vee \theta$, and $\psi \to \theta$.

Since φ has no repetitions of atoms, its subsentences ψ and θ have no atoms in common. Thus any truth-value assignment to the atoms in ψ can be combined with any truth-value assignment to the atoms in θ—with no 'clashes' of assigned truth values—to produce a single truth-value assignment to all the atoms involved in φ. Clearly also ψ and θ themselves contain no repetitions of atoms; whence the inductive hypothesis (IH) applies to them.

($\neg$): By IH ψ can be made false and can be made true; whence $\neg\psi$ can be made true and can be made false.

($\wedge$): By IH ψ can be made true and can be made false, and θ can be made true and can be made false.

To make $\psi \wedge \theta$ true, make ψ true and make θ true; and combine the two truth-value assignments involved.

To make $\psi \wedge \theta$ false, make either one of ψ and θ false.

($\vee$): To make $\psi \vee \theta$ true, make either one of ψ and θ true.

To make $\psi \vee \theta$ false, make ψ false and make θ false; and combine the two truth-value assignments involved.

($\rightarrow$): To make $\psi \rightarrow \theta$ true, make ψ false or make θ true.

To make $\psi \rightarrow \theta$ false, make ψ true and make θ false; and combine the two truth-value assignments involved. □

Lemma 2. *Every logical truth contains a repetition of atoms. So too does every logical falsehood.*

Proof. Suppose φ is a logical truth. Suppose further for *reductio ad absurdum* that φ contains no repetition of atoms. Then, by Lemma 1, φ can be made false. Contradiction. Thus φ *does* contain a repetition of atoms. Similar reasoning applies when φ is a logical falsehood. □

Here now is a meta-result of rather beguiling simplicity whose proof, which will be given in due course, is considerably more demanding than the proof by induction just given for Lemma 1 (whose immediate consequence is Lemma 2). We simply state the deeper meta-result here, and defer its proof to our intended sequel to this work.

Lemma 3. *Suppose that for every finite subset Γ of a (possibly infinite) set Δ of sentences some truth-value assignment makes every member of*

Γ *true. Then there is a truth-value assignment that makes every member of* Δ *true.*

We have already mentioned compactness of *logical consquence*, on p. 377. Lemma 3 states the compactness of *satisfiability.* It is a nice problem—but one which we shall likewise defer to the sequel—to prove (at the meta-level) that these two forms of compactness are equivalent.

5.6 Templates for some propositional arguments

Here are two recipes, or templates, for generating exercises in appraising validity of argument in propositional logic.

Let $[\neg]P$ have as its possible values either the sentence P itself, or its negation $\neg P$.

Let $P@Q$ have as its possible values

$$P \wedge Q,\ P \vee Q,\ P \rightarrow Q \text{ and } Q \rightarrow P.$$

The first template

Consider the argument-pattern

$$\frac{\begin{array}{l}[\neg]A \\ [\neg]([\neg]A@[\neg]B)\end{array}}{[\neg]B}$$

There are $2^5 \times 4 = 128$ distinct instantiatitions of this argument-pattern, given the substitution possibilities above.

Three examples of such instantiations are

$$\frac{\begin{array}{l}A \\ A \rightarrow B\end{array}}{B} \qquad \frac{\begin{array}{l}\neg A \\ \neg(\neg A \wedge B)\end{array}}{\neg B} \qquad \frac{\begin{array}{l}\neg A \\ \neg(B \rightarrow A)\end{array}}{B}$$

The second template

Now consider the argument-pattern

$$\begin{array}{c} [\neg]([\neg]A@[\neg]B) \\ [\neg]([\neg]B@[\neg]C) \\ \hline [\neg]([\neg]A@[\neg]C) \end{array}$$

There are $2^9 \times 4^3 = 32,768$ possible instantiations of this argument-pattern, given the substitution possibilities above.

Three examples of such instantiations are

$$\begin{array}{c} A \to B \\ B \to C \\ \hline A \to C \end{array} \qquad \begin{array}{c} \neg(\neg A \vee \neg B) \\ \neg(\neg B \vee \neg C) \\ \hline \neg(\neg A \vee \neg C) \end{array} \qquad \begin{array}{c} \neg(A \wedge \neg B) \\ \neg(B \wedge \neg C) \\ \hline \neg A \vee C \end{array}$$

Exercise 21. Using the translation manual
A : Apples are in season;
B : Bananas are in season;
C : Cucumbers are in season,
which of the following English sentences has the logical form

$$(A \wedge \neg B) \to \neg C\ ?$$

1. Apples are in season but if bananas aren't then cucumbers aren't.
2. If apples are in season but bananas aren't, then neither are cucumbers.
3. Apples are in season and bananas are not, even though cucumbers aren't.
4. Apples but not bananas are in season only if cucumbers are.
5. If apples are in season but bananas are not cucumbers, then bananas aren't in season.

Exercise 22. Construct a falsification for the sentence

$$\neg(A \to B) \to (\neg C \wedge \neg D)$$

relative to the truth-value assignment

$$\tau = \begin{cases} A & B & D \\ \downarrow & \downarrow & \downarrow \\ F & F & T \end{cases}.$$

Exercise 23. Determine of each of the following formal sentences whether it is logically true, or logically false, or contingent.

1. $\neg\neg(A \vee \neg A)$
2. $\neg\neg(A \wedge \neg A)$
3. $\neg(A \to B) \vee \neg(B \to A)$
4. $(A \vee \neg A) \to (B \wedge \neg B)$
5. $(A \wedge \neg A) \to (B \vee \neg B)$
6. $A \to (B \to C)$
7. $\neg A \to (\neg B \to \neg C)$
8. $A \vee \neg A$
9. $A \to \neg A$
10. $\neg A \to A$
11. $A \to A$
12. $(A \to A) \to A$
13. $A \to (A \to A)$
14. $A \to (B \to A)$
15. $B \to (A \to A)$
16. $(A \to A) \to B$
17. $A \wedge \neg A$
18. $(A \to B) \to B$
19. $(A \to B) \vee (A \to \neg B)$
20. $(A \to B) \wedge (A \to \neg B)$
21. $(A \to B) \vee (B \to \neg A)$
22. $(A \wedge B) \vee (\neg A \wedge \neg B)$
23. $\neg(A \vee \neg A)$
24. $(A \wedge B) \to A$
25. $(A \wedge B) \vee A$
26. $(A \to B) \vee A$
27. $\neg\neg A \to A$
28. $A \to \neg\neg\neg A$
29. $(\neg A \wedge B) \vee (\neg B \vee A)$
30. $A \to (\neg A \to \neg A)$
31. $\neg(A \wedge \neg A)$
32. $(A \to \neg A) \to A$
33. $(A \to \neg A) \to \neg A$
34. $(A \to B) \vee (B \to A)$
35. $(A \to B) \wedge (B \to A)$
36. $\neg((A \to B) \wedge (B \to \neg A))$
37. $(A \to B) \to (A \to \neg B)$

Exercise 24. Describe the respective advantages and disadvantages of inverted vs. upright tree analyses of the logico-grammatical forms of formal sentences.

Exercise 25. There are sixteen possible binary connectives, as characterized by their two-valued truth tables. We saw in §4.12 that the Sheffer stroke can be taken as a sole primitive connective in classical propositional logic, because all one-place and two-place connectives can be defined in terms of it. Can you find another binary connective that can serve the same purpose? If so, specify its truth table and provide definitions in terms of it for $\neg\varphi$, $\varphi \wedge \psi$, $\varphi \vee \psi$, and $\varphi \rightarrow \psi$.

Chapter 6

From Truth Tables to Rules of Evaluation

We aim to show how in propositional logic the truth tables for the connectives $\neg$, $\wedge$, $\vee$, and $\rightarrow$ lead to the formulation, in inferentialist fashion, of rules of *verification* and rules of *falsification* for sentences with those connectives dominant. We shall do so by concentrating on the row-by-row, left-to-right reading of each truth table. In Chapter 11 we shall extend the treatment of verifications and falsifications to quantified sentences.[1]

6.1 Truth-tabular computation as inferential

The inferentialist construal of attributions of truth and falsity

The inferentialist construes an attribution of *truth* to a sentence φ (under an assignment τ of truth values to atomic sentences) as '*concluding to* φ'. If φ is an *atomic* sentence A, and τ assigns T to A, then this 'concluding' is immediate:

$$\frac{}{A}\,\tau$$

If φ is *compound*, however, then justifying the attribution of truth to

[1]The present author put forward this evaluational semantics, involving model-relative verifications and falsifications, in Tennant [2010] and Tennant [2018].

it will involve some non-trivial evaluating, representing how one uses the truth tables to work out that φ *is true* under τ (i.e., how one can *verify* φ relative to τ):

$$\begin{array}{c} \vdots_{\tau} \\ \varphi \end{array}$$

The inferentialist construes an attribution of *falsity* to a sentence φ (under an assignment τ of truth values to atomic sentences) as '*concluding to* $\perp$ *(absurdity) from* φ'. If φ is an *atomic* sentence A, and τ assigns F to A, then this 'concluding' is immediate:

$$\frac{A}{\perp}\tau\ ,$$

If φ is *compound*, however, then the attribution of falsity to it will involve some non-trivial evaluating, representing how one uses the truth tables to work out that φ *is false* under τ (i.e., how one can *falsify* φ relative to τ):

$$\begin{array}{c} \varphi \\ \vdots_{\tau} \\ \perp \end{array}$$

Literals and factoids

A *literal* is an atomic sentence, or the negation of an atomic sentence. Atomic sentences are called *positive* literals; their negations are called *negative* literals.

In the propositional case, a truth-value assignment τ can be represented as a (consistent, or coherent) set of literals.

A is in the set if and only if $\tau(A) = T$.

$\neg A$ is in the set if and only if $\tau(A) = F$.

It follows that in the propositional case the simplest possible explicitly τ-relative verifications and falsifications are, respectively,

$$\frac{}{A}\tau \quad \text{and} \quad \frac{A}{\perp}\tau\ .$$

We can use the former 'inference *to* A' in place of A itself *as a positive literal*. And we can use the latter 'inference *from* A *to* $\bot$' in place of $\neg A$ itself *as a negative literal*. We can also call them *positive* or *negative* **factoids**, respectively.

We shall use Λ as a placeholder for sets of positive and negative factoids generated by the truth-value assignment in question. We choose purple as the coloration here to remind the reader that these sets Λ can in general contain both positive and negative factoids.

6.2 Rules of Verification and Falsification for the Connectives

Verifications and falsifications are two kinds of (semantic) *evaluation*. They are constructed according to rules. In the following graphic statements of our rules of evaluation, sets of factoids feature as the sets of 'undischarged side assumptions'. The rules are of the respective forms @-$\mathcal{V}$ and @-$\mathcal{F}$ for each logical operator @. @-$\mathcal{V}$ is the verification rule for @, and @-$\mathcal{F}$ is its falsification rule.

In our statement of these rules below, boxes ($\Box$) appended to discharge strokes indicate that the discharged 'assumptions' (*qua* inputs to the computation of truth values) *are required to have been used as indicated* in order for the evaluation rule in question (be it a verification rule or a falsification rule) to be applicable.

This is a requirement of *relevance*. It arises from the fact that one has to have certain truth-value inputs in order to obtain certain truth-value outputs. This requirement of relevance is *obviously justified*, given that the context is one of performing computations with truth values.

Rules of Verification and Falsification for $\neg$

$$(\neg\text{-}\mathcal{V}) \qquad \dfrac{\begin{array}{c} \Box\!\!\overline{}^{(i)} \\ \underbrace{\varphi\ ,\ \Lambda} \\ \vdots \\ \bot \end{array}}{\neg\varphi}\,{\scriptstyle (i)}$$

To verify a negation $\neg\varphi$ it suffices to falsify φ.

(¬-$\mathcal{F}$)

$$\dfrac{\neg\varphi \qquad \begin{array}{c}\Lambda\\ \vdots\\ \varphi\end{array}}{\bot}$$

To falsify a negation $\neg\varphi$ it suffices to verify φ.

Rules of Verification and Falsification for $\wedge$

($\wedge$-$\mathcal{V}$)

$$\dfrac{\begin{array}{c}\Lambda_1\\ \vdots\\ \varphi\end{array} \qquad \begin{array}{c}\Lambda_2\\ \vdots\\ \psi\end{array}}{\varphi\wedge\psi}$$

To verify a conjunction, one must verify both conjuncts.

($\wedge$-$\mathcal{F}$)

$$\dfrac{\varphi\wedge\psi \qquad \begin{array}{c}\Box\text{—}(i)\\ \underbrace{\varphi\ ,\ \Lambda}\\ \vdots\\ \bot\end{array}}{\bot}(i) \qquad\qquad \dfrac{\varphi\wedge\psi \qquad \begin{array}{c}\Box\text{—}(i)\\ \underbrace{\psi\ ,\ \Lambda}\\ \vdots\\ \bot\end{array}}{\bot}(i)$$

To falsify a conjunction, one has to falsify one of the conjuncts.

Rules of Verification and Falsification for $\vee$

($\vee$-$\mathcal{V}$)

$$\dfrac{\begin{array}{c}\Lambda\\ \vdots\\ \varphi\end{array}}{\varphi\vee\psi} \qquad \dfrac{\begin{array}{c}\Lambda\\ \vdots\\ \psi\end{array}}{\varphi\vee\psi}$$

To verify a disjunction, one has to verify one of the disjuncts.

($\vee$-$\mathcal{F}$)

$$\dfrac{\varphi\vee\psi \qquad \begin{array}{c}\Box\text{—}(i)\\ \underbrace{\varphi\ ,\ \Lambda_1}\\ \vdots\\ \bot\end{array} \qquad \begin{array}{c}\Box\text{—}(i)\\ \underbrace{\psi\ ,\ \Lambda_2}\\ \vdots\\ \bot\end{array}}{\bot}(i)$$

To falsify a disjunction, one must falsify both disjuncts.

Rules of Verification and Falsification for →

(→-$\mathcal{V}$)

$$\dfrac{\begin{array}{c}\overset{\square\text{---}(i)}{\underbrace{\varphi\ ,\ \Lambda}}\\ \vdots\\ \bot\end{array}}{\varphi\to\psi}\,(i) \qquad\qquad \dfrac{\begin{array}{c}\Lambda\\ \vdots\\ \psi\end{array}}{\varphi\to\psi}$$

To verify a conditional, one must either falsify its antecedent or verify its consequent.

(→-$\mathcal{F}$)

$$\dfrac{\varphi\to\psi \qquad \begin{array}{c}\Lambda_1\\ \vdots\\ \varphi\end{array} \qquad \begin{array}{c}\overset{\square\text{---}(i)}{\underbrace{\psi\ ,\ \Lambda_2}}\\ \vdots\\ \bot\end{array}}{\bot}\,(i)$$

To falsify a conditional, one must both verify its antecedent and falsify its consequent.

Remarks on Propositional Verifications and Falsifications

1. Any verification of a complex sentence φ has φ as the conclusion of its last step, which is an application of the $\mathcal{V}$-rule for the operator dominant in φ.

2. Any falsification of a complex sentence φ has φ as the major premise of its last step, which is an application of the $\mathcal{F}$-rule for the operator dominant in φ.

3. Verifications and falsifications are *relative to a truth-value assignment* τ. They 'depend' *only on* τ*-literals* as their 'undischarged assumptions'. (We use scare quotes here because these are assumptions for *evaluations*, not for *deductions*.)

4. Verifications and falsifications are, mathematically speaking, (labeled) trees all of whose branches are finite.

5. The 'premises' of a verification or falsification are 'atomic facts' of the truth-value assignment τ—positive and/or negative factoids.

6. In any verification or falsification, every major premise for the application of an $\mathcal{F}$-rule *stands proud.* It is not the conclusion of any verification. Moreover, any such major premise is *discharged* at the very next step. Hence all undischarged major premises φ are those of terminal applications of $\mathcal{F}$-rules in falsifications of φ.

7. The evaluation rules used in verifications and falsifications invite generalization to the case of formal deduction of a *possibly complex sentential* conclusion (not just: $\bot$) from a set of *possibly complex* premises (not just: literals).

8. The *verification rules* (@-$\mathcal{V}$) will generalize to become (or, as we shall say, *morph into*) *@-Introduction rules.*

9. The *falsification rules* (@-$\mathcal{F}$) will generalize to become (or, as we shall say, *morph into*) *@-Elimination rules.*

Remarks on Bivalence and the Law of Non-Contradiction

We can see from the rules of evaluation for negation that **Bivalence** will hold for evaluations with respect to any truth-value assignment τ : every sentence being evaluated will either come out true (by having a τ-relative verification), or come out false (by having a τ-relative falsification).

Moreover, the **Law of Non-Contradiction** will hold at the meta-level: no sentence will have both a τ-relative verification and a τ-relative falsification. This can be proved by induction on the complexity of sentences, by appeal to the basis fact that the truth-value assignment τ is both *atomically complete* and *coherent.*

6.3 Rules of evaluation as inductive clauses in a co-inductive definition of evaluations

A theoretically important way to construe the foregoing rules of verification and falsification is as inductive clauses in a co-inductive definition of evaluations (i.e., of verifications and falsifications). We need to state a basis clause, telling us what the simplest verifications and falsifications are. We need to make clear that verifications and falsifications are *relative to an interpretation* (here, a truth-value assignment). We shall call it τ, and the reader should understand that it is taken to be fixed throughout. We need to state how each rule of evaluation succeeds in constructing a new evaluation out of ones already constructed. And we need to ensure, through a closure clause, that these are the *only* ways to construct evaluations.

Basis. The simplest *verifications* are of the form

$$\frac{}{A}\tau$$

where A is an atom that τ makes true. Such a construct is a verification of A relative to τ. It uses the (singleton) set

$$\left\{\frac{}{A}\tau\right\}$$

of τ-factoids.

The simplest *falsifications* are of the form

$$\frac{A}{\bot}\tau$$

where A is an atom that τ makes false. Such a construct is a falsification of A relative to τ. It uses the (singleton) set

$$\left\{\frac{A}{\bot}\tau\right\}$$

of τ-factoids.

Inductive Clauses. We label each clause here with the same label as we used for the corresponding rule stated in §6.2.

($\neg$-$\mathcal{V}$) If F is a falsification of φ using the set Λ of τ-factoids, then

$$\frac{F}{\neg\varphi}$$

is a verification of $\neg\varphi$ using the same set Λ of τ-factoids.

($\neg$-$\mathcal{F}$) If V is a verification of φ using the set Λ of τ-factoids, then

$$\frac{\neg\varphi \quad V}{\bot}$$

is a falsification of $\neg\varphi$ using the same set Λ of τ-factoids.

($\wedge$-$\mathcal{V}$) If V_1 is a verification of φ using the set Λ_1 of τ-factoids and V_2 is a verification of ψ using the set Λ_2 of τ-factoids, then

$$\frac{V_1 \quad V_2}{\varphi \wedge \psi}$$

is a verification of $\varphi \wedge \psi$ using the set $\Lambda_1 \cup \Lambda_2$ of τ-factoids.

($\wedge$-$\mathcal{F}$) If F is a falsification of φ using the set Λ of τ-factoids, then

$$\frac{\varphi \wedge \psi \quad F}{\bot}$$

is a falsification of $\varphi \wedge \psi$ using the same set Λ of τ-factoids.

Likewise, if F is a falsification of ψ using the set Λ of τ-factoids, then

$$\frac{\varphi \wedge \psi \quad F}{\bot}$$

is a falsification of $\varphi \wedge \psi$ using the same set Λ of τ-factoids.

($\vee$-$\mathcal{V}$) If V is a verification of φ using the set Λ of τ-factoids, then

$$\frac{V}{\varphi \vee \psi}$$

is a verification of $\varphi \vee \psi$ using the same set Λ of τ-factoids.

Likewise, if V is a verification of ψ using the set Λ of τ-factoids, then

$$\frac{V}{\varphi \vee \psi}$$

is a verification of $\varphi \vee \psi$ using the same set Λ of τ-factoids.

($\vee$-$\mathcal{F}$) If F_1 is a falsification of φ using the set Λ_1 of τ-factoids and F_2 is a falsification of ψ using the set Λ_2 of τ-factoids, then

$$\frac{\varphi \vee \psi \quad F_1 \quad F_2}{\bot}$$

is a falsification of $\varphi \vee \psi$ using the set $\Lambda_1 \cup \Lambda_2$ of τ-factoids.

($\rightarrow$-$\mathcal{V}$) If F is a falsification of φ using the set Λ of τ-factoids, then

$$\frac{F}{\varphi \rightarrow \psi}$$

is a verification of $\varphi \rightarrow \psi$ using the same set Λ of τ-factoids.

Likewise, if V is a verification of ψ using the set Λ of τ-factoids, then

$$\frac{V}{\varphi \rightarrow \psi}$$

is a verification of $\varphi \rightarrow \psi$ using the same set Λ of τ-factoids.

($\rightarrow$-$\mathcal{F}$) If V is a verification of φ using the set Λ_1 of τ-factoids and F is a falsification of ψ using the set Λ_2 of τ-factoids, then

$$\frac{\varphi \rightarrow \psi \quad V \quad F}{\bot}$$

is a falsification of $\varphi \rightarrow \psi$ using the set $\Lambda_1 \cup \Lambda_2$ of τ-factoids.

Closure Clause. A construction is a verification or a falsification of a given sentence using a certain set of τ-factoids only if its being so follows from the preceding clauses.

The reader should now compare each of the foregoing inductive clauses with the earlier rule in §6.2 with which it has been paired by having the same label. The two members of each such pair are just *different ways of saying the same thing.*

The sayings in question are, of course, in the metalanguage. It can be convenient to abbreviate, symbolically, various important metalinguistic notions for the purpose of framing more succinct definitions of them. The reader will find convenient the following two abbreviations of the notions that we have just co-inductively defined, in order to re-write more succinctly the co-inductive definition of verifications and falsifications that has been furnished in this section.

$\mathcal{V}(V, \varphi, \Lambda, \tau)$ abbreviates 'V verifies φ using the set Λ of τ-factoids'.

$\mathcal{F}(F, \varphi, \Lambda, \tau)$ abbreviates 'F falsifies φ using the set Λ of τ-factoids'.

Here is the resulting re-writing, with the inductive clauses rendered as metalinguistic rules of inference using our symbolic abbreviations.

Basis. $\dfrac{}{\mathcal{V}\left(\dfrac{}{A}^{\tau}, A, \left\{\dfrac{}{A}^{\tau}\right\}, \tau\right)}$ $\qquad$ $\dfrac{}{\mathcal{F}\left(\dfrac{A}{\bot}^{\tau}, A, \left\{\dfrac{A}{\bot}^{\tau}\right\}, \tau\right)}$

Inductive Clauses. Once again we label each clause here with the same label as we used for the corresponding rule stated in §6.2.

$(\neg\text{-}\mathcal{V})$ $\dfrac{\mathcal{F}(F, \varphi, \Lambda, \tau)}{\mathcal{V}\left(\dfrac{F}{\neg\varphi}, \varphi, \Lambda, \tau\right)}$

$(\neg\text{-}\mathcal{F})$ $\dfrac{\mathcal{V}(V, \varphi, \Lambda, \tau)}{\mathcal{F}\left(\dfrac{\neg\varphi \quad V}{\bot}, \varphi, \Lambda, \tau\right)}$

$(\wedge\text{-}\mathcal{V})$ $\dfrac{\mathcal{V}(V_1, \varphi, \Lambda_1, \tau) \quad \mathcal{V}(V_2, \psi, \Lambda_2, \tau)}{\mathcal{V}\left(\dfrac{V_1 \quad V_2}{\varphi \wedge \psi}, \varphi, \Lambda_1 \cup \Lambda_2, \tau\right)}$

$(\wedge\text{-}\mathcal{F})$ $\dfrac{\mathcal{F}(F, \varphi, \Lambda, \tau)}{\mathcal{F}\left(\dfrac{\varphi \wedge \psi \quad F}{\bot}, \varphi \wedge \psi, \Lambda, \tau\right)}$ $\qquad$ $\dfrac{\mathcal{F}(F, \psi, \Lambda, \tau)}{\mathcal{F}\left(\dfrac{\varphi \wedge \psi \quad F}{\bot}, \varphi \wedge \psi, \Lambda, \tau\right)}$

$(\vee\text{-}\mathcal{V})$ $\dfrac{\mathcal{V}(V, \varphi, \Lambda, \tau)}{\mathcal{V}\left(\dfrac{V}{\varphi \vee \psi}, \varphi \vee \psi, \Lambda, \tau\right)}$ $\qquad$ $\dfrac{\mathcal{V}(V, \psi, \Lambda, \tau)}{\mathcal{V}\left(\dfrac{V}{\varphi \vee \psi}, \varphi \vee \psi, \Lambda, \tau\right)}$

$(\vee\text{-}\mathcal{F})$ $\dfrac{\mathcal{F}(F_1, \varphi, \Lambda_1, \tau) \quad \mathcal{F}(F_2, \psi, \Lambda_2, \tau)}{\mathcal{F}\left(\dfrac{\varphi \vee \psi \quad F_1 \quad F_2}{\bot}, \varphi \vee \psi, \Lambda_1 \cup \Lambda_2, \tau\right)}$

$(\rightarrow\text{-}\mathcal{V})$ $\dfrac{\mathcal{F}(F, \varphi, \Lambda, \tau)}{\mathcal{V}\left(\dfrac{F}{\varphi \rightarrow \psi}, \varphi \rightarrow \psi, \Lambda, \tau\right)}$ $\qquad$ $\dfrac{\mathcal{V}(V, \psi, \Lambda, \tau)}{\mathcal{V}\left(\dfrac{V}{\varphi \rightarrow \psi}, \varphi \rightarrow \psi, \Lambda, \tau\right)}$

$(\rightarrow\text{-}\mathcal{F})$ $\dfrac{\mathcal{V}(V, \varphi, \Lambda_1, \tau) \quad \mathcal{F}(F, \psi, \Lambda_2, \tau)}{\mathcal{F}\left(\dfrac{\varphi \rightarrow \psi \quad V \quad F}{\bot}, \varphi \rightarrow \psi, \Lambda_1 \cup \Lambda_2, \tau\right)}$

Closure Clause. A construction is a verification or a falsification of a given sentence using a certain set of τ-factoids only if its being so follows from the preceding clauses.

Chapter 7

From Rules of Evaluation to Rules of Deduction

The rules of evaluation that were stated in §6.2 serve, as we saw in §6.3, as inductive clauses in a co-inductive definition of verifications and falsifications. The basis clause stipulates that the simplest verifications and falsifications are of the respective forms

$$\frac{}{A}\,\tau \quad \text{and} \quad \frac{A}{\bot}\,\tau \;.$$

where τ stands for the truth-value assignment to atoms in question.

7.1 The method of morphing

Evaluations (verifications and falsifications) are *model relative.* In the propositional case, this amounts to their being relative to a particular *truth value assignment* τ. By contrast, *logical deductions* (built up by means of *logical rules of inference*) are *model invariant.* Our aim here is to show how to 'morph' our rules of evaluation into *rules of deduction.* For each logical operator @, its Verification rule (@-$\mathcal{V}$) will morph into its Introduction Rule in natural deduction, and its Falsification rule (@-$\mathcal{F}$) will morph into the corresponding Elimination Rule.

To effect such morphing, do the following.

1. Replace sets Λ of literals with sets Δ of arbitrary sentences.

2. In the $\mathcal{F}$-rules for $\wedge$, $\vee$, and $\rightarrow$, allow for an arbitrary sentence θ as the conclusion, as an alternative to $\bot$. One can still, of course, have $\bot$ as a conclusion, as a 'special case' of θ.

3. In step (2) above, when formulating $\vee$-E from $\vee$-$\mathcal{F}$, allow for θ to be the overall conclusion even when it is the conclusion of just one of the two case-proofs, the other case-proof concluding with $\bot$. This is called 'liberalized' $\vee$-E.[1]

4. **Keep the boxes** appended to discharge strokes when morphing. This ensures that the resulting deductions will be **relevant**.

5. **Keep major premises standing proud** when morphing. This ensures that the resulting deductions will be in **normal form**.

A natural deduction is said to be in normal form just in case no major premise for an application of an elimination rule for its dominant operator stands as the conclusion of an application of the introduction rule for that operator. If major premises of eliminations (henceforth: MPEs) always stand proud, this requirement of normality will trivially be met.

Remarks to help one understand the following Morphings

1. We shall use V (with or without numerical subscripts) as a placeholder for verifications.

2. Reminder: A verification has a sentence as its conclusion.

3. We shall use F (with or without numerical subscripts) as a placeholder for falsifications.

4. Reminder: A falsification has the absurdity symbol $\bot$ as its conclusion.

5. We shall use Λ (with or without numerical subscripts) as a placeholder for sets of literals of the truth-value assignment relative to which a sentence is being evaluated.

---

[1]This form of $\vee$-E was first put forward in Tennant [1979a], at p. 183 *supra*. In the same paper, at p. 175 *supra*, we put forward the 'no vacuous discharge' requirement on the rules $\vee$-E and $\exists$-E.

6. We shall use Π (with or without numerical subscripts) as a placeholder for proofs (natural deductions).

7. A proof can have either a sentence or the absurdity symbol $\bot$ as its conclusion.

8. We shall use Δ (with or without numerical subscripts) as a placeholder for sets of sentences (the set of premises of a proof).

9. We shall color in brown the changes wrought by the morphings, which will be indicated by the squiggly arrow $\rightsquigarrow$.

7.2 The morphings themselves

$$(\neg\text{-}\mathcal{V})\qquad \cfrac{\begin{array}{c}\square\text{---}(i)\\ \underbrace{\varphi\ ,\ \Lambda}\\ F\\ \bot\end{array}}{\neg\varphi}\,(i) \qquad\rightsquigarrow\qquad (\neg\text{-I})\qquad \cfrac{\begin{array}{c}\square\text{---}(i)\\ \underbrace{\varphi\ ,\ \Delta}\\ \Pi\\ \bot\end{array}}{\neg\varphi}\,(i)$$

$$(\wedge\text{-}\mathcal{V})\qquad \cfrac{\begin{array}{c}\Lambda_1\\ V_1\\ \varphi_1\end{array}\qquad\begin{array}{c}\Lambda_2\\ V_2\\ \varphi_2\end{array}}{\varphi_1\wedge\varphi_2} \qquad\rightsquigarrow\qquad (\wedge\text{-I})\qquad \cfrac{\begin{array}{c}\Delta_1\\ \Pi_1\\ \varphi_1\end{array}\qquad\begin{array}{c}\Delta_2\\ \Pi_2\\ \varphi_2\end{array}}{\varphi_1\wedge\varphi_2}$$

$$(\vee\text{-}\mathcal{V})\qquad \cfrac{\begin{array}{c}\Lambda\\ V\\ \varphi\end{array}}{\varphi\vee\psi}\qquad \cfrac{\begin{array}{c}\Lambda\\ V\\ \psi\end{array}}{\varphi\vee\psi} \qquad\rightsquigarrow\qquad (\vee\text{-I})\qquad \cfrac{\begin{array}{c}\Delta\\ \Pi\\ \varphi\end{array}}{\varphi\vee\psi}\qquad \cfrac{\begin{array}{c}\Delta\\ \Pi\\ \psi\end{array}}{\varphi\vee\psi}$$

$$(\rightarrow\text{-}\mathcal{V})\qquad \cfrac{\begin{array}{c}\square\text{---}(i)\\ \underbrace{\varphi\ ,\ \Lambda}\\ F\\ \bot\end{array}}{\varphi\rightarrow\psi}\,(i)\qquad \cfrac{\begin{array}{c}\Lambda\\ V\\ \psi\end{array}}{\varphi\rightarrow\psi} \qquad\rightsquigarrow\qquad (\rightarrow\text{-I})\qquad \cfrac{\begin{array}{c}\square\text{---}(i)\\ \underbrace{\varphi\ ,\ \Delta}\\ \Pi\\ \bot\end{array}}{\varphi\rightarrow\psi}\,(i)\qquad \cfrac{\begin{array}{c}\Delta\\ \Pi\\ \psi\end{array}}{\varphi\rightarrow\psi}$$

Note that these two parts of ($\rightarrow$-I) resulting from this direct morphing of the verification rule ($\rightarrow$-$\mathcal{V}$) do not yet furnish that form of ($\rightarrow$-I) that is conventionally known as 'Conditional Proof' (CP). This is the form that permits one to assume φ for the sake of argument, then to deduce ψ by (possibly, but not necessarily) using the assumption φ, and finally to discharge φ (if one *has* used it) when one infers the conclusion $\varphi \to \psi$. We shall see in due course how to get this 'missing part' (CP) of ($\to$-I).

So much, for the time being, for the Verification rules and how to morph them into Introduction rules. We turn now to the Falsification rules and how to morph *them* into *Elimination* rules.

$$(\neg\text{-}\mathcal{F}) \quad \dfrac{\neg\varphi \qquad \begin{array}{c}\Lambda\\ V\\ \varphi\end{array}}{\bot} \quad \rightsquigarrow \quad (\neg\text{-E}) \quad \dfrac{\neg\varphi \qquad \begin{array}{c}\Delta\\ \Pi\\ \varphi\end{array}}{\bot}$$

$$(\wedge\text{-}\mathcal{F}) \quad \dfrac{\varphi\wedge\psi \qquad \begin{array}{c}\overset{\square\text{---}(i)}{\underbrace{\varphi\ ,\ \Lambda}}\\ F\\ \bot\end{array}}{\bot}(i) \quad \dfrac{\varphi\wedge\psi \qquad \begin{array}{c}\overset{\square\text{---}(i)}{\underbrace{\psi\ ,\ \Lambda}}\\ F\\ \bot\end{array}}{\bot}(i) \quad \rightsquigarrow \quad (\wedge\text{-E}) \quad \dfrac{\varphi\wedge\psi \qquad \begin{array}{c}\overset{\square\text{---}(i)}{\underbrace{\varphi\ ,\ \Delta}}\\ \Pi\\ \theta/\bot\end{array}}{\theta/\bot}(i) \quad \dfrac{\varphi\wedge\psi \qquad \begin{array}{c}\overset{\square\text{---}(i)}{\underbrace{\psi\ ,\ \Delta}}\\ \Pi\\ \theta/\bot\end{array}}{\theta/\bot}(i)\ ;$$

hence, more efficiently:

$$\dfrac{\varphi\wedge\psi \qquad \begin{array}{c}\overset{(i)\text{---}\square\text{---}(i)}{\underbrace{\varphi\,,\psi\ ,\ \Delta}}\\ \Pi\\ \theta/\bot\end{array}}{\theta/\bot}(i)\ ,$$

where the box between the two discharge strokes indicates that at least one of φ or ψ must feature as an undischarged assumption of the subproof Π.

$$(\vee\text{-}\mathcal{F}) \quad \dfrac{\varphi_1\vee\varphi_2 \qquad \begin{array}{c}\overset{\square\text{---}(i)}{\underbrace{\varphi_1\ ,\ \Lambda_1}}\\ F_1\\ \bot\end{array} \qquad \begin{array}{c}\overset{\square\text{---}(i)}{\underbrace{\varphi_2\ ,\ \Lambda_2}}\\ F_2\\ \bot\end{array}}{\bot}(i) \quad \rightsquigarrow \quad (\vee\text{-E}) \quad \dfrac{\varphi_1\vee\varphi_2 \qquad \begin{array}{c}\overset{\square\text{---}(i)}{\underbrace{\varphi_1\ ,\ \Delta_1}}\\ \Pi_1\\ \theta/\bot\end{array} \qquad \begin{array}{c}\overset{\square\text{---}(i)}{\underbrace{\varphi_2\ ,\ \Delta_2}}\\ \Pi_2\\ \theta/\bot\end{array}}{\theta/\bot}(i)$$

($\vee$-E) says 'If either of the two case-proofs Π_1, Π_2 has $\bot$ as its conclusion, bring down as the main conclusion the conclusion of the other case-proof.'

$$(\to\text{-}\mathcal{F})\quad \dfrac{\varphi\to\psi \quad \begin{matrix}\Lambda_1\\ V\\ \varphi\end{matrix} \quad \begin{matrix}\square\!\!-\!(i)\\ \underbrace{\psi\,,\Lambda_2}\\ F\\ \bot\end{matrix}}{\bot}\,(i) \qquad\rightsquigarrow\qquad (\to\text{-E})\quad \dfrac{\varphi\to\psi \quad \begin{matrix}\Delta_1\\ \Pi_1\\ \varphi\end{matrix} \quad \begin{matrix}\square\!\!-\!(i)\\ \underbrace{\psi\,,\Delta_2}\\ \Pi_2\\ \theta/\bot\end{matrix}}{\theta/\bot}\,(i)$$

Here, Π_1 is called the *minor* subproof, and Π_2 is called the *major* subproof, for ($\to$-E).

7.3 The effect of Bivalence

Because **Bivalence** holds, one can appeal, when making *deductions*, to the classical Law of Excluded Middle:

$$\textbf{LEM}\qquad \dfrac{}{\varphi\vee\neg\varphi}$$

Recall 'liberalized' $\vee$-E in step 3 above: *when formulating* $\vee$*-E from* $\vee$*-*$\mathcal{F}$*, allow for* θ *to be the overall conclusion even when it is the conclusion of just one of the two case-proofs, the other case-proof concluding with* $\bot$.

When **LEM** is the major premise for $\vee$-E, we can therefore reason as follows:

$$\dfrac{\dfrac{}{\varphi\vee\neg\varphi} \quad \begin{matrix}(i)\overline{}\\ \varphi\\ \vdots\\ \psi/\bot\end{matrix} \quad \begin{matrix}\overline{}(i)\\ \neg\varphi\\ \vdots\\ \psi/\bot\end{matrix}}{\psi/\bot}\,(i)$$

The allowable forms of Dilemma

Since **LEM** is an axiom, we can suppress it. It has allowed us to derive

$$\textbf{Dilemma}\qquad \dfrac{\begin{matrix}(i)\overline{}\\ \varphi\\ \vdots\\ \psi/\bot\end{matrix} \quad \begin{matrix}\overline{}(i)\\ \neg\varphi\\ \vdots\\ \psi/\bot\end{matrix}}{\psi/\bot}\,(i)$$

This is graphic portmanteau for the following four possibilities:

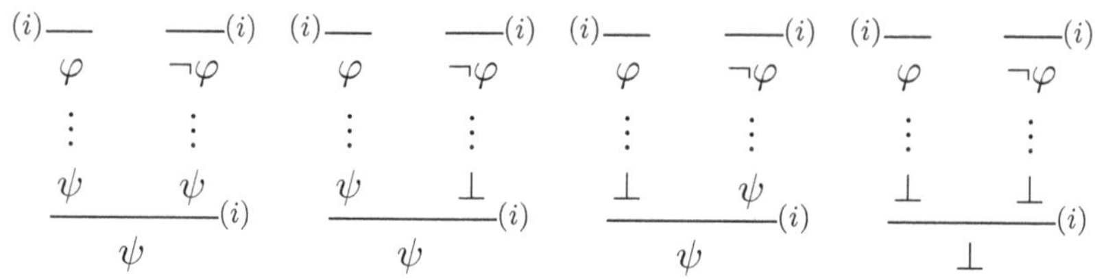

Only the first two of these, however, are strictly classical; the third and fourth can be derived constructively. (Apply $\neg$-I to their positive horns.)

The actually needed forms of Dilemma

In order to ensure full classical strength for the resulting system of natural deduction, we therefore need Dilemma in only the following two forms, where now ψ cannot be instantiated by $\bot$.

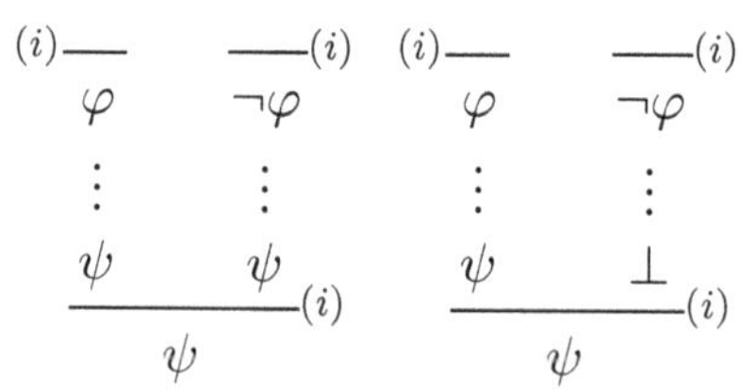

Allowing $\rightarrow$-Introduction to *discharge* an assumption

Note that we have thus far obtained an introduction rule for $\rightarrow$ in only the two parts that correspond to the respective considerations that it suffices, for the truth of a conditional, to have its antecedent false, or to have its consequent true. But as already explained, this leaves out of the picture the most familiar form of $\rightarrow$-Introduction, known as Conditional Proof (CP), which allows for the discharge of the antecedent φ as an assumption *if* (but not: *only* if) it has been used in the derivation of the consequent ψ as the conclusion of the subordinate proof. We shall use a diamond $\diamond$ affixed to the discharge stroke to encode the liberal discharge condition that we have just explained:

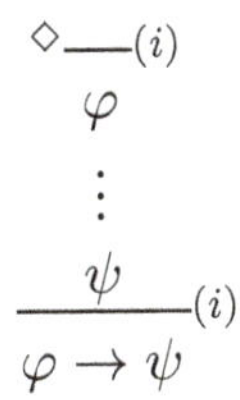

Rules of Deduction: An extra part for →-Introduction

The first two parts of →-Introduction immediately below were obtained by morphing the rule of →-Verification. The third part (CP), allowing discharge of assumptions, can now be added to them, in light of the derivations below.

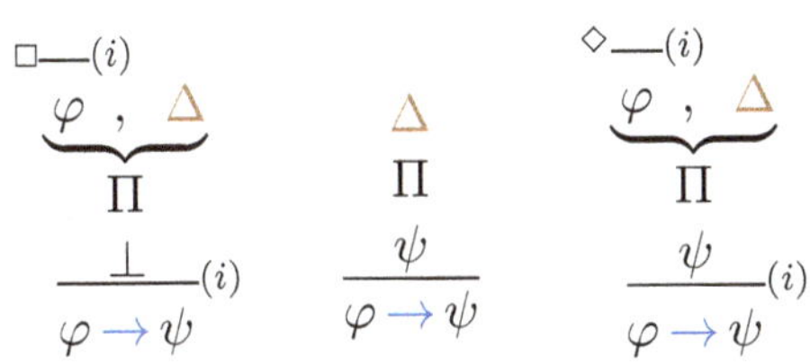

The following two derivations of (CP) employ **Bivalence** in the form of the classical rule of Dilemma. The first derivation uses the conditional's antecedent A as its positive horn-assumption; the second derivation uses the conditional's consequent B as its positive horn-assumption. The respective derivations have applications of Dilemma as their final steps, which are labeled with '(2)':

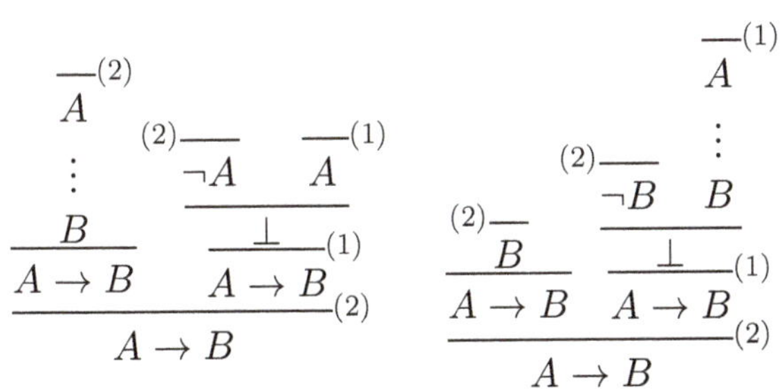

Succinct graphic statement of →-Introduction

We now have the above-displayed three 'parts' of the rule of →-Introduction, which we reprise here for visual convenience:

$$\dfrac{\begin{array}{c}\overset{\Box\text{---}(i)}{\underbrace{\varphi\ ,\ \Delta}}\\ \Pi\\ \bot\end{array}}{\varphi\rightarrow\psi}\,(i)\qquad\qquad \dfrac{\begin{array}{c}\Delta\\ \Pi\\ \psi\end{array}}{\varphi\rightarrow\psi}\qquad\qquad \dfrac{\begin{array}{c}\overset{\Diamond\text{---}(i)}{\underbrace{\varphi\ ,\ \Delta}}\\ \Pi\\ \psi\end{array}}{\varphi\rightarrow\psi}\,(i)$$

But a moment's thought makes one realize that the second part is now *covered* by the third part. So the third part subsumes the second part. Accordingly the final display for (→-I) becomes

$$\dfrac{\begin{array}{c}\overset{\Box\text{---}(i)}{\underbrace{\varphi\ ,\ \Delta}}\\ \Pi\\ \bot\end{array}}{\varphi\rightarrow\psi}\,(i)\qquad\qquad \dfrac{\begin{array}{c}\overset{\Diamond\text{---}(i)}{\underbrace{\varphi\ ,\ \Delta}}\\ \Pi\\ \psi\end{array}}{\varphi\rightarrow\psi}\,(i)$$

The diamond in the graphic rule-part on the *right* says 'vacuous' discharge is permitted: φ need not be an undischarged assumption of the subordinate proof Π. This is exactly the rule of (→-I) in the Gentzen–Prawitz tradition. The graphic rule-part on the *left* is a 'new part' of (→-I), arising naturally via the foregoing method of morphing. It directly respects the third and fourth lines of the truth table for →, according to which the falsity of the antecedent φ renders a conditional $\varphi \rightarrow \psi$ true. To be sure, the graphic rule-part on the left is 'non-trivially' *derivable* in the Gentzen–Prawitz system:

$$\dfrac{\dfrac{\begin{array}{c}\overset{\Box\text{---}(i)}{\underbrace{\varphi\ ,\ \Delta}}\\ \Pi\\ \bot\end{array}}{\psi}\,\text{EFQ}}{\varphi\rightarrow\psi}\,(i)$$

but only by dint of *their* appeal to the rule *Ex Falso Quodlibet* (EFQ), *which has not raised its head anywhere among our own rules thus far.*

To be sure, we *mentioned* EFQ in passing, on p. 54, as part of our explanation of *Gentzen's* choice of natural-deduction rules in the snippet displayed there from his pioneering paper. But we have not *used* or *adopted* EFQ ourselves. Moreover it will emerge that we can keep it that way. It will emerge that *we have no use for EFQ in our successful project of furnishing a formal logical system for the regimentation of mathematical reasoning.*

7.4 Graphic statement of the classical rules of inference

We are now in a position to collect together all the natural-deduction rules for propositional logic that we have discovered by the method of morphing our 'proof-theoretically' *semantic* rules of verification and falsification.

We shall revert to the usual way of stating such rules. We shall use three vertically descending dots to gesture at possible proof-work implicitly there. We shall suppress explicit reminders of sets of undischarged side-assumptions. *But* we shall be meticulous in annotating discharge rules with their boxes or diamonds.

Note that all elimination rules are now in *parallelized* form, and their major premises *stand proud*; hence, all proofs are in *normal form.*

Graphic Deductive Rules for the Connectives

First, Introduction rules:

$$\dfrac{\begin{array}{c}\square\!\!-\!\!(i)\\ \varphi\\ \vdots\\ \bot\end{array}}{\neg\varphi}{\scriptstyle(i)} \qquad \dfrac{\varphi \quad \psi}{\varphi\wedge\psi} \qquad \dfrac{\varphi}{\varphi\vee\psi} \quad \dfrac{\psi}{\varphi\vee\psi} \qquad \dfrac{\begin{array}{c}\square\!\!-\!\!(i)\\ \varphi\\ \vdots\\ \bot\end{array}}{\varphi\rightarrow\psi}{\scriptstyle(i)} \quad \dfrac{\begin{array}{c}\diamond\!\!-\!\!(i)\\ \varphi\\ \vdots\\ \psi\end{array}}{\varphi\rightarrow\psi}{\scriptstyle(i)}$$

Second, Elimination rules:

$$\frac{\neg\varphi \quad \varphi}{\bot} \qquad \frac{\varphi\wedge\psi \quad \begin{array}{c}{}^{(i)}\square^{(i)}\\ \underbrace{\varphi,\psi}\\ \vdots\\ \theta\end{array}}{\theta}{\scriptstyle(i)} \qquad \frac{\varphi\vee\psi \quad \begin{array}{c}\square^{(i)}\\ \varphi\\ \vdots\\ \theta/\bot\end{array} \quad \begin{array}{c}\square^{(i)}\\ \psi\\ \vdots\\ \theta/\bot\end{array}}{\theta/\bot}{\scriptstyle(i)} \qquad \frac{\varphi\rightarrow\psi \quad \begin{array}{c}\\ \\ \vdots\\ \varphi\end{array} \quad \begin{array}{c}\square^{(i)}\\ \psi\\ \vdots\\ \theta\end{array}}{\theta}{\scriptstyle(i)}$$

... and finally a rule that makes the logic **classical**, such as:

Dilemma

$$\frac{\begin{array}{c}{}^{(i)}\square\\ \varphi\\ \vdots\\ \psi\end{array} \quad \begin{array}{c}\square^{(i)}\\ \neg\varphi\\ \vdots\\ \psi\end{array}}{\psi}{\scriptstyle(i)} \qquad \frac{\begin{array}{c}{}^{(i)}\square\\ \varphi\\ \vdots\\ \psi\end{array} \quad \begin{array}{c}\square^{(i)}\\ \neg\varphi\\ \vdots\\ \bot\end{array}}{\psi}{\scriptstyle(i)}$$

7.5 Serial forms of elimination for $\wedge$ and $\rightarrow$

The Elimination Rules for $\wedge$ and $\rightarrow$ that have been obtained above by our method of morphing are called *parallelized* forms. This is because the inferential steps they immediately afford are to conclusions θ in general, which do not always have to occur as immediate constituents (φ or ψ) of the respective major premises (namely, the conjunction $\varphi\wedge\psi$, or the conditional $\varphi\rightarrow\psi$) whose dominant operator is being eliminated.

These parallelized forms happen to be interderivable with the *serial* forms of elimination for $\wedge$ and $\rightarrow$ that were furnished by Gentzen when he invented the method of natural deduction. These serial forms afford inferential steps only to conclusions that *do* occur as immediate constituents of the major premise whose dominant operator is being eliminated.

The serial forms of elimination for $\wedge$ and $\rightarrow$ are these:

$$\frac{\varphi\wedge\psi}{\varphi} \quad \frac{\varphi\wedge\psi}{\psi} \qquad \frac{\varphi\rightarrow\psi \quad \varphi}{\psi}$$

Here are *inter*derivations of these Gentzenian serial forms of elimination with our parallelized forms.

Derivations of parallelized forms using serial forms

$$\begin{array}{c}\underbrace{\dfrac{\varphi\wedge\psi}{\varphi}\;,\;\dfrac{\varphi\wedge\psi}{\psi}}\\ \vdots\\ \theta\end{array}\qquad\qquad\begin{array}{c}\vdots\\ \dfrac{\varphi\to\psi\quad\varphi}{\psi}\\ \vdots\\ \theta\end{array}$$

Derivations of serial forms using parallelized forms

$$\dfrac{\varphi\wedge\psi\quad\overline{\varphi}^{(1)}}{\varphi}\qquad\dfrac{\varphi\wedge\psi\quad\overline{\varphi}^{(1)}}{\varphi}\qquad\dfrac{\varphi\to\psi\quad\varphi\quad\overline{\psi}^{(1)}}{\psi}$$

Exercise 26. We refer again to Exercise 23 on page 164.

(i) You will have identified which sentences in the given list are logical truths. For each such sentence φ, use the primitive deductive rules of classical logic that we have just displayed to prove φ *from no assumptions.* You may of course make certain assumptions 'for the sake of argument'; but, by the stage at which you infer φ as the overall conclusion of your proof, all such assumptions must be discharged. (If you need to see more proofs to 'get the hang' of how they can be put together, you might wish to take a look at Chapter 21, 'Proofs Galore'. You could also consult §13.4, 'A recipe for proof construction'.)

(ii) You will have identified which sentences in the given list are logical falsehoods. For each such sentence φ, use the primitive deductive rules of **constructive** logic to disprove φ—that is, to derive $\bot$ from φ as the sole premise. This means that you cannot use any of the strictly classical rules, such as Dilemma, Classical Reductio, Double Negation Elimination, or the Law of Excluded Middle. You may of course make certain assumptions other than φ 'for the sake of argument'; but, by the stage at which you infer $\bot$ as the conclusion of your (dis)proof, all such assumptions must be discharged, leaving φ as the sole undischarged assumption.

Chapter 8

Strictly Classical Rules of Negation

8.1 Strictly classical negation rules

There are four of these rules. One of them—**DNE**—we saw 'in action' in two of the deductive problems in §3.17. Another two of them—**LEM** and **Dilemma**—we have already seen in §7.3. It remains to acquaint the reader with Classical Reductio (**CR**). Here, for the record and for some further investigation, is a complete statement of the classical quartet.

Classical Reductio ad Absurdum (CR):

$$\begin{array}{c} \square\overline{\quad}^{(i)} \\ \neg\varphi \\ \vdots \\ \dfrac{\bot}{\varphi}(i) \end{array}$$

Law of Excluded Middle (LEM) :

$$\frac{}{\varphi \vee \neg\varphi}$$

Dilemma (Dil):

$$\dfrac{\begin{array}{cc} \square\underline{\quad}(i) & \square\underline{\quad}(i) \\ \varphi & \neg\varphi \\ \vdots & \vdots \\ \psi & \psi \end{array}}{\psi}(i) \qquad \dfrac{\begin{array}{cc} \square\underline{\quad}(i) & \square\underline{\quad}(i) \\ \varphi & \neg\varphi \\ \vdots & \vdots \\ \psi & \bot \end{array}}{\psi}(i)$$

Double Negation Elimination (DNE):

$$\dfrac{\neg\neg\varphi}{\varphi}$$

We shall call the sentence φ in this statement of the four classical rules the *litmus sentence*. It is the sentence for which **Bivalence** is being presupposed by the deductive reasoner who applies the rule in question.

The litmus sentences occur in these rules in idiosyncratic ways, which do not conform to the constructivist's pattern of introduction or elimination rules. First, the dominant operator of φ is never mentioned. The litmus sentence does not occur as the conclusion of an introduction, nor as the major premise of an elimination. And the strictly classical rules 'for negation' themselves defy any accommodation within the scheme of introduction and elimination rules. The latter rules always focus on just a single, dominant, occurrence of the logical operator in question, be it in the conclusion (of an introduction) or in the major premise (of an elimination). But look at the aberrant form of each of the strictly classical rules! In **CR** we have $\neg\varphi$ occurring as an assumption of the *reductio* subproof, and being discharged by the application of the rule. In **LEM** we have an occurrences of *yet another* logical operator, namely $\vee$, alongside $\neg$. In **Dil** we once again have $\neg\varphi$ occurring as an assumption of a subproof, and being discharged by the application of the rule. Finally, in **DNE** we have *two* occurrences of $\neg$, not just one.

Despite these idiosyncratic features of each of the four strictly classical rules 'for negation', they are all on a par in their net effect upon being adjoined to our (constructive) introduction and elimination rules. For, *modulo* our constructive logic for the logical operators, the strictly classical rules are *interderivable*. Using any one of them, one can derive any one of the remaining three. We shall now demonstrate that this is the case.

The system that results from adjoining any one of the strictly classical rules to Constructive Logic is Classical Logic. The classicist may as well use *any* of them, subject only to considerations of convenience.

8.2 Interderiving the classical negation rules

Using **CR** one can derive **LEM** in 6 steps:

$$\dfrac{\dfrac{\dfrac{\dfrac{\dfrac{\overline{\varphi}^{\,(2)}}{\varphi\vee\neg\varphi} \qquad \overline{\neg(\varphi\vee\neg\varphi)}^{\,(1)}}{\bot}}{\neg\varphi}{\scriptstyle(2)}}{\varphi\vee\neg\varphi} \qquad \overline{\neg(\varphi\vee\neg\varphi)}^{\,(1)}}{\dfrac{\bot}{\varphi\vee\neg\varphi}{\scriptstyle(1)}}$$

Using **CR** one can derive **Dil** in 4 steps:[1]

$$\dfrac{\dfrac{\dfrac{\begin{array}{c}\overline{\varphi}^{\,(2)}\\ \vdots \\ \psi\end{array} \qquad \overline{\neg\psi}^{\,(1)}}{\bot}}{\begin{array}{c}\neg\varphi\\ \vdots \\ \psi\end{array}}{\scriptstyle(2)} \qquad \overline{\neg\psi}^{\,(1)}}{\dfrac{\bot}{\psi}{\scriptstyle(1)}}$$

Using **CR** one can derive **DNE** in 2 steps:

$$\dfrac{\neg\neg\varphi \qquad \overline{\neg\varphi}^{\,(1)}}{\dfrac{\bot}{\varphi}{\scriptstyle(1)}}$$

[1]For the part of **Dil** that uses a disproof of the negative horn, the derivation will obviously take only 3 steps.

Using **DNE** one can derive **CR** in 2 steps:

$$\dfrac{\dfrac{\begin{array}{c}\overline{\neg\varphi}^{\,(1)}\\ \vdots\\ \bot\end{array}}{\neg\neg\varphi}\,(1)}{\varphi}$$

Using **DNE** one can derive **LEM** in 7 steps:

$$\dfrac{\dfrac{\dfrac{\dfrac{\dfrac{\dfrac{\dfrac{\overline{\varphi}^{\,(2)}}{\varphi\vee\neg\varphi}\quad \overline{\neg(\varphi\vee\neg\varphi)}^{\,(1)}}{\bot}}{\neg\varphi}\,(2)}{\varphi\vee\neg\varphi}\quad \overline{\neg(\varphi\vee\neg\varphi)}^{\,(1)}}{\bot}}{\neg\neg(\varphi\vee\neg\varphi)}\,(1)}{\varphi\vee\neg\varphi}$$

Using **DNE** one can derive **Dil** in 5 steps:[2]

$$\dfrac{\dfrac{\dfrac{\begin{array}{c}\dfrac{\dfrac{\begin{array}{c}\overline{\varphi}^{\,(2)}\\ \vdots\\ \psi\end{array}\quad \overline{\neg\psi}^{\,(1)}}{\bot}}{\neg\varphi}\,(2)\\ \vdots\\ \psi\end{array}\quad \overline{\neg\psi}^{\,(1)}}{\bot}}{\neg\neg\psi}\,(1)}{\psi}$$

[2]For the part of **Dil** that uses a disproof of the negative horn, the derivation will obviously take only 4 steps.

Using **LEM** one can derive **Dil** in 1 step:

$$\dfrac{\overline{\varphi \vee \neg\varphi} \qquad \begin{matrix}\overline{\varphi}^{\,(1)}\\ \vdots \\ \psi\end{matrix} \qquad \begin{matrix}\overline{\neg\varphi}^{\,(1)}\\ \vdots \\ \psi/\bot\end{matrix}}{\psi}\,(1)$$

Using **LEM** one can derive **CR** in 1 step:

$$\dfrac{\overline{\varphi \vee \neg\varphi} \qquad \overline{\varphi}^{\,(1)} \qquad \begin{matrix}\overline{\neg\varphi}^{\,(1)}\\ \vdots \\ \bot\end{matrix}}{\varphi}\,(1)$$

Using **LEM** one can derive **DNE** in 2 steps:

$$\dfrac{\overline{\varphi \vee \neg\varphi} \qquad \overline{\varphi}^{\,(1)} \qquad \dfrac{\neg\neg\varphi \qquad \overline{\neg\varphi}^{\,(1)}}{\bot}}{\varphi}\,(1)$$

Using **Dil** one can derive **LEM** in 3 steps:

$$\dfrac{\dfrac{\overline{\varphi}^{\,(1)}}{\varphi \vee \neg\varphi} \qquad \dfrac{\overline{\neg\varphi}^{\,(1)}}{\varphi \vee \neg\varphi}}{\varphi \vee \neg\varphi}\,(1)$$

Using **Dil** one can derive **CR** in 1 step:

$$\dfrac{\overline{\varphi}^{\,(1)} \qquad \begin{matrix}\overline{\neg\varphi}^{\,(1)}\\ \vdots \\ \bot\end{matrix}}{\varphi}\,(1)$$

Using **Dil** one can derive **DNE** in 2 steps:

$$\dfrac{\overline{\varphi}^{\,(1)} \qquad \dfrac{\neg\neg\varphi \qquad \overline{\neg\varphi}^{\,(1)}}{\bot}}{\varphi}\,(1)$$

8.3 Interderiving the Classical Negation Rules: Lengths of Derivations

The following diagram affords easy comparison of the numbers of steps involved in each of the derivations undertaken above. To establish that all these classical rules are interderivable, all one needs is a single 'cycle' of rules $\rho_1 \longrightarrow \rho_2 \longrightarrow \rho_3 \longrightarrow \rho_4 \longrightarrow \rho_1$.[3]

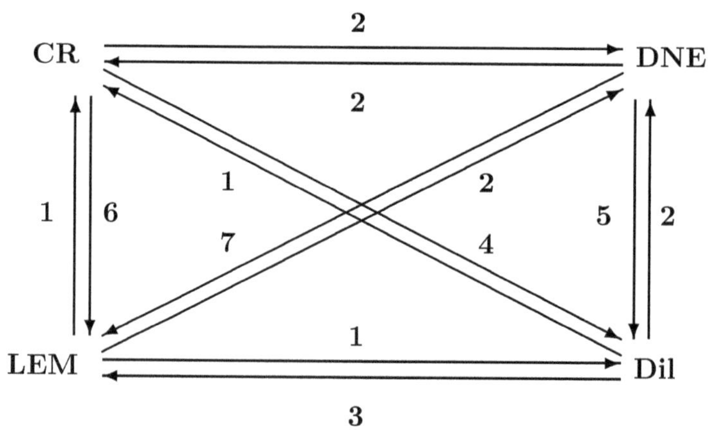

The shortest cycle of derivations

Each cycle consists of four segments. There are six distinct cycles. Each number in the following table indicates how many steps are taken in deriving the rule to the right of the number, using the rule to its left. The total numbers of steps for the six cycles are given in the final column.

[3]These solid-headed arrows are short versions of the longer arrows in the diagram. They are not to be confused with the conditional connective $\rightarrow$.

	→		→		→		→		Total
LEM	**1**	CR	**4**	Dil	3	DNE	**7**	LEM	15
LEM	**2**	DNE	**5**	Dil	2	CR	**6**	LEM	15
LEM	2	CR	2	DNE	**5**	Dil	3	LEM	12
LEM	3	DNE	2	CR	4	Dil	3	LEM	12
LEM	1	Dil	**1**	CR	2	DNE	**7**	LEM	11
LEM	1	Dil	**2**	DNE	2	CR	**6**	LEM	**11**

The bottommost cycle minimizes the **total number of steps (11)** *and* minimizes the **maximum segment (6)**.

The Most Economical Way to Interderive the Classical Negation Rules

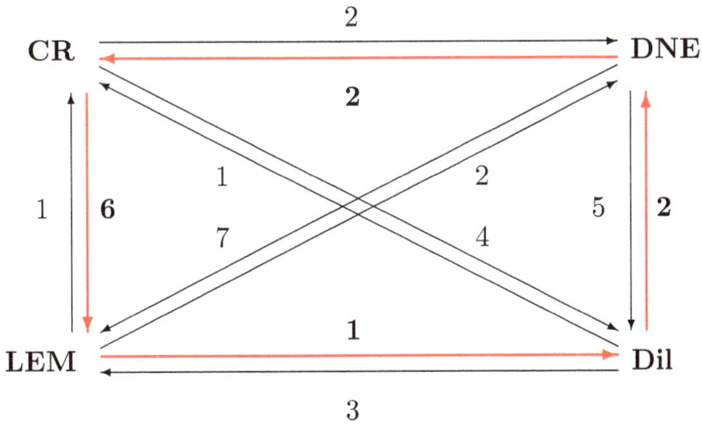

Take the anti-clockwise circuit.

The Most Economical Way to Interderive the Classical Negation Rules, with Derivations Interpolated

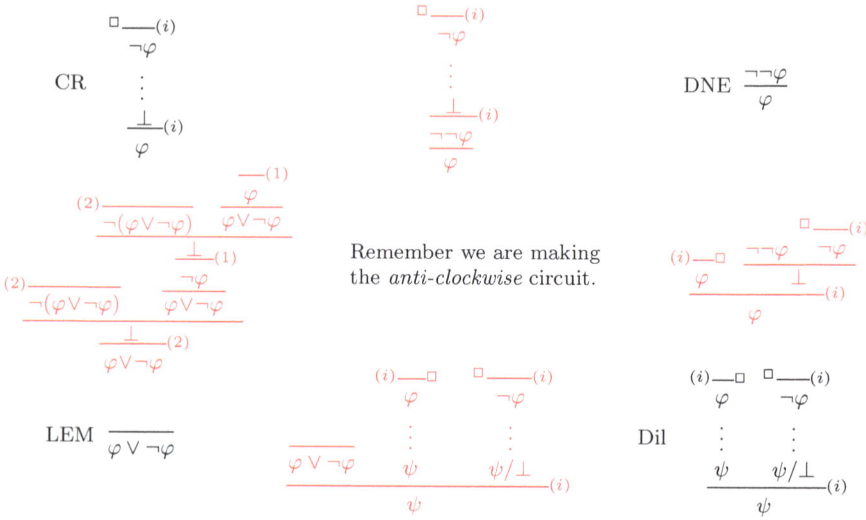

Exercise 27. Give a classical natural deduction for the one 'De Morgan' argument at the propositional level that is *not* constructively provable, namely the argument

$$\neg(A \wedge B) : \neg A \vee \neg B.$$

In your proof use only Classical Reductio as the required strictly classical rule of negation. Then give a classical proof of the same argument that uses only Dilemma as its classical rule of negation. Which proof has fewer steps?

Chapter 9

Classical Interdefinability of Logical Operators

9.1 What it is to define one expression in terms of others

In Classical Logic, taking $\neg$ and any one of the three binary connectives $\wedge, \vee, \rightarrow$ as primitives, one can define the other two binary connectives. When we write

$$\theta \equiv_{df} \chi$$

we mean that the definiendum θ is being given the definiens χ. The definiendum is the expression to be defined, while the definiens is the expression that 'does the defining'. The definiens contains, as logical expressions, only ones that are being taken as primitive; while the definiendum contains the non-primitive logical expression that is being defined in terms of those primitives.

The definitions to follow hold only within classical logic. In classical logic with all the logical operators primitive, it will be the case, for any of our definitions below of the form

$$\theta \equiv_{df} \chi ,$$

that

$$\theta \vdash \chi \quad \text{and} \quad \chi \vdash \theta$$

9.2 Taking $\neg$ and $\wedge$ as primitives

Defining $\vee$ in terms of $\neg$ and $\wedge$

$$\varphi \vee \psi \ \equiv_{df} \ \neg(\neg\varphi \wedge \neg\psi)$$

This definition does not work in constructive logic, because in constructive logic

$$\neg(\neg\varphi \wedge \neg\psi) \nvdash \varphi \vee \psi.$$

Defining $\rightarrow$ in terms of $\neg$ and $\wedge$

$$\varphi \rightarrow \psi \ \equiv_{df} \ \neg(\varphi \wedge \neg\psi)$$

This definition does not work in constructive logic, because in constructive logic

$$\neg(\varphi \wedge \neg\psi) \nvdash \varphi \rightarrow \psi$$

9.3 Taking $\neg$ and $\vee$ as primitives

Defining $\wedge$ in terms of $\neg$ and $\vee$

$$\varphi \wedge \psi \ \equiv_{df} \ \neg(\neg\varphi \vee \neg\psi)$$

This definition does not work in constructive logic, because in constructive logic

$$\neg(\neg\varphi \vee \neg\psi) \nvdash \varphi \wedge \psi$$

Defining $\rightarrow$ in terms of $\neg$ and $\vee$

$$\varphi \rightarrow \psi \equiv_{df} \neg\varphi \vee \psi$$

This definition does not work in constructive logic, because in constructive logic

$$\varphi \rightarrow \psi \nvdash \neg\varphi \vee \psi$$

9.4 Taking $\neg$ and $\to$ as primitives

Defining $\wedge$ in terms of $\neg$ and $\to$

$$\varphi \wedge \psi \;\equiv_{df}\; \neg(\varphi \to \neg\psi)$$

This definition does not work in constructive logic, because in constructive logic

$$\neg(\varphi \to \neg\psi) \;\nvdash\; \varphi \wedge \psi$$

Defining $\vee$ in terms of $\neg$ and $\to$

$$\varphi \vee \psi \;\equiv_{df}\; \neg\varphi \to \psi$$

This definition does not work in constructive logic, because in constructive logic

$$\neg\varphi \to \psi \;\nvdash\; \varphi \vee \psi$$

9.4.1 Defining $\leftrightarrow$ in terms of $\wedge$ and $\to$.

This definition works in constructive logic, hence also in classical logic:

$$\varphi \leftrightarrow \psi \;\equiv_{df}\; (\varphi \to \psi) \wedge (\psi \to \varphi)$$

9.5 The Sheffer stroke

To reprise part of §4.12: The Sheffer stroke | is a single binary connective in terms of which one can define all the other connectives of Classical Propositional Logic. It forms a new sentence from any two given sentences φ and ψ, just as do the binary connectives $\wedge$, $\vee$, and $\to$, with which we are already familiar.

$$(\varphi|\psi)$$

can be read as 'φ stroke ψ'.

It means 'Not both φ and ψ'.

Here are the definitions, in terms of the stroke, of all the usual connectives of classical propositional logic:

$$\neg\varphi \equiv_{df} \varphi|\varphi$$
$$\varphi \wedge \psi \equiv_{df} (\varphi|\psi)|(\varphi|\psi)$$
$$\varphi \vee \psi \equiv_{df} (\varphi|\varphi)|(\psi|\psi)$$
$$\varphi \to \psi \equiv_{df} \varphi|(\psi|\psi)$$

Exercise 28. Formulate a classically sound and complete set of inference rules for propositional logic based solely on the Sheffer stroke. Try to make your choice of rules for the stroke look laterally symmetric. There are rules 'out there', in Plato's heaven, that really are beautiful. (Hint: think up an introduction rule; an elimination rule; and a strictly classical rule for 'negation'. For the aesthetic appeal sought, use infix notation.)

Exercise 29. Now do the same for the *other* binary operator which, like the Sheffer stroke, suffices for defining all the other connectives of classical logic. You will have identified this other binary operator if you did Exercise 25.

Chapter 10

Propagating Truth Values and Validities

10.1 Atomic sequents and truth-value assignments

An *atomic sequent* is a sequent of the form ' $: A$' or of the form '$A :$ ', where A is an atomic sentence, and where the space on the other side of the colon could be filled by the symbol for the empty set:

$$\emptyset : A \quad ; \quad A : \emptyset.$$

A truth-value assignment to atomic sentences can be represented as a set of atomic sequents. Those of the form $\emptyset : A$ say that A is true; those of the form $A : \emptyset$ say that A is false. No truth-value assignment to atomic sentences can say both, on pain of incoherence.

Note that sequents of the form $A : A$, where A is an atom, are not, on our definition, atomic sequents. Rather, they would be special cases of what will be called *initial* sequents—that is, sequents of the form $\varphi : \varphi$. Initial sequents are important because they are the starting points for sequent *proofs* in the sequent *calculus*—of which more in due course, before chapter's end.

10.2 Propagating truth values

By 'truth value propagation' we mean the determination of the truth value of a compound sentence from the truth values of its immediate constituents. For the sake of simplicity in an exposition of new ideas we confine ourselves at the outset to propositional logic.

Let the atomic sequents codifying a truth-value assignment be called Λ ('Lambda', with or without numerical subscripts). A good way to remind oneself what kind of object Λ is, is to think of the logician's term *literal*, meaning an atom or the negation of an atom. The atomic sequent $\emptyset : A$ is a positive literal; the atomic sequent $A : \emptyset$ is a negative literal.

We shall now offer a calculus for calculating the truth value of a compound sentence under a given truth-value assignment Λ that deals with its atoms. The rules of this calculus can be understood as *rules of propagation* of truth values, *to* sentential compounds, *from* their immediate constituents. If Λ makes a sentence θ true, we shall write $\Lambda : \theta$; but if Λ makes θ false, we shall write $\Lambda, \theta : \emptyset$.

Note that these are simple cases of *valid sequents*. This is because the only truth-value assignment that makes all of Λ true is the very one that Λ represents. So, trivially, if $\Lambda : \theta$ then every truth-value assignment that makes all of Λ true makes θ true also. Likewise, if $\Lambda, \theta : \emptyset$, then every truth-value assignment that makes all of Λ true makes θ false, whence $\Lambda \cup \{\theta\}$ is unsatisfiable.

We shall state the rules of truth-value propagation connective by connective. The reader will see that the rules simply recapitulate the content of the familiar truth tables. And this primrose path will lead to our relevantized sequent calculus.

Propagating truth values: negations

If φ is made false (by Λ), then $\neg\varphi$ is made true (by Λ):

$$\frac{\Lambda, \varphi : \emptyset}{\Lambda : \neg\varphi}$$

Henceforth we shall omit the qualification 'by Λ', since it will be clear in context that all makings-true and makings-false will be by whatever truth-value assignment Λ is in question.

If φ is made true, then $\neg\varphi$ is made false:

$$\frac{\Lambda : \varphi}{\Lambda, \neg\varphi : \emptyset}$$

Propagating truth values: conjunctions

If φ and ψ are both made true, then so is $\varphi \wedge \psi$:

$$\frac{\Lambda_1 : \varphi \quad \Lambda_2 : \psi}{\Lambda_1, \Lambda_2 : \varphi \wedge \psi}$$

If φ is made false, then so is $\varphi \wedge \psi$:

$$\frac{\Lambda, \varphi : \emptyset}{\Lambda, \varphi \wedge \psi : \emptyset}$$

If ψ is made false, then so is $\varphi \wedge \psi$:

$$\frac{\Lambda, \psi : \emptyset}{\Lambda, \varphi \wedge \psi : \emptyset}$$

Propagating truth values: disjunctions

If φ is made true, then so is $\varphi \vee \psi$:

$$\frac{\Lambda : \varphi}{\Lambda : \varphi \vee \psi}$$

If ψ is made true, then so is $\varphi \vee \psi$:

$$\frac{\Lambda : \psi}{\Lambda : \varphi \vee \psi}$$

If φ and ψ are both made false, then so is $\varphi \vee \psi$:

$$\frac{\Lambda_1, \varphi : \emptyset \quad \Lambda_2, \psi : \emptyset}{\Lambda_1, \Lambda_2, \varphi \vee \psi : \emptyset}$$

Propagating truth values: conditionals

If ψ is made true, then so is $\varphi \rightarrow \psi$:

$$\frac{\Lambda : \psi}{\Lambda : \varphi \rightarrow \psi}$$

If φ is made false, then $\varphi \rightarrow \psi$ is made true:

$$\frac{\Lambda, \varphi : \emptyset}{\Lambda : \varphi \rightarrow \psi}$$

If φ is made true and ψ is made false, then $\varphi \rightarrow \psi$ is made false:

$$\frac{\Lambda_1 : \varphi \quad \Lambda_2, \psi : \emptyset}{\Lambda_1, \Lambda_2, \varphi \rightarrow \psi : \emptyset}$$

10.3 Propagating validities

We turn now to the propagation of *validities*. These are valid arguments in general, of one of the two forms

$$\Delta : \theta \quad \text{or} \quad \Delta : \emptyset$$

where Δ is an arbitrary set of sentences, and θ is a sentence. The case of a set Λ of literals is just a special case of such Δ in general. An argument of the form $\Delta : \emptyset$ is stating that Δ is not satisfiable.

Alternative and equivalent ways of saying that *the argument* $\Delta : \theta$ *is valid* are:

> Δ logically implies θ;

and

> θ is a logical consequence of Δ.

The following forms of validity-propagation are intimately derived from those for truth-value propagation. Deltas take over from Lambdas; and thetas can now feature as conclusions of rules that have their connectives appearing on the left.

Propagating validities: negations

If $\Delta \cup \{\varphi\}$ is unsatisfiable, then $\neg\varphi$ follows logically from Δ:

$$\frac{\Delta, \varphi : \emptyset}{\Delta : \neg\varphi}$$

If φ follows logically from Δ, then $\Delta \cup \{\neg\varphi\}$ is unsatisfiable:

$$\frac{\Delta : \varphi}{\Delta, \neg\varphi : \emptyset}$$

Propagating validities: conjunctions

If φ and ψ follow logically from Δ_1 and Δ_2 respectively, then $\varphi \wedge \psi$ follows logically from $\Delta_1 \cup \Delta_2$:

$$\frac{\Delta_1 : \varphi \quad \Delta_2 : \psi}{\Delta_1, \Delta_2 : \varphi \wedge \psi}$$

If θ follows logically from $\Delta \cup \{\varphi\}$, then θ follows logically from $\Delta \cup \{\varphi \wedge \psi\}$:

$$\frac{\Delta, \varphi : \theta}{\Delta, \varphi \wedge \psi : \theta}$$

If θ follows logically from $\Delta \cup \{\psi\}$, then θ follows logically from $\Delta \cup \{\varphi \wedge \psi\}$:

$$\frac{\Delta, \psi : \theta}{\Delta, \varphi \wedge \psi : \theta}$$

Propagating validities: disjunctions

If φ follows logically from Δ, then so does $\varphi \vee \psi$: $$\frac{\Delta : \varphi}{\Delta : \varphi \vee \psi}$$

If ψ follows logically from Δ, then so does $\varphi \vee \psi$: $$\frac{\Delta : \varphi}{\Delta : \varphi \vee \psi}$$

If θ follows logically from $\Delta_1 \cup \{\varphi\}$ and also from $\Delta_2 \cup \{\psi\}$, then θ follows logically from $\Delta_1 \cup \Delta_2 \cup \{\varphi \vee \psi\}$: $$\frac{\Delta_1, \varphi : \theta \quad \Delta_2, \psi : \theta}{\Delta_1, \Delta_2, \varphi \vee \psi : \theta}$$

If θ follows logically from $\Delta_1 \cup \{\varphi\}$ but $\Delta_2 \cup \{\psi\}$ is unsatisfiable, then θ follows logically from $\Delta_1 \cup \Delta_2 \cup \{\varphi \vee \psi\}$: $$\frac{\Delta_1, \varphi : \theta \quad \Delta_2, \psi : \emptyset}{\Delta_1, \Delta_2, \varphi \vee \psi : \theta}$$

If θ follows logically from $\Delta_2 \cup \{\psi\}$ but $\Delta_1 \cup \{\varphi\}$ is unsatisfiable, then θ follows logically from $\Delta_1 \cup \Delta_2 \cup \{\varphi \vee \psi\}$: $$\frac{\Delta_1, \varphi : \emptyset \quad \Delta_2, \psi : \theta}{\Delta_1, \Delta_2, \varphi \vee \psi : \theta}$$

If both $\Delta_1 \cup \{\varphi\}$ and $\Delta_2 \cup \{\psi\}$ are unsatisfiable, then so too is $\Delta_1 \cup \Delta_2 \cup \{\varphi \vee \psi\}$: $$\frac{\Delta_1, \varphi : \emptyset \quad \Delta_2, \psi : \emptyset}{\Delta_1, \Delta_2, \varphi \vee \psi : \emptyset}$$

Propagating validities: conditionals

If ψ follows logically from Δ, then $\varphi \rightarrow \psi$ follows logically from $\Delta \backslash \{\varphi\}$:

$$\frac{\Delta : \psi}{\Delta \backslash \{\varphi\} : \varphi \rightarrow \psi}$$

If $\Delta \cup \{\varphi\}$ is unsatisfiable, then $\varphi \rightarrow \psi$ follows logically from Δ:

$$\frac{\Delta, \varphi : \emptyset}{\Delta : \varphi \rightarrow \psi}$$

If φ follows logically from Δ_1 and $\Delta_2 \cup \{\psi\}$ logically implies θ, then $\Delta_1 \cup \Delta_2 \cup \{\varphi \rightarrow \psi\}$ logically implies θ:

$$\frac{\Delta_1 : \varphi \quad \Delta_2, \psi : \theta}{\Delta_1, \Delta_2, \varphi \rightarrow \psi : \theta}$$

10.4 The whole set of rules

$$\neg\text{R}\ \frac{\Delta, \varphi : \emptyset}{\Delta : \neg\varphi}$$

$$\neg\text{L}\ \frac{\Delta : \varphi}{\Delta, \neg\varphi : \emptyset}$$

$$\wedge\text{R}\ \frac{\Delta : \varphi \quad \Gamma : \psi}{\Delta, \Gamma : \varphi \wedge \psi}$$

$$\wedge\text{L}\ \frac{\Delta, \varphi : \theta}{\Delta, \varphi \wedge \psi : \theta} \quad \frac{\Delta, \psi : \theta}{\Delta, \varphi \wedge \psi : \theta}$$

$$\vee\text{R}\ \frac{\Delta : \varphi}{\Delta : \varphi \vee \psi} \quad \frac{\Delta : \varphi}{\Delta : \varphi \vee \psi}$$

$$\vee\text{L}\ \frac{\Delta, \varphi{:}\theta \quad \Gamma, \psi{:}\theta}{\Delta, \Gamma, \varphi \vee \psi{:}\theta} \quad \frac{\Delta, \varphi{:}\theta \quad \Gamma, \psi{:}\emptyset}{\Delta, \Gamma, \varphi \vee \psi{:}\theta} \quad \frac{\Delta, \varphi{:}\emptyset \quad \Gamma, \psi{:}\theta}{\Delta, \Gamma, \varphi \vee \psi{:}\theta} \quad \frac{\Delta, \varphi{:}\emptyset \quad \Gamma, \psi{:}\emptyset}{\Delta, \Gamma, \varphi \vee \psi{:}\emptyset}$$

$$\rightarrow\text{R}\ \frac{\Delta : \psi}{\Delta \backslash \{\varphi\} : \varphi \rightarrow \psi} \quad \frac{\Delta, \varphi : \emptyset}{\Delta : \varphi \rightarrow \psi}$$

$$\rightarrow\text{L}\ \frac{\Delta : \varphi \quad \Gamma, \psi : \theta}{\Delta, \Gamma, \varphi \rightarrow \psi : \theta}$$

The whole set of rules, with empty space replacing $\emptyset$

$$\neg\text{R}\ \frac{\Delta,\varphi :}{\Delta : \neg\varphi}$$

$$\neg\text{L}\ \frac{\Delta : \varphi}{\Delta, \neg\varphi :}$$

$$\wedge\text{R}\ \frac{\Delta : \varphi \quad \Gamma : \psi}{\Delta, \Gamma : \varphi \wedge \psi}$$

$$\wedge\text{L}\ \frac{\Delta, \varphi : \theta}{\Delta, \varphi \wedge \psi : \theta} \quad \frac{\Delta, \psi : \theta}{\Delta, \varphi \wedge \psi : \theta}$$

$$\vee\text{R}\ \frac{\Delta : \varphi}{\Delta : \varphi \vee \psi} \quad \frac{\Delta : \varphi}{\Delta : \varphi \vee \psi}$$

$$\vee\text{L}\ \frac{\Delta, \varphi : \theta \ \ \Gamma, \psi : \theta}{\Delta, \Gamma, \varphi \vee \psi : \theta} \quad \frac{\Delta, \varphi : \theta \ \ \Gamma, \psi :}{\Delta, \Gamma, \varphi \vee \psi : \theta} \quad \frac{\Delta, \varphi : \ \ \Gamma, \psi : \theta}{\Delta, \Gamma, \varphi \vee \psi : \theta} \quad \frac{\Delta, \varphi : \ \ \Gamma, \psi :}{\Delta, \Gamma, \varphi \vee \psi :}$$

$$\rightarrow\text{R}\ \frac{\Delta : \psi}{\Delta \backslash \{\varphi\} : \varphi \rightarrow \psi} \quad \frac{\Delta, \varphi :}{\Delta : \varphi \rightarrow \psi}$$

$$\rightarrow\text{L}\ \frac{\Delta : \varphi \quad \Gamma, \psi : \theta}{\Delta, \Gamma, \varphi \rightarrow \psi : \theta}$$

Finally, let us not forget (for classical logic) the rule of

$$\textbf{Dilemma:}\ \frac{\Delta, \varphi : \psi \quad \Gamma, \neg\psi : \psi/\emptyset}{\Delta, \Gamma : \psi}$$

10.5 A remarkable isomorphism

The reader has probably been aware, since the beginning of this chapter, that we seem to be plowing the same furrow that
(i) began with the left-to-right inferentialist readings of the truth tables;
(ii) formulated truth-tabular computations via the rules of evaluation (i.e., the rules of verification and falsification); and
(iii) morphed the latter rules into the introduction and elimination rules, respectively, of natural deduction.

In this chapter, it should seem (because it is indeed the case) that we are re-plowing that familiar route, but with a slightly different notation for rules and proofs. And indeed we are. This re-tracing of that route, however, finds the terrain softer underfoot, and yields some rather unexpected and special benefits. These we shall now expound.

The deductive rules for propositional logic that have been developed in the previous section are so-called *sequent* rules. These have one or two premise sequents, and a conclusion sequent. Every sequent has one of the following two forms:
(1) '$\Omega : \chi$' where Ω is a finite (possibly empty) set of sentences and χ is a sentence; or
(2) '$\Omega :$ ' (i.e., '$\Omega : \emptyset$') where Ω is a finite *non-empty* set of sentences.

For each connective @ there is a pair of sequent rules. These tell one how to introduce a dominant occurrence of @ in a sentence either on the left or on the right side of the colon in the conclusion sequent.

There is only one kind of sequents that need no derivational justification. These are the so-called *initial* sequents, which are of the form $\chi : \chi$ (i.e., to be captious, of the form $\{\chi\} : \chi$). They correspond, in the sequent calculus, to the natural-deductive basis fact that χ is a proof of χ from $\{\chi\}$. Natural deductions, and sequent proofs, have to get started, and are allowed to do so only with the blindingly obvious.

The way to understand the validity claim represented by a sequent '$\Omega : \chi$' or '$\Omega :$ ' is as follows:

> Every model or interpretation that makes all of the sentences before the colon true (i.e. that makes every member of Ω true) makes at least one sentence after the colon true.

When there is no sentence after the colon this entails that there is no model or interpretation that makes all of Ω true. Thus a sequent of the form '$\Omega :$ ' is saying that Ω is not satisfiable. Ω is incoherent. Ω has no model.

On the other hand, a sequent of the form '$\Omega : \chi$' is saying that truth is preserved from Ω to χ. There is no model that makes all of Ω true but makes χ false. Moreover, a sequent of the form '$\emptyset : \chi$' is saying that χ is true under *every* interpretation *simpliciter*—i.e., χ is *logically true.*

One can now understand why *there is no sequent of the form* ' : ' (i.e., of the form $\emptyset : \emptyset$). For, if there were, it would be making the absurd claim

> Every model or interpretation that makes all of the sentences in $\emptyset$ true makes at least one sentence in $\emptyset$ true.

This is absurd because every model or interpretation *does* make true all of the sentences in the empty set $\emptyset$ (because $\emptyset$ has no sentences at all in it); yet must fail to make at least one sentence in $\emptyset$ true (because, again, $\emptyset$ has no sentences at all in it). This is why certain logicians (including Gentzen) take the task of a consistency proof for a sequent system (whether for logic alone, or for some mathematical theory such as number theory) to be that of showing that the 'sequent' $\emptyset : \emptyset$ is not derivable in the system.

Sequent proofs are constructed by using initial sequents and then deriving yet other sequents by means of the left- and right-rules for the logical operators. (The reader will have plenty of practice constructing sequent proofs, by doing the exercises in Chapter 13.) The left- and right-rules for the logical operators have either one or two premise sequents. A sequent proof is therefore in general tree-like, with monofurcations and bifurcations. Its leaf nodes are labeled by initial sequents, and its root node is labeled by the sequent that is proved. One useful feature of a sequent proof is that every node in it is labeled by the sequent whose validity has been established by the subtree above and rooted in that node. There is no need to indicate discharge of assumptions. This is because with a left rule @L the assumptions in question just get replaced on the left by the new sentence with @ dominant that is being introduced on the left, while with a right rule @R the assumption to be discharged is separated off from the other assumptions by a comma, and gets absorbed into the new conclusion—which has @ dominant—that is the newcomer on the right. (Remember here that 'left' ['right'] means 'left of the colon' ['right of the colon'].)

The reader will find that every natural deduction constructed in accordance with the introduction and (parallelized) elimination rules, and the rule of **Dilemma**, corresponds *directly* to a sequent proof of the same result; and *vice versa*. One can therefore, as a logical theorist, flit effortlessly back and forth between natural deductions and sequent proofs. This reveals that proofs as abstract objects have distinct but equivalent forms of representation. One can therefore choose one's favorite form of representation in accordance with one's strategic interests and purposes.

If one is searching for a proof of a given argument, it may be that thinking in terms of introduction and elimination rules produces the best results. Indeed, it may be that using the *serial* elimination rules for $\wedge$, $\rightarrow$, and $\forall$ gives an extra little bit of help in this regard. But that is compatible with the fact that when it comes to programming powerful and efficient automated theorem-provers it is the *parallelized* forms of elimination rules that one would be well advised to adhere to.

Using parallelized elimination rules enables one to avoid having to repeat chunks of formal proof on the way to the overall conclusion. For example, consider the greater economy achieved from parallelizing $\wedge$-E in the following two deductions for the same argument, namely

$$\frac{(A\vee D)\rightarrow(B\wedge C) \quad B\rightarrow E \quad C\rightarrow F}{A\rightarrow(E\wedge F)}\;.$$

The *first* deduction uses the serial forms of $\wedge$-E and of $\rightarrow$-E, and has to use them *twice*, thereby incurring unnecessary repetitions of the derivations of the major premises involved. That makes the proof 10 steps long. Here now is the longer, 10-step natural deduction that uses serial $\wedge$-E and serial $\rightarrow$-E (*modus ponens*):

$$\dfrac{\dfrac{\dfrac{\dfrac{\dfrac{\overline{A}^{(1)}}{A\vee D} \quad (A\vee D)\rightarrow(B\wedge C)}{\dfrac{B\wedge C}{B}} \quad B\rightarrow E}{E} \qquad \dfrac{\dfrac{\dfrac{\overline{A}^{(1)}}{A\vee D} \quad (A\vee D)\rightarrow(B\wedge C)}{\dfrac{B\wedge C}{C}} \quad C\rightarrow F}{F}}{E\wedge F}}{A\rightarrow(E\wedge F)}{\scriptstyle (1)}$$

The *second* deduction uses the parallelized forms of $\wedge$-E and of $\rightarrow$-E, and needs to use each of them *only once*, respectively on the consequent of the main premise, and on the main premise itself, as major premises for those eliminations. The application of $\wedge$-E affords discharge of both the conjuncts (of the major premise) where they appear as assumptions for the subproof. The resulting proof is only 7 steps long—exactly as many steps as there are occurrences of logical operators in the argument being proved. Here now is the shorter, 7-step natural deduction that uses parallelized $\wedge$-E and parallelized $\to$-E:

$$\dfrac{\dfrac{(A\vee D)\to(B\wedge C) \quad \dfrac{\overline{A}^{\,(5)}}{A\vee D} \quad \dfrac{{}^{(4)}\overline{B\wedge C} \qquad \dfrac{\dfrac{B\to E \quad \overline{B}^{\,(3)} \quad \overline{E}^{\,(1)}}{E}{\scriptstyle(1)} \quad \dfrac{C\to F \quad \overline{C}^{\,(3)} \quad \overline{F}^{\,(2)}}{F}{\scriptstyle(2)}}{E\wedge F}}{E\wedge F}{\scriptstyle(3)}}{E\wedge F}{\scriptstyle(4)}}{A\to(E\wedge F)}{\scriptstyle(5)}$$

Let us turn now to the construction of a sequent proof of the same argument. It has the same macro-structure as the natural deduction that uses only parallelized elimination rules:

$$\dfrac{\dfrac{\dfrac{A:A}{A:A\vee D} \qquad \dfrac{\dfrac{\dfrac{B:B \quad E:E}{B\to E,\ B:E} \quad \dfrac{C:C \quad F:F}{C\to F,\ C:F}}{B\to E,\ B,\ C\to F,\ C:E\wedge F}}{B\to E,\ C\to F,\ B\wedge C:E\wedge F}}{(A\vee D)\to(B\wedge C),\ A,\ B\to E,\ C\to F:E\wedge F}}{(A\vee D)\to(B\wedge C),\ B\to E,\ C\to F:A\to(E\wedge F)}$$

Having found the sought proof, there is then the consideration of how best to *present* it. Treating it as a natural deduction, in which nodes are labeled by sentences, can be more economical than treating it as a sequent proof, in which every node is labeled by the whole sequent that has been established by that stage.

There is a trade-off to consider, however: natural deductions tend to become overly wide quite quickly, when—rather ironically—the corresponding sequent proof, for all its lengthier labels on nodes (whole sequents instead of sentences) does not spill off the sides of the page. This, however, is what Bertrand Russell would have called a merely medical limitation for creatures with insufficiently panoramic fields of vision.

The reader simply needs to be aware of this pervasive but illuminating correspondence:

Natural deduction with parallelized E-rules		**Sequent proof**
$\begin{array}{c}\Omega\\ \Pi\\ \chi\end{array}$	$\leftrightsquigarrow$	$\begin{array}{c}\Pi\\ \Omega : \chi\end{array}$

One can transform a natural deduction into its corresponding sequent proof by iteratively rotating clockwise through 90 degrees and annotating nodes with overall sequents proved. And the inverse operation is just as effortless, from sequent proofs to their corresponding natural deductions. A word of caution, however: this simple and direct relationship between natural deductions and their corresponding sequent proofs holds only courtesy of the requirement, adopted here, that all elimination rules in natural deduction be *parallelized*, **and** that major premises for eliminations *stand proud*, with no non-trivial proof-work above them. Gentzen did not impose such a requirement; and that is why Gentzenian natural deductions and Gentzenian sequent proofs 'enjoyed' such a fraught and complicated interrelationship over several decades. In our treatment, however, the one is the yin to the other's yang.

Exercise 30. *We refer to Exercise 26 on p. 185, which in turn referred to Exercise 23 on page 164. For each logical truth φ that you identified, provide now a* sequent proof *of $\emptyset : \varphi$. For each logical falsehood φ that you identified, provide now a* sequent proof *of $\varphi : \emptyset$.*

Exercise 31. Give a classical sequent proof of the argument

$$\neg(A \wedge B) : \neg A \vee \neg B$$

that uses Classical Reductio as its only classical rule of negation.

Chapter 11

Evaluations of Sentences with Quantifiers

This chapter aims to extend the treatment of verification and falsification rules to the quantifiers $\exists$ and $\forall$, and to show how they morph into the respective pairs of introduction and elimination rules in natural deduction. We are therefore building on the treatment of the connectives of propositional logic given in Chapter 6.

We begin by reprising some general remarks made in Chapter 6, where the notion of an interpretation of the language was that of an assignment of truth values to propositional atoms. Here, at first order, we have to deal with reference to individuals, and with predications involving them. And we have to deal with existential and universal generalizations. So the notion of an interpretation will no longer be the simple one of an assignment of truth values to propositional atoms. It will be the more general and slightly more complicated one of a *model* for a first-order language.

A model (call it M) has a domain of individuals (call it D). M provides extensions in D for names, function signs and primitive predicates, as follows. M furnishes individuals in the domain as denotations for names; n-place mappings from D into D for n-place function symbols; and set of n-tuples of individuals in D for n-place predicates. A first-order model thereby affords a more 'structured' representation of the atomic facts than is available in the propositional case. The reader has already seen, in Chapter 3, how we represented 'little models' as boxes, with their individuals being the bullets located inside them, and with monadic predicate-extensions indicated Venn–Euler style, as circles or as yet further (and smaller) boxes within the 'big box' (for the whole domain), and with any dyadic predicate-extension being represented by arrows of a specific kind. We shall be engaging in more of this box-picturing of models in due course, to help the student become well-versed in how the quantifiers work.

11.1 Rules of evaluation: verification and falsification

1. As in the propositional case, our rules of evaluation at first order determine whether a sentence is **verified** or **falsified** in a given interpretation.

2. An *interpretation*, or *model*, (in the **first-order** case) consists of a *domain of individuals*, and a specification of *extensions of predicates* within that domain. In the **propositional** case, as we have seen, an interpretation is a mere assignment of truth values to the atomic sentences.

3. As before, a **verification** of a sentence φ (but relative now to an interpretation M) establishes φ as its *conclusion*:

$$\begin{array}{c} \vdots \\ \varphi \end{array}$$

4. As before, a **falsification** of a sentence φ (but relative now to an interpretation M) establishes absurdity ($\bot$) from φ as its *major premise*:

$$\begin{array}{c} \varphi \\ \vdots \\ \bot \end{array}$$

The simplest verifications (relative to a **first-order interpretation** M) are of the form

$$\frac{}{P(\alpha_1, \ldots, \alpha_n)}\,{}^{M}$$

where in M the primitive predicate P holds of the individuals $\alpha_1, \ldots, \alpha_n$. This is a verification of $P(\alpha_1, \ldots, \alpha_n)$ relative to M.

The simplest falsifications (relative to an interpretation M) are of the form

$$\frac{P(\alpha_1, \ldots, \alpha_n)}{\bot}\,M$$

where in M the primitive predicate P fails to hold of the individuals $\alpha_1, \ldots, \alpha_n$. This is a falsification of $P(\alpha_1, \ldots, \alpha_n)$ relative to M.

More complex verifications and falsifications (relative to M) are built up in accordance with the rules (@-$\mathcal{V}$) and (@-$\mathcal{F}$) for logical operators. We are already acquainted with those rules for the connectives. In this chapter we shall provide such rules for the quantifiers.

In the **propositional** case, as we have already seen, the simplest verifications and falsifications relative to a truth-value assignment τ took the respective forms

$$\frac{}{A}\,\tau \qquad \text{and} \qquad \frac{A}{\bot}\,\tau\ ,$$

for atomic sentences A.

11.2 Saturated formulae

Note that we are dealing here, in the first-order case, with *saturated* formulae. These are created by taking a formula $\varphi(x_1, \ldots, x_n)$ (with the indicated variables free) and replacing each free occurrence of a variable x_i $(1 \leq i \leq n)$ with an individual α_i:

$$\varphi^{x_1}_{\alpha_1} \cdots {}^{x_n}_{\alpha_n} = \varphi(\alpha_1, \ldots, \alpha_n).$$

Saturated formulae do not have any free variables. That is why they are able to be true, or false, in an interpretation. They can be thought of as *sentences*, liable to come out true, or come out false, in any model that interprets their extra-logical vocabulary. They can be thought of as 'Russellian propositions'—the individuals Brutus and Caesar are 'embedded within' the proposition that Brutus killed Caesar.

11.3 Literals

A *literal* is an atomic sentence, or the negation of an atomic sentence. Thus atomic saturated formulae and their negations are literals. Construed inferentially, literals are our simplest possible evaluations.

This constitutes the *basis clause* in our **co-inductive definition of** (the tree-like arrays that we are calling) **verifications** and **falsifications**.

As we saw in §6.3, in the **propositional case** the simplest possible verifications and falsifications were, respectively,

$$\frac{}{A}\,\tau \qquad \text{and} \qquad \frac{A}{\bot}\,\tau \ .$$

In the **first-order** case, however, they are to be

$$\frac{}{P(\alpha_1,\ldots,\alpha_n)}\,M \qquad \text{and} \qquad \frac{P(\alpha_1,\ldots,\alpha_n)}{\bot}\,M \ .$$

So: the simplest possible verifications and falsifications express, respectively, *positive* or *negative* **factoids**. These can be given the obvious label '*predicational* factoids'.

We intend to keep matters as simple as possible in the first-order case, as an extension of what was done in the propositional case. We are therefore omitting a treatment—reasonably straightforward though it would turn out to be—of '*functional* factoids'. These would be of the form

$$\frac{}{\alpha = f(\alpha_1,\ldots,\alpha_n)}\,M \qquad \text{and} \qquad \frac{\alpha = f(\alpha_1,\ldots,\alpha_n)}{\bot}\,M \ .$$

To keep the analogy between the propositional and first-order cases as tight as possible, we shall treat only *predicates* in the first-order case. We shall not treat names; and, as already intimated, we shall not treat function-symbols either. Nor shall we treat any of the variable-binding term-forming operators. The need for a venture into *free* logic is thereby obviated. The reader is hereby assured, however, that such treatments and the accompanying venture into free logic *can* be undertaken—but only at the expense of the very smooth expository analogy that we are seeking to sustain between the propositional case and the first-order case.

We shall continue to use Λ as a placeholder for sets of literals of the interpretation in question. Every such Λ is a bunch of (predicational) **factoids**. We choose purple here for Lambda once again because that coloration serves to remind the reader that these sets Λ can in general contain both positive and negative factoids.

Such sets Λ of literals feature (as they did in the propositional case) as the sets of 'undischarged side assumptions' in the following graphic statements of our rules of evaluation for the quantifiers. These rules are of the respective forms $@\text{-}\mathcal{V}$ and $@\text{-}\mathcal{F}$ for each quantifier @. Because the evaluation rules for the connectives have already been given in §6.2, we shall not reprise them here. We need only to point out that the evaluation rules for the connectives are now to be construed as M-relative rather than τ-relative, since we have progressed from the propositional case to the first-order case. We shall proceed here to state only the new rules that are required for the first-order case. These are the rules of evaluation for the two quantifiers.

Each such rule can be construed (in the now familiar way) as an inductive clause in our co-inductive definition of the notions

$$\mathcal{V}(\Pi, \varphi, M, D, \Lambda) \text{ and } \mathcal{F}(\Pi, \varphi, M, D, \Lambda).$$

These metalinguistic predications can be read, respectively, as

> 'Π is a verification of φ relative to model M with domain D, using the set Λ of M-literals'

and

> 'Π is a falsification of φ relative to model M with domain D, using the set Λ of M-literals'.

The positive and negative factoids of a model M constitute what we shall call its *atomic diagram*.[1] The *positive* factoids within the atomic diagram of a model M are the ones that get 'pictured' or depicted in

[1]Some authors define an atomic diagram in terms of literals, i.e. atoms and negations of atoms. Our treatment of factoids is equivalent to this, but makes for a smoother inferentialist treatment of first-order evaluations.

the box-diagram for M. One has to bear in mind, though, that the negative factoids 'are there', even if not depicted within the box-diagram. The negative factoids are present by virtue of their depictions being absent! This is why we must always be able to resort to the *inferential* representation of the atomic diagram, in which the negative factoids are *explicitly represented*, as inferences with the conclusion $\bot$.

11.4 Rules of verification and falsification for the quantifiers

We have to bear in mind that in the first-order case we are dealing with a model M with domain D, rather than (as in the propositional case) with just a truth-value assignment τ. The positive and negative M-factoids in the various Λs will involve individuals from the domain D. The graphic forms of the evaluation rules for $\forall$ and $\exists$ are as follows.

Rules of Verification and Falsification for $\exists$

$(\exists\text{-}\mathcal{V})$

$$\frac{\begin{matrix}\Lambda\\ \vdots\\ \varphi^x_{\boldsymbol{\alpha}}\end{matrix}}{\exists x\varphi}$$

where $\boldsymbol{\alpha}$ is a particular individual in the domain D

To verify an existential in M, one must verify **some** instance of it from D.

$(\exists\text{-}\mathcal{F})$

$$\frac{\exists x\varphi \qquad \begin{matrix}\overset{\square\text{---}(i)}{\underbrace{\varphi^x_{\boldsymbol{\alpha}_1}\ ,\ \Lambda_1}}\\ \vdots\\ \bot\end{matrix} \quad \ldots \quad \begin{matrix}\overset{\square\text{---}(i)}{\underbrace{\varphi^x_{\boldsymbol{\alpha}_n}\ ,\ \Lambda_n}}\\ \vdots\\ \bot\end{matrix} \quad \ldots}{\bot}\ (i)\ M$$

where $\boldsymbol{\alpha}_1, \ldots, \boldsymbol{\alpha}_n, \ldots$ are all the individuals in the domain D

To falsify an existential in M, one must falsify **every** instance of it from D. Note that we add the subscript M at the right-hand end of the horizontal stroke, to emphasize the *M-relativity* of this evaluation rule.

Rules of Verification and Falsification for ∀

$$(\forall\text{-}\mathcal{V}) \qquad \frac{\begin{matrix}\Lambda_1 & & \Lambda_n & \\ \vdots & \ldots & \vdots & \ldots \\ \varphi(\boldsymbol{\alpha_1}) & & \varphi(\boldsymbol{\alpha_n}) & \end{matrix}}{\forall x\varphi(x)}\,{\scriptstyle M}$$

where $\boldsymbol{\alpha_1, \ldots, \alpha_n, \ldots}$ are all the individuals in the domain D

To verify a universal in M, one must verify **every** instance of it from the domain D. Note that here too we add the subscript M at the right-hand end of the horizontal stroke, to emphasize the *M-relativity* of this evaluation rule.

$$(\forall\text{-}\mathcal{F}) \qquad \frac{\forall x\varphi(x) \qquad \begin{matrix}\overset{\square\!-\!-(i)}{\underbrace{\varphi(\boldsymbol{\alpha})\ ,\ \Lambda}} \\ \vdots \\ \bot\end{matrix}}{\bot}\,{\scriptstyle (i)}$$

where $\boldsymbol{\alpha}$ is a particular individual in the domain D

To falsify a universal in M, one must falsify **some** instance of it from the domain D.

Having thus stated graphically the evaluation rules for the quantifiers, shall now re-formulate them as inductive clauses in the co-inductive definition of verifications and falsifications for sentences of first-order logic. The reader will recall how we did this in the propositional case, in §6.3. In the first-order case, we shall use the symbolic abbreviations in the metalanguage that were introduced on p. 215, and of which we shall remind the reader here:

$\mathcal{V}(V, \varphi, \Lambda, M, D)$ abbreviates 'V verifies φ using the set Λ of M-factoids involving individuals from D'.

$\mathcal{F}(F, \varphi, \Lambda, M, D)$ abbreviates 'F falsifies φ using the set Λ of M-factoids involving individuals from D'.

We can now re-write our evaluation rules for $\forall$ and $\exists$ as the following inductive clauses, in the form of metalinguistic rules of inference (albeit potentially infinitary ones, when the domain D is infinite), in a co-inductive definition of first-order M-relative verifications and falsifications.

$$(\exists\text{-}\mathcal{V}) \quad \frac{\mathcal{V}(V, \varphi^x_{\boldsymbol{\alpha}}, \Lambda, M, D) \quad \boldsymbol{\alpha} \in D}{\mathcal{V}\left(\frac{V}{\exists x \varphi}, \exists x \varphi, \Lambda, M, D\right)}$$

$$(\exists\text{-}\mathcal{F}) \quad \frac{\mathcal{F}(F_{\boldsymbol{\alpha_1}}, \varphi^x_{\boldsymbol{\alpha_1}}, \Lambda_1, M, D) \ldots \mathcal{F}(F_{\boldsymbol{\alpha_n}}, \varphi^x_{\boldsymbol{\alpha_n}}, \Lambda_2, M, D) \ldots \quad D = \{\boldsymbol{\alpha_1}, \ldots, \boldsymbol{\alpha_n}, \ldots\}}{\mathcal{F}\left(\frac{\exists x \varphi \;\; F_{\boldsymbol{\alpha_1}} \ldots F_{\boldsymbol{\alpha_n}} \;\; \cdots}{\bot}, \exists x \varphi, \bigcup_i \Lambda_i, M, D\right)}$$

$$(\forall\text{-}\mathcal{V}) \quad \frac{\mathcal{V}(V_{\boldsymbol{\alpha_1}}, \varphi^x_{\boldsymbol{\alpha_1}}, \Lambda_1, M, D) \ldots \mathcal{V}(V_{\boldsymbol{\alpha_n}}, \varphi^x_{\boldsymbol{\alpha_n}}, \Lambda_2, M, D) \ldots \quad D = \{\boldsymbol{\alpha_1}, \ldots, \boldsymbol{\alpha_n}, \ldots\}}{\mathcal{V}\left(\frac{V_{\boldsymbol{\alpha_1}} \ldots V_{\boldsymbol{\alpha_n}} \;\; \cdots}{\forall x \varphi}, \forall x \varphi, \Lambda, M, D\right)}$$

$$(\forall\text{-}\mathcal{F}) \quad \frac{\mathcal{F}(F, \varphi^x_{\boldsymbol{\alpha}}, \Lambda, M, D) \quad \boldsymbol{\alpha} \in D}{\mathcal{F}\left(\frac{\forall x \varphi \;\; F}{\bot}, \forall x \varphi, \Lambda, M, D\right)}$$

Remarks on First-Order Verifications and Falsifications

Because we are dealing in the first-order case with a model M rather than a propositional truth-value assignment τ, some of the following remarks—namely 3, 4, and 5 —update the corresponding remarks in §6.2 concerning the propositional case. The other remarks remain the same.

1. Any verification of a complex sentence φ has φ as the conclusion of its last step, which is an application of the $\mathcal{V}$-rule for the operator dominant in φ.

2. Any falsification of a complex sentence φ has φ as the major premise of its last step, which is an application of the $\mathcal{F}$-rule for the operator dominant in φ.

3. Verifications and falsifications are *relative to a model M*. They 'depend' *only on certain M-literals* (forming the set Λ) as their 'undischarged assumptions'. (We use scare quotes here because these are assumptions for *evaluations*, not for *deductions*.)

4. Verifications and falsifications are, mathematically speaking, (labeled) trees all of whose branches are finite. But, if the domain of M is infinite, these trees have infinite branchings at applications of the M-relative rules ($\forall$-$\mathcal{V}$) and/or ($\exists$-$\mathcal{F}$).

5. The 'premises' of a verification or falsification are 'atomic facts' of M—positive or negative M-factoids.

6. The evaluation rules used in verifications and falsifications invite generalization to the case of formal deduction of a *possibly complex sentential* conclusion (or of $\bot$) from a set of *possibly complex* premises.

7. The *verification rules* (@-$\mathcal{V}$) will generalize to become (or, as we shall say, *morph into*) *@-Introduction rules*.

8. The *falsification rules* (@-$\mathcal{F}$) will generalize to become (or, as we shall say, *morph into*) *@-Elimination rules*.

9. Note that in any verification or falsification, every major premise for the application of an $\mathcal{F}$-rule *stands proud*, in the sense that it is not the conclusion of any verification. Moreover, any such major premise is *discharged* at the very next step. Hence all undischarged major premises φ are those of terminal applications of $\mathcal{F}$-rules in falsifications of φ.

11.5 Extending the new turnstile to the first-order case

The rules of verification and falsification for the connectives, in §6.2, along with the rules of verification and falsification for the quantifiers that we have stated in §11.4, are the inductive clauses of a co-inductive definition of the two notions 'Π is an M-verification of φ' and 'Π is an M-falsification of φ'. The basis step of the definition was given in §11.3. We shall abbreviate these two notions as $\mathcal{V}(\Pi, \varphi, M)$ and $\mathcal{F}(\Pi, \varphi, M)$ respectively. These are ternary metalinguistic predications. We shall now see what we can do by existentially quantifying into the argument-places occupied by Π.

Let M be a model (which of course has a consistent set of literals).

Definition 8.

(i) $M \Vdash \varphi$ *(read:* M verifies φ*) if and only if there is an* M*-relative verification of* φ*, i.e.* $\exists\Pi\, \mathcal{V}(\Pi, \varphi, M)$.

(ii) $M, \varphi \Vdash \bot$ *(read:* M falsifies φ*) if and only if there is an* M*-relative falsification of* φ*, i.e.* $\exists\Pi\, \mathcal{F}(\Pi, \varphi, M)$.

The turnstile statement $M \Vdash \varphi$ can also be read as 'M makes φ true', or 'there is an M-relative truthmaker for φ'.

Likewise, the turnstile statement $M, \varphi \Vdash \bot$ can also be read as 'M makes φ false', or 'there is an M-relative falsitymaker for φ'.

On our analysis, the truth-in-M of a sentence *consists in* the *existence* of an M-relative truthmaker for it. Likewise, the falsity-in-M of a sentence *consists in* the *existence* of an M-relative falsitymaker for it.

By introducing model-relative truthmakers (and falsitymakers) we go beyond the standard Tarskian account of model-relative truth (and model-relative falsity). The Tarskian account provided, in effect, only an inductive definition of the *binary* metalinguistic predications 'φ is true-in-M' and 'φ is false-in-M', without defining any *complex abstract objects* that could function as M-relative truthmakers or falsitymakers respectively.[2] An advantage of our account is that an M-relative truthmaker for φ represents a distinct *way in which* φ is true in M. And with any given sentence, and any given model M interpreting its descriptive vocabulary, there will in general be a *multitude of different such ways* in which φ is made true (or made false) in M.

[2] Tarski's original article on the concept of truth in formalized languages did not treat of models. It was easily adapted to do so, however, after the introduction to formal semantics of the set-theoretically definable notion of a model, due to Kemeny [1948].

11.6 Equivalence with Tarskian truth

Let $\overline{\overline{D}}$ be the cardinality of the domain D of a given model M.

Theorem 1.
Modulo *a metatheory which contains the mathematics of* $\overline{\overline{D}}$*-furcating trees of finite depth, we have, for all models* M *with domain* D,

$$\exists \Pi\ \mathcal{V}(\Pi, \varphi, M, D) \quad \Leftrightarrow \quad M \Vdash \varphi, \ \textit{i.e.}, \ \varphi \textit{ is true in } M$$

where the right-hand side is in the sense of Tarski.

We are explaining here what Tarskian truth (in a model M) *consists in.* It consists in the *existence of an* M*-relative verification* (in our sense) of the sentence in question.

Proof. A fully detailed proof will be given in the sequel. It proceeds by the obvious induction on the complexity of the sentence (or saturated formula) φ. Note that the proof is constructive, provided only that the M-relative truthmakers for universals and the M-relative falsitymakers for existentials, in the case where D is infinite, can be assumed to exist, courtesy of the background metamathematics. This observation affects the right-to-left direction in the relevant cases of the inductive step. □

11.7 Some model-relative first-order evaluations

Consider the model M whose domain is $\{\alpha, \beta, \gamma\}$ and whose atomic diagram $\mathcal{B}$ is the complete set

$$\left\{ \frac{L\alpha\alpha}{\bot}, \frac{}{L\alpha\beta}, \frac{L\alpha\gamma}{\bot}, \frac{L\beta\alpha}{\bot}, \frac{L\beta\beta}{\bot}, \frac{L\beta\gamma}{\bot}, \frac{}{L\gamma\alpha}, \frac{}{L\gamma\beta}, \frac{L\gamma\gamma}{\bot}, \frac{\alpha=\beta}{\bot}, \frac{\alpha=\gamma}{\bot}, \frac{\beta=\gamma}{\bot} \right\}$$

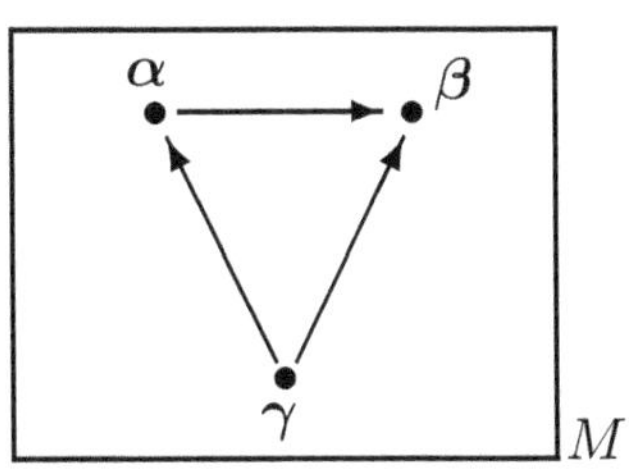

Now consider each of the following eight sentences (as φ):

$\forall x \exists y Lxy$	Everyone loves someone
$\forall x \exists y Lyx$	Everyone is loved by someone
$\forall x \exists y \neg Lxy$	Everyone is unloving toward someone
$\forall x \exists y \neg Lyx$	Everyone is unloved by someone
$\exists y \forall x Lxy$	Someone is loved by everyone
$\exists y \forall x Lyx$	Someone loves everyone
$\exists y \forall x \neg Lxy$	Someone is unloved by everyone
$\exists y \forall x \neg Lyx$	Someone is unloving toward everyone

For each of these eight sentences we can provide either a verification or a falsification relative to M (but not both!).

These will just be *illustrations*. The method, of course, is completely general. For *any* model M (even if infinite), and *any* sentence φ interpreted by M, there exists *either* an M-relative verification of φ, *or* an M-relative falsification of φ (but not both), depending on whether φ is true, or false, respectively, in the classical sense of Tarski. These abstract tree-like structures (i.e., verifications and falsifications) are finite when M is finite. But for infinite M they can involve 'infinite sideways branching' (with the *lengths* of their branches nevertheless all remaining finite). Still, the structures are well-defined, being built up in a recursive manner.

Let us proceed now with our (very finite) illustrations. At every stage we shall repeat the list of sentences being evaluated, so that the reader can keep track of our progression down the list. Moreover, the diagram for the model M will also be repeated at every stage, so that the reader can more easily check the choices of instances when applying the rules of evaluation for the quantifiers.

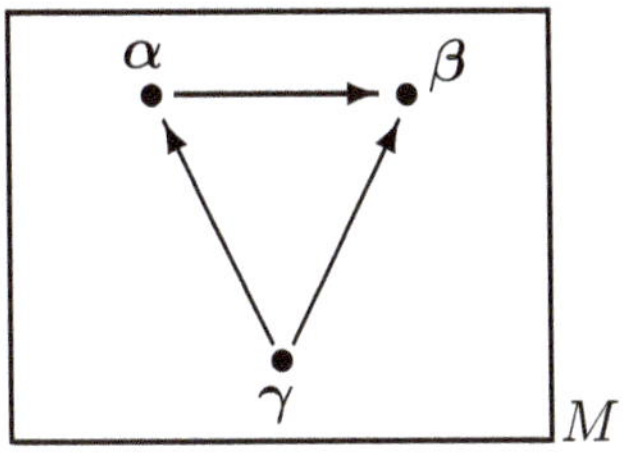

$\forall x\exists y Lxy$ False
$\forall x\exists y Lyx$
$\forall x\exists y\neg Lxy$
$\forall x\exists y\neg Lyx$
$\exists y\forall x Lxy$
$\exists y\forall x Lyx$
$\exists y\forall x\neg Lxy$
$\exists y\forall x\neg Lyx$

$$\dfrac{\forall x\exists y Lxy \qquad \dfrac{\dfrac{}{\exists y L\beta y}(2) \quad \dfrac{\dfrac{}{L\beta\alpha}(1)}{\bot}M \quad \dfrac{\dfrac{}{L\beta\beta}(1)}{\bot}M \quad \dfrac{\dfrac{}{L\beta\gamma}(1)}{\bot}M}{\bot}(1)\,M}{\bot}(2)$$

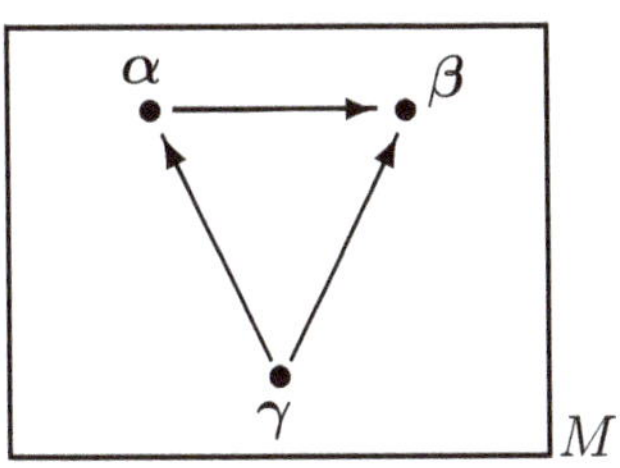

$\forall x\exists y Lxy$
$\forall x\exists y Lyx$ False
$\forall x\exists y\neg Lxy$
$\forall x\exists y\neg Lyx$
$\exists y\forall x Lxy$
$\exists y\forall x Lyx$
$\exists y\forall x\neg Lxy$
$\exists y\forall x\neg Lyx$

$$\dfrac{\forall x\exists y Lyx \qquad \dfrac{\dfrac{}{\exists y Ly\gamma}(2) \quad \dfrac{\dfrac{}{L\alpha\gamma}(1)}{\bot}M \quad \dfrac{\dfrac{}{L\beta\gamma}(1)}{\bot}M \quad \dfrac{\dfrac{}{L\gamma\gamma}(1)}{\bot}M}{\bot}(1)\,M}{\bot}(2)$$

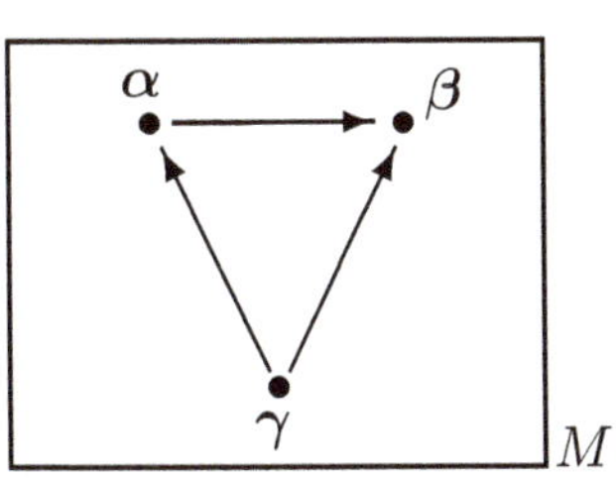

$\forall x\exists y Lxy$
$\forall x\exists y Lyx$
$\forall x\exists y\neg Lxy$ True
$\forall x\exists y\neg Lyx$
$\exists y\forall x Lxy$
$\exists y\forall x Lyx$
$\exists y\forall x\neg Lxy$
$\exists y\forall x\neg Lyx$

$$\dfrac{\dfrac{\dfrac{\dfrac{\dfrac{}{L\alpha\alpha}(1)}{\bot}M}{\neg L\alpha\alpha}(1)}{\exists y\neg L\alpha y}\qquad \dfrac{\dfrac{\dfrac{\dfrac{}{L\beta\alpha}(2)}{\bot}M}{\neg L\beta\alpha}(2)}{\exists y\neg L\beta y}\qquad \dfrac{\dfrac{\dfrac{\dfrac{}{L\gamma\gamma}(3)}{\bot}M}{\neg L\gamma\gamma}(3)}{\exists y\neg L\gamma y}}{\forall x\exists y\neg Lxy}M$$

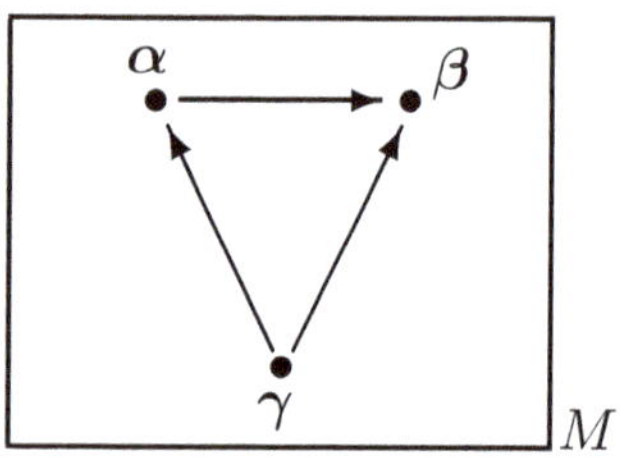

$\forall x\exists y Lxy$
$\forall x\exists y Lyx$
$\forall x\exists y\neg Lxy$
$\forall x\exists y\neg Lyx$ True
$\exists y\forall x Lxy$
$\exists y\forall x Lyx$
$\exists y\forall x\neg Lxy$
$\exists y\forall x\neg Lyx$

$$\dfrac{\dfrac{\dfrac{\dfrac{\dfrac{}{L\alpha\alpha}(1)}{\bot}M}{\neg L\alpha\alpha}(1)}{\exists y\neg Ly\alpha}\qquad \dfrac{\dfrac{\dfrac{\dfrac{}{L\beta\beta}(2)}{\bot}M}{\neg L\beta\beta}(2)}{\exists y\neg Ly\beta}\qquad \dfrac{\dfrac{\dfrac{\dfrac{}{L\beta\gamma}(3)}{\bot}M}{\neg L\beta\gamma}(3)}{\exists y\neg Ly\gamma}}{\forall x\exists y\neg Lyx}M$$

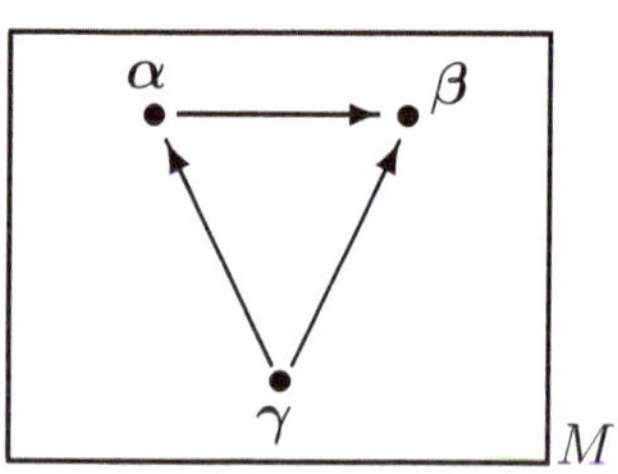

$\forall x\exists y Lxy$
$\forall x\exists y Lyx$
$\forall x\exists y\neg Lxy$
$\forall x\exists y\neg Lyx$
$\exists y\forall x Lxy$ False
$\exists y\forall x Lyx$
$\exists y\forall x\neg Lxy$
$\exists y\forall x\neg Lyx$

$$\dfrac{\exists y\forall x Lxy\qquad \dfrac{\dfrac{}{\forall x Lx\alpha}(4)\quad \dfrac{\dfrac{}{L\alpha\alpha}(1)}{\bot}M}{\bot}(1)\qquad \dfrac{\dfrac{}{\forall x Lx\beta}(4)\quad \dfrac{\dfrac{}{L\beta\beta}(2)}{\bot}M}{\bot}(2)\qquad \dfrac{\dfrac{}{\forall x Lx\gamma}(4)\quad \dfrac{\dfrac{}{L\alpha\gamma}(3)}{\bot}M}{\bot}(3)}{\bot}(4)\ M$$

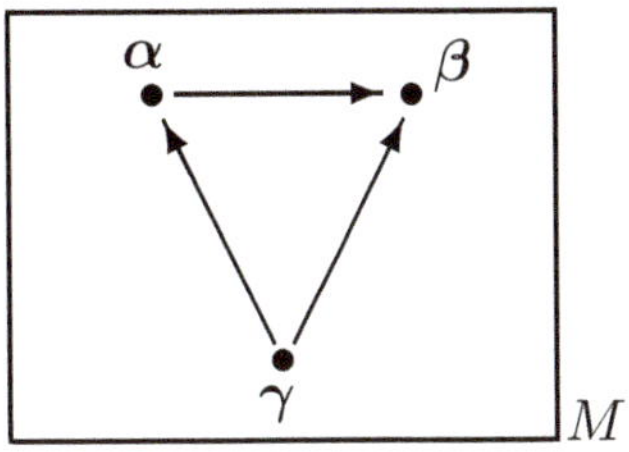

$\forall x\exists yLxy$
$\forall x\exists yLyx$
$\forall x\exists y\neg Lxy$
$\forall x\exists y\neg Lyx$
$\exists y\forall xLxy$
$\exists y\forall xLyx$ False
$\exists y\forall x\neg Lxy$
$\exists y\forall x\neg Lyx$

$$\dfrac{\exists y\forall xLyx \qquad \dfrac{\overline{\forall xL\alpha x}^{(4)} \quad \dfrac{\overline{L\alpha\alpha}^{(1)}}{\bot}{\scriptstyle M}}{\bot}{\scriptstyle (1)} \qquad \dfrac{\overline{\forall xL\beta x}^{(4)} \quad \dfrac{\overline{L\beta\alpha}^{(2)}}{\bot}{\scriptstyle M}}{\bot}{\scriptstyle (2)} \qquad \dfrac{\overline{\forall xL\gamma x}^{(4)} \quad \dfrac{\overline{L\gamma\gamma}^{(3)}}{\bot}{\scriptstyle M}}{\bot}{\scriptstyle (3)}}{\bot}{\scriptstyle (4)\ M}$$

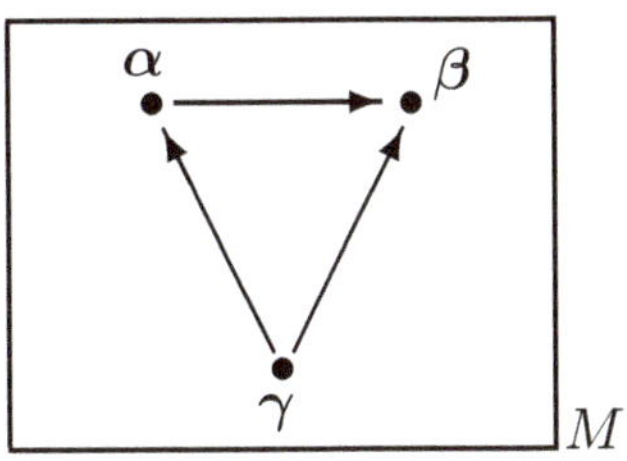

$\forall x\exists yLxy$
$\forall x\exists yLyx$
$\forall x\exists y\neg Lxy$
$\forall x\exists y\neg Lyx$
$\exists y\forall xLxy$
$\exists y\forall xLyx$
$\exists y\forall x\neg Lxy$ True
$\exists y\forall x\neg Lyx$

$$\dfrac{\dfrac{\dfrac{\dfrac{\overline{L\alpha\gamma}^{(1)}}{\bot}{\scriptstyle M}}{\neg L\alpha\gamma}{\scriptstyle (1)} \qquad \dfrac{\dfrac{\overline{L\beta\gamma}^{(2)}}{\bot}{\scriptstyle M}}{\neg L\beta\gamma}{\scriptstyle (2)} \qquad \dfrac{\dfrac{\overline{L\gamma\gamma}^{(3)}}{\bot}{\scriptstyle M}}{\neg L\gamma\gamma}{\scriptstyle (3)}}{\forall x\neg Lx\gamma}{\scriptstyle M}}{\exists y\forall x\neg Lxy}$$

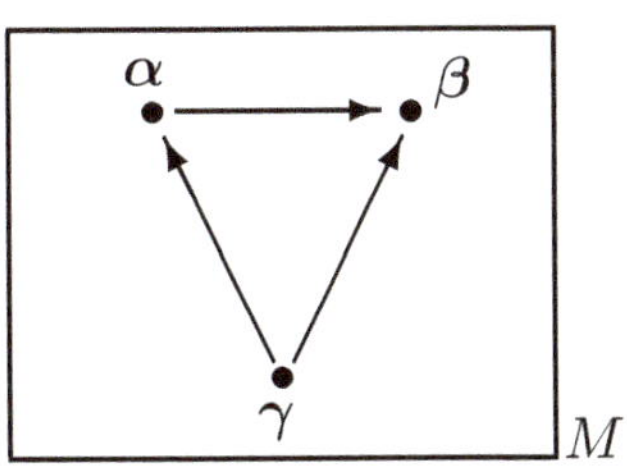

$\forall x\exists yLxy$
$\forall x\exists yLyx$
$\forall x\exists y\neg Lxy$
$\forall x\exists y\neg Lyx$
$\exists y\forall xLxy$
$\exists y\forall xLyx$
$\exists y\forall x\neg Lxy$
$\exists y\forall x\neg Lyx$ True

$$\dfrac{\dfrac{\dfrac{\dfrac{}{L\beta\alpha}{\scriptstyle(1)}}{\bot}{\scriptstyle M}}{\neg L\beta\alpha}{\scriptstyle(1)} \qquad \dfrac{\dfrac{\dfrac{}{L\beta\beta}{\scriptstyle(2)}}{\bot}{\scriptstyle M}}{\neg L\beta\beta}{\scriptstyle(2)} \qquad \dfrac{\dfrac{\dfrac{}{L\beta\gamma}{\scriptstyle(3)}}{\bot}{\scriptstyle M}}{\neg L\beta\gamma}{\scriptstyle(3)}}{\dfrac{\forall x\neg L\beta x}{\exists y\forall x\neg Lyx}}{\scriptstyle M}$$

Exercise 32. *Identify, in the foregoing octet of M-relative evaluations (beginning on p. 222), which, if any, of the eight different sentences involved might enjoy* other *M-relative evaluations to the same effect.*

11.8 Bivalence and Law of Non-Contradiction

Any interpretation M (like the particular example M that we have been working with above) is *atomically complete*: every atomic sentence features either in a positive M-factoid, or in a negative M-factoid. Any interpretation M is also *coherent*: no atomic sentence features both in a positive M-factoid, and in a negative M-factoid.

We can see from the rules of evaluation for negation that **Bivalence** will hold for evaluations with respect to any interpretation M: every sentence being evaluated will either come out true (by having an M-relative verification), or come out false (by having an M-relative falsification).

Moreover, the **Law of Non-Contradiction** will hold at the meta-level: no sentence will have both an M-relative verification and an M-relative falsification. This can be proved by induction on the complexity of sentences, by appeal to the basis fact that the interpretation M is both *atomically complete* and *coherent.*

Exercise 33. Give each of the following invalid single-premise arguments a counterexample. That is, provide in each case, relative to your counterexample M, an M-relative verification of each premise, and an M-relative falsification of the conclusion.

1. $\dfrac{\forall xFx}{\forall x\neg Fx}$
2. $\dfrac{\forall x\neg Fx}{\exists xFx}$
3. $\dfrac{\forall xFx}{\exists x\neg Fx}$
4. $\dfrac{\exists x\neg\neg Fx}{\forall xFx}$
5. $\dfrac{\neg\forall xFx}{\neg\exists xFx}$
6. $\dfrac{\exists xFx}{\forall xFx}$

7. $\dfrac{\exists x(Fx \wedge Gx)}{\forall x(Fx \vee Gx)}$

8. $\dfrac{\exists x(Fx \vee Gx)}{\exists xFx \wedge \exists xGx}$

9. $\dfrac{\exists xFx \wedge \exists xGx}{\exists x(Fx \wedge Gx)}$

10. $\dfrac{\forall xFx}{\forall x(Fx \wedge Gx)}$

11. $\dfrac{\forall xFx \to \forall xGx}{\forall x(Fx \to Gx)}$

12. $\dfrac{\forall x(Fx \to Gx)}{\forall x(Gx \to Fx)}$

13. $\dfrac{\forall y\exists xLxy}{\exists x\forall yLxy}$

14. $\dfrac{\forall xLxx}{\forall x\forall yLxy}$

15. $\dfrac{\exists x\exists yLxy}{\exists xLxx}$

16. $\dfrac{\exists x\neg Fx}{\forall x\neg Fx}$

17. $\dfrac{\neg\forall xFx}{\forall x\neg Fx}$

18. $\dfrac{\exists x\neg Fx}{\neg\exists xFx}$

19. $\dfrac{\begin{array}{l}\exists x\neg Gx\\ \forall x(Fx \to Gx)\end{array}}{\neg\exists xFx}$

20. $\dfrac{\begin{array}{l}\exists x\neg Fx\\ \exists x(Fx \vee Gx)\end{array}}{\exists x\neg Gx}$

21. $\dfrac{\forall x(Fx \vee Gx)}{\forall xFx}$

22. $\dfrac{\forall x(Fx \to Gx)}{\neg\forall xFx \to \neg\forall xGx}$

23. $\dfrac{\exists xFx \to \exists xGx}{\forall xFx \to \forall xGx}$

24. $\dfrac{\exists x(Fx \wedge \neg Gx)}{\forall x(Fx \to \neg Gx)}$

Chapter 12

Logic With Only The Quantifiers

In §4.4 we examined an important fragment of the language of first-order logic, namely the pure first-order language of identity. In §4.8 we mentioned the Horn fragment, on which the high-level programming language Prolog is based. In §4.11 we presented the language of propositional logic as the fragment of the language of first-order logic that is obtained by omitting the quantifiers from the language, and ignoring the predicate-argument structure of atomic sentences. We also made the point that in the classical case we can make do with any fragment based on negation, one of the binary connectives $\wedge$, $\vee$, and $\rightarrow$, and one of the quantifiers $\exists$ and $\forall$.

In this chapter we consider an unusual fragment, which has not yet been made the object of any systematic study. We believe, however, that it has great *pedagogical* value for *thorough* mastery of the existential and universal quantifiers (both semantically and inferentially); and that this thorough mastery can be achieved from within an *extremely simple* setting. This is the fragment whose *only* logical expressions are the quantifiers; and it is circumscribed further by containing only a single primitive predicate, let us say L. If the predicate L is n-place, then we call the language fragment $\mathbf{L}_n$. Note that the predicate L is in italics, denoting a relation among individuals, while the boldface $\mathbf{L}_n$, with numerical subscripts n, will denote the *language* that contains the predicate L of arity n.

The extraordinary thing about $\mathbf{L}_n$ is that it arguably contains all the complexities whose mastery would credit anyone displaying it with a full understanding of the two quantifiers $\exists$ and $\forall$. They can be fully understood without needing to take into account how they interact with the connectives. Moreover, all the quantificational behavior to be explored in this Chapter is fully constructive. There is no issue that can arise here on which the classicist could disagree with the constructivist, or *vice versa*.

12.1 The language $\mathbf{L}_1$

We work at the outset in a first-order language without identity, and with a single *unary* (monadic) predicate L as its only primitive extralogical expression. We eschew all logical connectives. We have only one individual variable, x (because with a single unary predicate that is all that we shall need). We consider only sentences containing only the usual quantifiers ($\exists$ or $\forall$) as their logical expressions. These sentences form the language we shall call $\mathbf{L}_1$. There are *only two* such sentences:

$$\exists xLx, \ \forall xLx.$$

There are therefore only two possible non-trivial single-premise arguments in $\mathbf{L}_1$ (non-trivial because their premises are distinct from their conclusions). They are

$$(1)\quad \frac{\exists xLx}{\forall xLx} \qquad (2)\quad \frac{\forall xLx}{\exists xLx}\ .$$

The argument

$$(1)\quad \frac{\exists xLx}{\forall xLx}$$

is invalid. It is the fallacy of over-generalization. A simple 2-element counterexample to it is

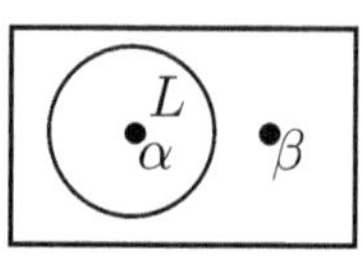

In this model, the individual α witnesses the truth of the existential premise, and the individual β witnesses the falsity of the universal conclusion.

The argument

$$(2)\quad \frac{\forall x Lx}{\exists x Lx}$$

has the empty model as its only counterexample. It therefore matters whether we allow ourselves to consider the empty model when classifying arguments as valid or invalid. For the rest of this study we shall rule the empty model out of consideration, thereby rendering argument (2) valid. Argument (2) is provable in standard logic, with its background assumption of non-empty domains. It has the proof

$$\dfrac{\dfrac{\forall x Lx}{La}}{\exists x Lx}$$

in standard logic. In what is known as universally free logic—the logic that takes into account the possibility that the domain is empty—the foregoing argument has the proof

$$\dfrac{\exists x \mathbf{E}x \qquad \dfrac{\dfrac{{}^{(1)}\overline{\mathbf{E}a}}{\exists ! a} \qquad \dfrac{\dfrac{{}^{(1)}\overline{\mathbf{E}a}}{\exists ! a} \quad \forall x Lx}{La}}{\exists x Lx}}{\exists x Lx}{\scriptstyle (1)}$$

Note that in the latter proof one appeals to the fact that there is at least one individual in the domain. That needs to be made explicit when working in universally free logic. Because the language does not contain identity, we need to use a monadic existence predicate $\mathbf{E}$ to express the assumption that there is at least one thing ($\exists x\mathbf{E}x$). Had the language contained identity, we could have used $x = x$ for this purpose, instead of $\mathbf{E}x$. Note also that the universal eliminations in the last two displayed proofs are of serial form. Indeed, throughout this chapter universal eliminations within proofs will be in serial form.

12.2 The language $\mathbf{L}_2$

Let us ratchet up from 1 to 2. We continue to work in a first-order language without identity, but now with a single *binary* (dyadic) predicate L as its only primitive extralogical expression. Once again we eschew all logical connectives. We require now only *two* distinct individual variables, x and y (because with a single binary predicate that is all we shall need). We consider only sentences containing only the usual quantifiers $\exists$ and $\forall$ as their logical expressions. They form the language $\mathbf{L}_2$. As the reader can check, there are *exactly ten* such sentences (disregarding different possible alphabetical choices of bound variables):

$$\exists x\exists yLxy,\ \exists x\forall yLxy,\ \forall x\exists yLxy,\ \forall x\forall yLxy,\ \exists x\exists yLyx,$$
$$\exists x\forall yLyx,\ \forall x\exists yLyx,\ \forall x\forall yLyx,\ \exists xLxx,\ \forall xLxx.$$

Note that we have used the variables x and y in their alphabetical order within quantifier prefixes. We have generated the necessary variety of sentences by ordering occurrences of x and y in all possible ways as arguments for the predicate L. We now consider all possible single-premise arguments in $\mathbf{L}_2$ whose premises are distinct from their conclusions. There are ninety such arguments. They are listed below. (Thirty three of them are valid; fifty seven of them are invalid.) The six arguments involving only the existential quantifier $\exists$ are in green; while the six involving only the universal quantifier $\forall$ are in orange. Among these 'single-quantifier' arguments the valid ones are square-bracketed: four of the green; and four of the orange.

(1) $\frac{\exists x \exists y Lxy}{\exists x \forall y Lxy}$	(2) $\frac{\exists x \exists y Lxy}{\forall x \exists y Lxy}$	(3) $\frac{\exists x \exists y Lxy}{\forall x \forall y Lxy}$	(4) $\left[\frac{\exists x \exists y Lxy}{\exists x \exists y Lyx}\right]$	(5) $\frac{\exists x \exists y Lxy}{\exists x \forall y Lyx}$
(6) $\frac{\exists x \exists y Lxy}{\forall x \exists y Lyx}$	(7) $\frac{\exists x \exists y Lxy}{\forall x \forall y Lyx}$	(8) $\frac{\exists x \exists y Lxy}{\exists x Lxx}$	(9) $\frac{\exists x \exists y Lxy}{\forall x Lxx}$	(10) $\frac{\exists x \forall y Lxy}{\exists x \exists y Lxy}$
(11) $\frac{\exists x \forall y Lxy}{\forall x \exists y Lxy}$	(12) $\frac{\exists x \forall y Lxy}{\forall x \forall y Lxy}$	(13) $\frac{\exists x \forall y Lxy}{\exists x \exists y Lyx}$	(14) $\frac{\exists x \forall y Lxy}{\exists x \forall y Lyx}$	(15) $\frac{\exists x \forall y Lxy}{\forall x \exists y Lyx}$
(16) $\frac{\exists x \forall y Lxy}{\forall x \forall y Lyx}$	(17) $\frac{\exists x \forall y Lxy}{\exists x Lxx}$	(18) $\frac{\exists x \forall y Lxy}{\forall x Lxx}$	(19) $\frac{\forall x \exists y Lxy}{\exists x \exists y Lxy}$	(20) $\frac{\forall x \exists y Lxy}{\exists x \forall y Lxy}$
(21) $\frac{\forall x \exists y Lxy}{\forall x \forall y Lxy}$	(22) $\frac{\forall x \exists y Lxy}{\exists x \exists y Lyx}$	(23) $\frac{\forall x \exists y Lxy}{\exists x \forall y Lyx}$	(24) $\frac{\forall x \exists y Lxy}{\forall x \exists y Lyx}$	(25) $\frac{\forall x \exists y Lxy}{\forall x \forall y Lyx}$
(26) $\frac{\forall x \exists y Lxy}{\exists x Lxx}$	(27) $\frac{\forall x \exists y Lxy}{\forall x Lxx}$	(28) $\frac{\forall x \forall y Lxy}{\exists x \exists y Lxy}$	(29) $\frac{\forall x \forall y Lxy}{\exists x \forall y Lxy}$	(30) $\frac{\forall x \forall y Lxy}{\forall x \exists y Lxy}$
(31) $\frac{\forall x \forall y Lxy}{\exists x \exists y Lyx}$	(32) $\frac{\forall x \forall y Lxy}{\exists x \forall y Lyx}$	(33) $\frac{\forall x \forall y Lxy}{\forall x \exists y Lyx}$	(34) $\left[\frac{\forall x \forall y Lxy}{\forall x \forall y Lyx}\right]$	(35) $\frac{\forall x \forall y Lxy}{\exists x Lxx}$
(36) $\left[\frac{\forall x \forall y Lxy}{\forall x Lxx}\right]$	(37) $\left[\frac{\exists x \exists y Lyx}{\exists x \exists y Lxy}\right]$	(38) $\frac{\exists x \exists y Lyx}{\exists x \forall y Lxy}$	(39) $\frac{\exists x \exists y Lyx}{\forall x \exists y Lxy}$	(40) $\frac{\exists x \exists y Lyx}{\forall x \forall y Lxy}$
(41) $\frac{\exists x \exists y Lyx}{\exists x \forall y Lyx}$	(42) $\frac{\exists x \exists y Lyx}{\forall x \exists y Lyx}$	(43) $\frac{\exists x \exists y Lyx}{\forall x \forall y Lyx}$	(44) $\frac{\exists x \exists y Lyx}{\exists x Lxx}$	(45) $\frac{\exists x \exists y Lyx}{\forall x Lxx}$
(46) $\frac{\exists x \forall y Lyx}{\exists x \exists y Lxy}$	(47) $\frac{\exists x \forall y Lyx}{\exists x \forall y Lxy}$	(48) $\frac{\exists x \forall y Lyx}{\forall x \exists y Lxy}$	(49) $\frac{\exists x \forall y Lyx}{\forall x \forall y Lxy}$	(50) $\frac{\exists x \forall y Lyx}{\exists x \exists y Lyx}$
(51) $\frac{\exists x \forall y Lyx}{\forall x \exists y Lyx}$	(52) $\frac{\exists x \forall y Lyx}{\forall x \forall y Lyx}$	(53) $\frac{\exists x \forall y Lyx}{\exists x Lxx}$	(54) $\frac{\exists x \forall y Lyx}{\forall x Lxx}$	(55) $\frac{\forall x \exists y Lyx}{\exists x \exists y Lxy}$
(56) $\frac{\forall x \exists y Lyx}{\exists x \forall y Lxy}$	(57) $\frac{\forall x \exists y Lyx}{\forall x \exists y Lxy}$	(58) $\frac{\forall x \exists y Lyx}{\forall x \forall y Lxy}$	(59) $\frac{\forall x \exists y Lyx}{\exists x \exists y Lyx}$	(60) $\frac{\forall x \exists y Lyx}{\exists x \forall y Lyx}$
(61) $\frac{\forall x \exists y Lyx}{\forall x \forall y Lyx}$	(62) $\frac{\forall x \exists y Lyx}{\exists x Lxx}$	(63) $\frac{\forall x \exists y Lyx}{\forall x Lxx}$	(64) $\frac{\forall x \forall y Lyx}{\exists x \exists y Lxy}$	(65) $\frac{\forall x \forall y Lyx}{\exists x \forall y Lxy}$
(66) $\left[\frac{\forall x \forall y Lyx}{\forall x \forall y Lxy}\right]$	(67) $\frac{\forall x \forall y Lyx}{\forall x \exists y Lxy}$	(68) $\frac{\forall x \forall y Lyx}{\exists x \exists y Lyx}$	(69) $\frac{\forall x \forall y Lyx}{\exists x \forall y Lyx}$	(70) $\frac{\forall x \forall y Lyx}{\forall x \exists y Lyx}$
(71) $\frac{\forall x \forall y Lyx}{\exists x Lxx}$	(72) $\left[\frac{\forall x \forall y Lyx}{\forall x Lxx}\right]$	(73) $\left[\frac{\exists x Lxx}{\exists x \exists y Lxy}\right]$	(74) $\frac{\exists x Lxx}{\exists x \forall y Lxy}$	(75) $\frac{\exists x Lxx}{\forall x \exists y Lxy}$
(76) $\frac{\exists x Lxx}{\forall x \forall y Lxy}$	(77) $\left[\frac{\exists x Lxx}{\exists x \exists y Lyx}\right]$	(78) $\frac{\exists x Lxx}{\exists x \forall y Lyx}$	(79) $\frac{\exists x Lxx}{\forall x \exists y Lyx}$	(80) $\frac{\exists x Lxx}{\forall x \forall y Lyx}$
(81) $\frac{\exists x Lxx}{\forall x Lxx}$	(82) $\frac{\forall x Lxx}{\exists x \exists y Lxy}$	(83) $\frac{\forall x Lxx}{\exists x \forall y Lxy}$	(84) $\frac{\forall x Lxx}{\forall x \exists y Lxy}$	(85) $\frac{\forall x Lxx}{\forall x \forall y Lxy}$
(86) $\frac{\forall x Lxx}{\exists x \exists y Lyx}$	(87) $\frac{\forall x Lxx}{\exists x \forall y Lyx}$	(88) $\frac{\forall x Lxx}{\forall x \exists y Lyx}$	(89) $\frac{\forall x Lxx}{\forall x \forall y Lyx}$	(90) $\frac{\forall x Lxx}{\exists x Lxx}$

The argument

$$(90) \qquad \frac{\forall x Lxx}{\exists x Lxx}$$

has the empty model as its only counterexample. It therefore matters, as it did in the unary case, whether we allow ourselves to consider the empty model when classifying arguments as valid or invalid. Recall that we are ruling the empty model out of consideration. Given that decision, argument (90) is to be regarded as valid. It has the proof

$$\dfrac{\dfrac{\forall x Lxx}{Laa}}{\exists x Lxx}$$

in standard logic, and in universally free logic it has the proof

$$\dfrac{\exists x \mathbf{E}x \qquad \dfrac{\dfrac{{}^{(1)}\overline{\mathbf{E}a}}{\exists! a} \qquad \dfrac{\dfrac{{}^{(1)}\overline{\mathbf{E}a}}{\exists! a} \quad \forall x Lxx}{Laa}}{\exists x Lxx}}{\exists x Lxx}\,{\scriptstyle(1)}$$

Note that in the latter proof one appeals once again to the fact that there is at least one individual in the domain.

Exercise 34. Of each remaining argument in the list of ninety above, determine whether it is valid or invalid. If valid, give a formal natural deduction of its conclusion from its premise as the only undischarged assumption, using only primitive rules of inference for the quantifiers. Otherwise, give a counterexample—a model that has at least one individual in its domain and that makes the premise true but makes the conclusion false.

Solutions to this exercise follow. Each *valid* argument receives a formal proof, in the form of a natural deduction. Universal eliminations, as heralded earlier, will be of serial form.

Each *invalid* argument receives *a* counterexample; and could well have others. We seek, however, to give in every case a *smallest possible* counterexample—that is, one with as few individuals as possible in its domain, and with as meager an extension as possible for L (i.e., as few arrows in the diagram as possible). In the diagrams to follow, the extension of L (as a collection of ordered pairs of individuals) is pictured by arrows.

Note that the L-'denuded' model

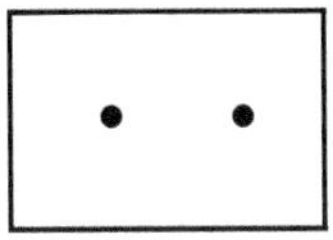

cannot make any $\mathbf{L}^2$-sentence true. And the L-'replete' model

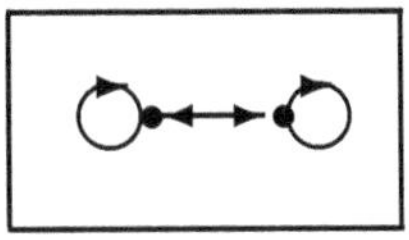

cannot make any $\mathbf{L}^2$-sentence false. Thus neither of these models can serve as a counterexample to any invalid single-premise $\mathbf{L}^2$-argument. But these two models are important extremal cases: every $\mathbf{L}^2$-sentence is true in the L-replete model; and every $\mathbf{L}^2$-sentence is false in the L-denuded model. $\mathbf{L}^2$-sentences are not only individually contingent; they are co-satisfiable and they are co-falsifiable.

There is only one possible two-element model that is not needed for the counterexemplification of any of the invalid single-premise arguments in $\mathbf{L}^2$, and that is the 'almost replete' model

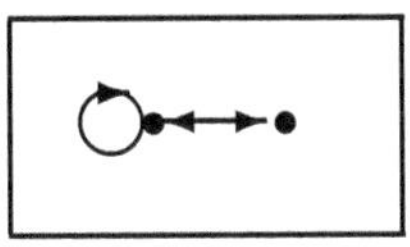

The arguments that this two-element model could be used to counterexemplify can be counterexemplified by simpler, less 'replete' two-element models.

Note that the last model just depicted, because its individuals are nameless, is isomorphic to the model depicted this way:

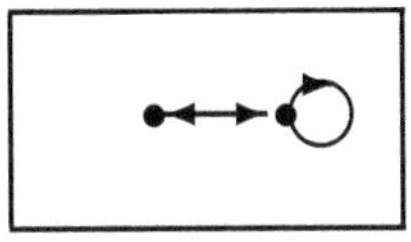

When we depict a model with more than one nameless individual, we must therefore bear in mind that 'the model' in question is to be thought of as a representative of *the class of all models with exactly the same structure*—its so-called *isomorphism class*.

It is interesting that all the invalid single-premise arguments in $\mathbf{L}^2$ admit of two-element counterexamples. And at least two elements are *required* for counterexemplification. There is no one-element counterexample to any invalid single-premise argument in $\mathbf{L}^2$. This is because any model with but a single element is either denuded or replete. The denuded one makes all $\mathbf{L}^2$-sentences false; the replete one makes them all true.

Here now are the ninety single-premise arguments in $\mathbf{L}^2$ once again. Each is given along with a proof or a counterexample as appropriate. No argument can have *both* a proof *and* a counterexample! (The reader should by now be able to explain why this is so.)

Each proof uses only the introduction and elimination rules for the quantifiers.

Each counterexample is given as a diagram, or model, in which dots are the individuals and, as already mentioned, arrows represent the extension of the binary predicate L. Along with each counterexemplifying model, we provide a model-relative *verification* of the argument's *premise*, and a model-relative *falsification* of its *conclusion*, in the senses explained in Chapter 11.

Note that our labeling of the two dots as α and β is a *meta*-linguistic device. There are no names in the object language that denote the dots (i.e., denote the individuals in the domain).

1. $\dfrac{\exists x\exists yLxy}{\exists x\forall yLxy}$

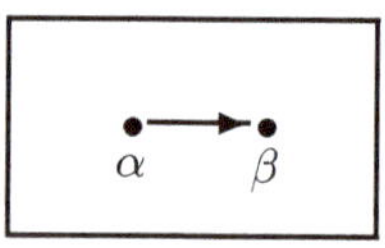

$$\dfrac{\dfrac{\dfrac{}{L\alpha\beta}}{\exists yL\alpha y}}{\exists x\exists yLxy}
\qquad
\dfrac{\exists x\forall yLxy \quad \dfrac{{}^{(3)}\overline{\forall yL\alpha y} \quad \dfrac{\overline{L\alpha\alpha}^{(1)}}{\bot}}{\bot}{\scriptstyle (1)} \quad \dfrac{{}^{(3)}\overline{\forall yL\beta y} \quad \dfrac{\overline{L\beta\alpha}^{(2)}}{\bot}}{\bot}{\scriptstyle (2)}}{\bot}{\scriptstyle (3)}$$

2. $\dfrac{\exists x\exists yLxy}{\forall x\exists yLxy}$

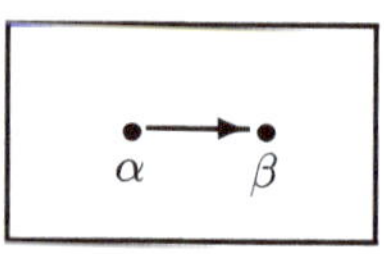

$$\dfrac{\dfrac{\dfrac{}{L\alpha\beta}}{\exists yL\alpha y}}{\exists x\exists yLxy}
\qquad
\dfrac{\forall x\exists yLxy \quad \dfrac{{}^{(2)}\overline{\exists yL\beta y} \quad \dfrac{\overline{L\beta\alpha}^{(1)}}{\bot} \quad \dfrac{\overline{L\beta\beta}^{(1)}}{\bot}}{\bot}{\scriptstyle (1)}}{\bot}{\scriptstyle (2)}$$

3. $\dfrac{\exists x \exists y Lxy}{\forall x \forall y Lxy}$

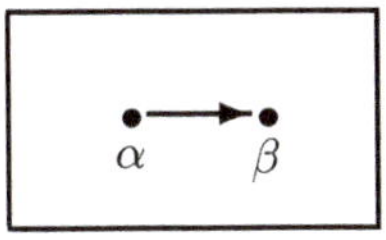

$$\dfrac{\dfrac{\overline{L\alpha\beta}}{\exists y L\alpha y}}{\exists x \exists y Lxy} \qquad \dfrac{\forall x \forall y Lxy \quad \dfrac{{}^{(2)}\overline{\forall y L\alpha y} \quad \dfrac{\overline{L\alpha\alpha}^{(1)}}{\bot}}{\bot}{\scriptstyle (1)}}{\bot}{\scriptstyle (2)}$$

4. $\dfrac{\exists x \exists y Lxy}{\exists x \exists y Lyx}$

$$\dfrac{\exists x \exists y Lxy \quad \dfrac{{}^{(1)}\overline{\exists y Lay} \quad \dfrac{\dfrac{\overline{Lab}^{(2)}}{\exists y Lyb}}{\exists x \exists y Lyx}}{\exists x \exists y Lyx}{\scriptstyle (2)}}{\exists x \exists y Lyx}{\scriptstyle (1)}$$

5. $\dfrac{\exists x \exists y Lxy}{\exists x \forall y Lyx}$

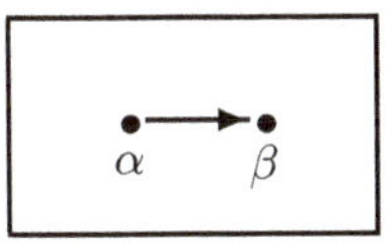

$$\dfrac{\dfrac{\overline{L\alpha\beta}}{\exists y L\alpha y}}{\exists x \exists y Lxy} \qquad \dfrac{\exists x \forall y Lyx \quad \dfrac{{}^{(3)}\overline{\forall y Ly\alpha} \quad \dfrac{\overline{L\alpha\alpha}^{(1)}}{\bot}}{\bot}{\scriptstyle (1)} \quad \dfrac{{}^{(3)}\overline{\forall y Ly\beta} \quad \dfrac{\overline{L\beta\beta}^{(2)}}{\bot}}{\bot}{\scriptstyle (2)}}{\bot}{\scriptstyle (3)}$$

6. $\dfrac{\exists x \exists y Lxy}{\forall x \exists y Lyx}$

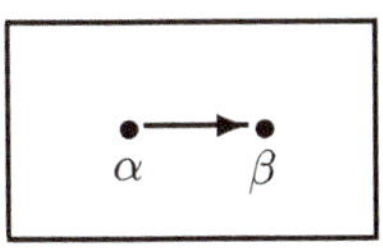

$$\dfrac{\dfrac{\overline{L\alpha\beta}}{\exists y L\alpha y}}{\exists x \exists y Lxy} \qquad \dfrac{\forall x \exists y Lyx \quad \dfrac{{}^{(2)}\overline{\exists y Ly\alpha} \quad \dfrac{\overline{L\alpha\alpha}^{(1)}}{\bot} \quad \dfrac{\overline{L\beta\alpha}^{(1)}}{\bot}}{\bot}{\scriptstyle (1)}}{\bot}{\scriptstyle (2)}$$

7. $\dfrac{\exists x \exists y Lxy}{\forall x \forall y Lyx}$

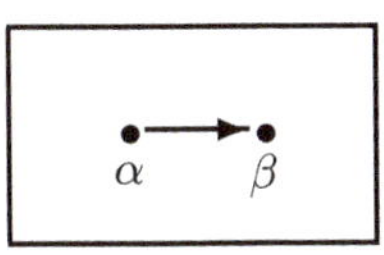

$$\cfrac{\cfrac{\overline{L\alpha\beta}}{\exists y L\alpha y}}{\exists x\exists y Lxy} \qquad \cfrac{\forall x\forall y Lyx \quad \cfrac{\overline{\forall y Ly\alpha}^{(2)} \quad \cfrac{\overline{L\alpha\alpha}^{(1)}}{\bot}}{\bot}(1)}{\bot}(2)$$

8. $\dfrac{\exists x\exists y Lxy}{\exists x Lxx}$

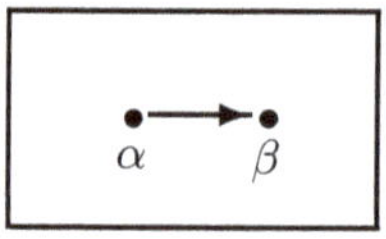

$$\cfrac{\cfrac{\overline{L\alpha\beta}}{\exists y L\alpha y}}{\exists x\exists y Lxy} \qquad \cfrac{\exists x Lxx \quad \cfrac{\overline{L\alpha\alpha}^{(1)}}{\bot} \quad \cfrac{\overline{L\beta\beta}^{(1)}}{\bot}}{\bot}(1)$$

9. $\dfrac{\exists x\exists y Lxy}{\forall x Lxx}$

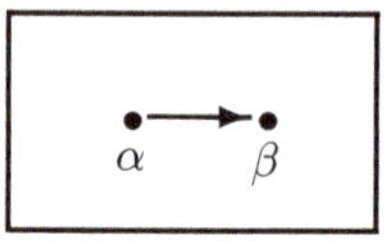

$$\cfrac{\cfrac{\overline{L\alpha\beta}}{\exists y L\alpha y}}{\exists x\exists y Lxy} \qquad \cfrac{\forall x Lxx \quad \cfrac{\overline{L\alpha\alpha}^{(1)}}{\bot}}{\bot}(1)$$

10. $\dfrac{\exists x\forall y Lxy}{\exists x\exists y Lxy}$

$$\cfrac{\exists x\forall y Lxy \quad \cfrac{\cfrac{\cfrac{\overline{\forall y Lay}^{(1)}}{Laa}}{\exists y Lay}}{\exists x\exists y Lxy}}{\exists x\exists y Lxy}(1)$$

11. $\dfrac{\exists x\forall y Lxy}{\forall x\exists y Lxy}$

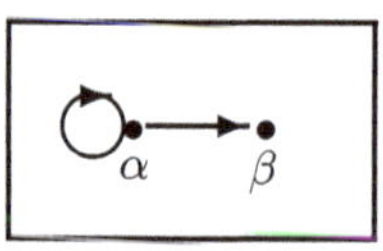

$$\cfrac{\cfrac{\overline{L\alpha\alpha} \quad \overline{L\alpha\beta}}{\forall y L\alpha y}}{\exists x\forall y Lxy} \qquad \cfrac{\forall x\exists y Lxy \quad \cfrac{\overline{\exists y L\beta y}^{(2)} \quad \cfrac{\overline{L\beta\alpha}^{(1)}}{\bot} \quad \cfrac{\overline{L\beta\beta}^{(1)}}{\bot}}{\bot}(1)}{\bot}(2)$$

12. $\dfrac{\exists x\forall y Lxy}{\forall x\forall y Lxy}$

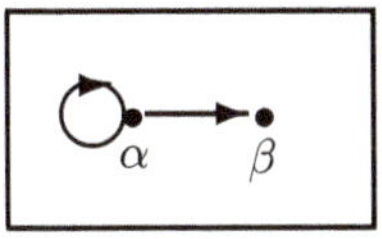

$$\dfrac{\dfrac{\overline{L\alpha\alpha}\quad\overline{L\alpha\beta}}{\forall y L\alpha y}}{\exists x\forall y Lxy} \qquad \dfrac{\forall x\forall y Lxy \quad \dfrac{{}^{(2)}\overline{\forall y L\beta y} \quad \dfrac{\overline{L\beta\alpha}^{(1)}}{\bot}}{\bot}{(1)}}{\bot}{(2)}$$

13. $\dfrac{\exists x\forall y Lxy}{\exists x\exists y Lyx}$

$$\dfrac{\exists x\forall y Lxy \quad \dfrac{\dfrac{\dfrac{\overline{\forall y Lay}^{(1)}}{Lab}}{\exists y Lay}}{\exists x\exists y Lxy}}{\exists x\exists y Lxy}{(1)}$$

14. $\dfrac{\exists x\forall y Lxy}{\exists x\forall y Lyx}$

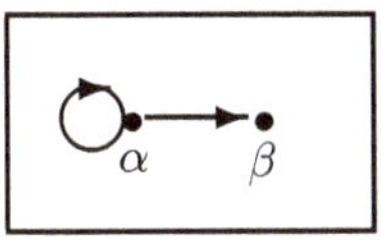

$$\dfrac{\dfrac{\overline{L\alpha\alpha}\quad\overline{L\alpha\beta}}{\forall y L\alpha y}}{\exists x\forall y Lxy} \qquad \dfrac{\exists x\forall y Lyx \quad \dfrac{{}^{(3)}\overline{\forall y Ly\alpha} \quad \dfrac{\overline{L\beta\alpha}^{(1)}}{\bot}}{\bot}{(1)} \quad \dfrac{{}^{(3)}\overline{\forall y Ly\beta} \quad \dfrac{\overline{L\beta\alpha}^{(2)}}{\bot}}{\bot}{(2)}}{\bot}{(3)}$$

15. $\dfrac{\exists x\forall y Lxy}{\forall x\exists y Lyx}$

$$\dfrac{\exists x\forall y Lxy \quad \dfrac{\dfrac{\dfrac{\overline{\forall y Lay}^{(1)}}{Lab}}{\exists y Lay}}{\forall x\exists y Lyx}}{\forall x\exists y Lyx}{(1)}$$

16. $\dfrac{\exists x\forall y Lxy}{\forall x\forall y Lyx}$

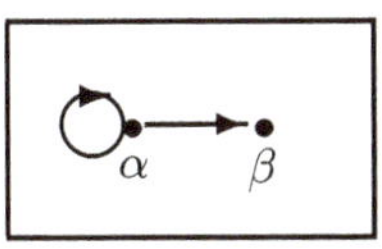

$$\dfrac{\dfrac{\overline{L\alpha\alpha}\quad\overline{L\alpha\beta}}{\forall y L\alpha y}}{\exists x\forall y Lxy} \qquad \dfrac{\forall x\forall y Lyx \quad \dfrac{{}^{(2)}\overline{\forall y Ly\beta} \quad \dfrac{\overline{L\beta\beta}^{(1)}}{\bot}}{\bot}{(1)}}{\bot}{(2)}$$

17. $\dfrac{\exists x\forall yLxy}{\exists xLxx}$

$$\dfrac{\exists x\forall yLxy \qquad \dfrac{\dfrac{\dfrac{}{\forall yLay}{\scriptstyle (1)}}{Laa}}{\exists xLxx}}{\exists xLxx}{\scriptstyle (1)}$$

18. $\dfrac{\exists x\forall yLxy}{\forall xLxx}$

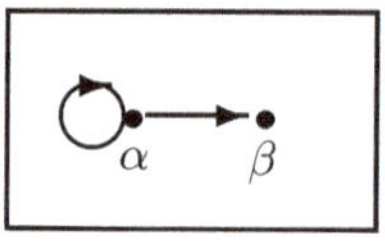

$$\dfrac{\dfrac{\dfrac{}{L\alpha\alpha} \quad \dfrac{}{L\alpha\beta}}{\forall yL\alpha y}}{\exists x\forall yLxy} \qquad \dfrac{\forall xLxx \qquad \dfrac{\dfrac{}{L\alpha\alpha}{\scriptstyle (1)}}{\bot}}{\bot}{\scriptstyle (1)}$$

19. $\dfrac{\forall x\exists yLxy}{\exists x\exists yLxy}$

$$\dfrac{\dfrac{\forall x\exists yLxy}{\exists yLay}}{\exists x\exists yLxy}$$

20. $\dfrac{\forall x\exists yLxy}{\exists x\forall yLxy}$

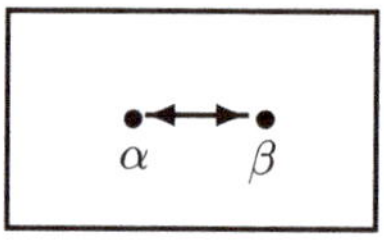

$$\dfrac{\dfrac{\dfrac{}{L\alpha\beta}}{\exists yL\alpha y} \quad \dfrac{\dfrac{}{L\beta\alpha}}{\exists yL\beta y}}{\forall x\exists yLxy} \qquad \dfrac{\exists x\forall yLxy \qquad \dfrac{{\scriptstyle (3)}\dfrac{}{\forall yL\alpha y} \quad \dfrac{\dfrac{}{L\alpha\alpha}{\scriptstyle (1)}}{\bot}}{\bot}{\scriptstyle (1)} \qquad \dfrac{{\scriptstyle (3)}\dfrac{}{\forall yL\beta y} \quad \dfrac{\dfrac{}{L\beta\beta}{\scriptstyle (2)}}{\bot}}{\bot}{\scriptstyle (2)}}{\bot}{\scriptstyle (3)}$$

21. $\dfrac{\forall x\exists yLxy}{\forall x\forall yLxy}$

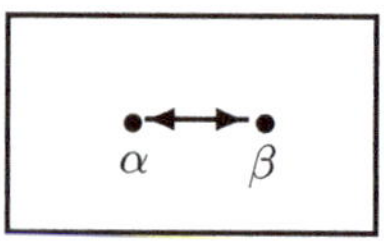

$$\dfrac{\dfrac{\dfrac{}{L\alpha\beta}}{\exists yL\alpha y} \quad \dfrac{\dfrac{}{L\beta\alpha}}{\exists yL\beta y}}{\forall x\exists yLxy} \qquad \dfrac{\forall x\forall yLxy \qquad \dfrac{{\scriptstyle (2)}\dfrac{}{\forall yL\alpha y} \quad \dfrac{\dfrac{}{L\alpha\alpha}{\scriptstyle (1)}}{\bot}}{\bot}{\scriptstyle (1)}}{\bot}{\scriptstyle (2)}$$

22. $\dfrac{\forall x\exists yLxy}{\exists x\exists yLyx}$

$$\dfrac{\dfrac{\forall x\exists yLxy}{\exists yLay} \qquad \dfrac{\dfrac{\dfrac{}{Lab}{\scriptstyle (1)}}{\exists yLay}}{\exists x\exists yLxy}}{\exists x\exists yLxy}{\scriptstyle (1)}$$

23. $\dfrac{\forall x\exists y Lxy}{\exists x\forall y Lyx}$

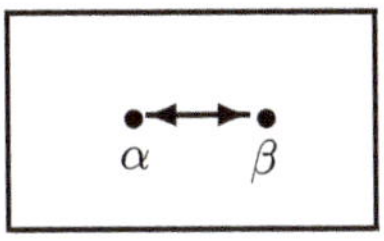

$$\dfrac{\dfrac{\dfrac{}{L\alpha\beta}}{\exists y L\alpha y} \qquad \dfrac{\dfrac{}{L\beta\alpha}}{\exists y L\beta y}}{\forall x\exists y Lxy}$$

$$\dfrac{\exists x\forall y Lyx \qquad \dfrac{{}^{(3)}\dfrac{}{\forall y Ly\alpha} \quad \dfrac{\dfrac{}{L\alpha\alpha}{}^{(1)}}{\bot}}{\bot}{}^{(1)} \qquad \dfrac{{}^{(3)}\dfrac{}{\forall y Ly\beta} \quad \dfrac{\dfrac{}{L\beta\beta}{}^{(2)}}{\bot}}{\bot}{}^{(2)}}{\bot}{}^{(3)}$$

24. $\dfrac{\forall x\exists y Lxy}{\forall x\exists y Lyx}$

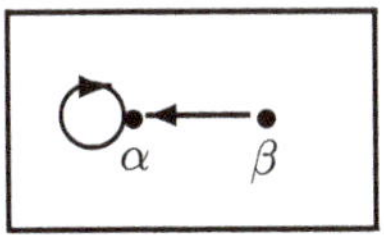

$$\dfrac{\dfrac{\dfrac{}{L\alpha\alpha}}{\exists y L\alpha y} \qquad \dfrac{\dfrac{}{L\beta\alpha}}{\exists y L\beta y}}{\forall x\exists y Lxy}$$

$$\dfrac{\forall x\exists y Lyx \qquad \dfrac{{}^{(2)}\dfrac{}{\exists y Ly\beta} \quad \dfrac{\dfrac{}{L\alpha\beta}{}^{(1)}}{\bot} \quad \dfrac{\dfrac{}{L\beta\beta}{}^{(1)}}{\bot}}{\bot}{}^{(1)}}{\bot}{}^{(2)}$$

25. $\dfrac{\forall x\exists y Lxy}{\forall x\forall y Lyx}$

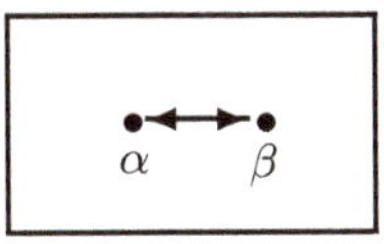

$$\dfrac{\dfrac{\dfrac{}{L\alpha\beta}}{\exists y L\alpha y} \qquad \dfrac{\dfrac{}{L\beta\alpha}}{\exists y L\beta y}}{\forall x\exists y Lxy}$$

$$\dfrac{\forall x\forall y Lyx \qquad \dfrac{{}^{(2)}\dfrac{}{\forall y Ly\alpha} \quad \dfrac{\dfrac{}{L\alpha\alpha}{}^{(1)}}{\bot}}{\bot}{}^{(1)}}{\bot}{}^{(2)}$$

26. $\dfrac{\forall x\exists y Lxy}{\exists x Lxx}$

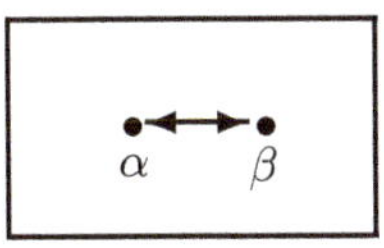

$$\dfrac{\dfrac{\dfrac{}{L\alpha\beta}}{\exists y L\alpha y} \qquad \dfrac{\dfrac{}{L\beta\alpha}}{\exists y L\beta y}}{\forall x\exists y Lxy}$$

$$\dfrac{\exists x Lxx \qquad \dfrac{\dfrac{}{L\alpha\alpha}{}^{(1)}}{\bot} \qquad \dfrac{\dfrac{}{L\beta\beta}{}^{(1)}}{\bot}}{\bot}{}^{(1)}$$

27. $\dfrac{\forall x\exists y Lxy}{\forall x Lxx}$

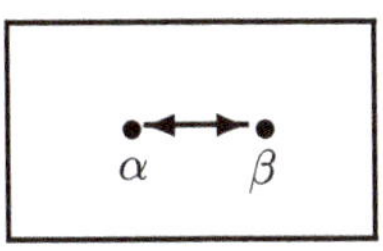

$$\dfrac{\dfrac{\overline{L\alpha\beta}}{\exists y L\alpha y} \qquad \dfrac{\overline{L\beta\alpha}}{\exists y L\beta y}}{\forall x\exists y Lxy} \qquad\qquad \dfrac{\forall x Lxx \qquad \dfrac{\overline{L\alpha\alpha}^{(1)}}{\bot}}{\bot}{\scriptstyle (1)}$$

28. $\dfrac{\forall x\forall y Lxy}{\exists x\exists y Lxy}$ $\qquad \dfrac{\dfrac{\dfrac{\dfrac{\forall x\forall y Lxy}{\forall y Lay}}{Lab}}{\exists y Lay}}{\exists x\exists y Lxy}$

29. $\dfrac{\forall x\forall y Lxy}{\exists x\forall y Lxy}$ $\qquad \dfrac{\dfrac{\forall x\forall y Lxy}{\forall y Lay}}{\exists x\forall y Lxy}$

30. $\dfrac{\forall x\forall y Lxy}{\forall x\exists y Lxy}$ $\qquad \dfrac{\dfrac{\dfrac{\dfrac{\forall x\forall y Lxy}{\forall y Lay}}{Laa}}{\exists y Lay}}{\forall x\exists y Lxy}$

31. $\dfrac{\forall x\forall y Lxy}{\exists x\exists y Lyx}$ $\qquad \dfrac{\dfrac{\dfrac{\dfrac{\forall x\forall y Lxy}{\forall y Lay}}{Laa}}{\exists y Lya}}{\exists x\exists y Lyx}$

32. $\dfrac{\forall x\forall y Lxy}{\exists x\forall y Lyx}$ $\qquad \dfrac{\dfrac{\dfrac{\dfrac{\forall x\forall y Lxy}{\forall y Lay}}{Laa}}{\forall y Lya}}{\exists x\forall y Lyx}$

33. $\dfrac{\forall x\forall y Lxy}{\forall x\exists y Lyx}$ $\qquad \dfrac{\dfrac{\dfrac{\dfrac{\forall x\forall y Lxy}{\forall y Lay}}{Laa}}{\exists y Lya}}{\forall x\exists y Lyx}$

34. $\dfrac{\forall x\forall y Lxy}{\forall x\forall y Lyx}$ $\qquad \dfrac{\dfrac{\dfrac{\dfrac{\forall x\forall y Lxy}{\forall y Lay}}{Laa}}{\forall y Lya}}{\forall x\forall y Lyx}$

35. $\dfrac{\forall x\forall y Lxy}{\exists x Lxx}$ $\qquad \dfrac{\dfrac{\dfrac{\forall x\forall y Lxy}{\forall y Lay}}{Laa}}{\exists x Lxx}$

36. $\dfrac{\forall x \forall y Lxy}{\forall x Lxx}$ $\qquad \dfrac{\dfrac{\dfrac{\forall x \forall y Lxy}{\forall y Lay}}{Laa}}{\forall x Lxx}$

37. $\dfrac{\exists x \exists y Lyx}{\exists x \exists y Lxy}$

$$\dfrac{\exists x \exists y Lyx \qquad \dfrac{\dfrac{}{\exists y Lya}{\scriptstyle (1)} \qquad \dfrac{\dfrac{\dfrac{}{Lba}{\scriptstyle (2)}}{\exists y Lby}}{\exists x \exists y Lxy}}{\exists x \exists y Lxy}{\scriptstyle (2)}}{\exists x \exists y Lxy}{\scriptstyle (1)}$$

38. $\dfrac{\exists x \exists y Lyx}{\exists x \forall y Lxy}$

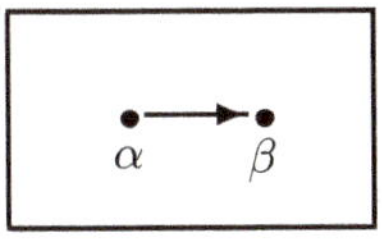

$\dfrac{\dfrac{L\alpha\beta}{\exists y Ly\beta}}{\exists x \exists y Lyx}$

$$\dfrac{\exists x \forall y Lxy \qquad \dfrac{\dfrac{}{\forall y L\alpha y}{\scriptstyle (3)} \qquad \dfrac{\dfrac{}{L\alpha\alpha}{\scriptstyle (1)}}{\bot}}{\bot}{\scriptstyle (1)} \qquad \dfrac{\dfrac{}{\forall y L\beta y}{\scriptstyle (3)} \qquad \dfrac{\dfrac{}{L\beta\beta}{\scriptstyle (2)}}{\bot}}{\bot}{\scriptstyle (2)}}{\bot}{\scriptstyle (3)}$$

39. $\dfrac{\exists x \exists y Lyx}{\forall x \exists y Lxy}$

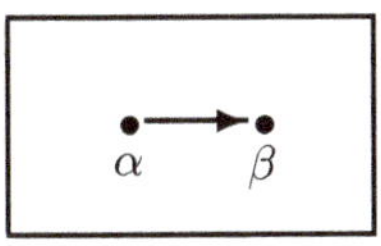

$\dfrac{\dfrac{L\alpha\beta}{\exists y Ly\beta}}{\exists x \exists y Lyx}$

$$\dfrac{\forall x \exists y Lxy \qquad \dfrac{\dfrac{}{\exists y L\beta y}{\scriptstyle (2)} \qquad \dfrac{\dfrac{}{L\beta\alpha}{\scriptstyle (1)}}{\bot} \qquad \dfrac{\dfrac{}{L\beta\beta}{\scriptstyle (1)}}{\bot}}{\bot}{\scriptstyle (1)}}{\bot}{\scriptstyle (2)}$$

40. $\dfrac{\exists x \exists y Lyx}{\forall x \forall y Lxy}$

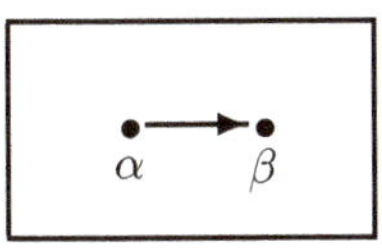

$\dfrac{\dfrac{L\alpha\beta}{\exists y Ly\beta}}{\exists x \exists y Lyx}$

$$\dfrac{\forall x \forall y Lxy \qquad \dfrac{\dfrac{}{\forall y L\alpha y}{\scriptstyle (2)} \qquad \dfrac{\dfrac{}{L\alpha\alpha}{\scriptstyle (1)}}{\bot}}{\bot}{\scriptstyle (1)}}{\bot}{\scriptstyle (2)}$$

41. $\dfrac{\exists x\exists y Lyx}{\exists x\forall y Lyx}$

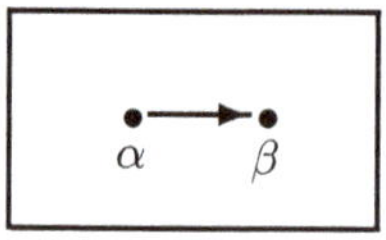

$$\cfrac{\cfrac{L\alpha\beta}{\exists y Ly\beta}}{\exists x\exists y Lyx}$$

$$\cfrac{\exists x\forall y Lyx \qquad \cfrac{\overline{\forall y Ly\alpha}^{\,(3)} \qquad \cfrac{\overline{L\alpha\alpha}^{\,(1)}}{\bot}}{\bot}{\scriptstyle(1)} \qquad \cfrac{\overline{\forall y Ly\beta}^{\,(3)} \qquad \cfrac{\overline{L\beta\beta}^{\,(2)}}{\bot}}{\bot}{\scriptstyle(2)}}{\bot}{\scriptstyle(3)}$$

42. $\dfrac{\exists x\exists y Lyx}{\forall x\exists y Lyx}$

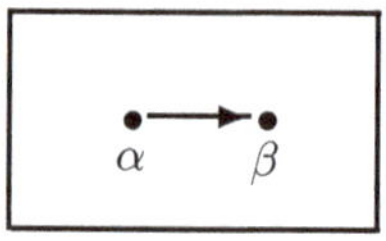

$$\cfrac{\cfrac{L\alpha\beta}{\exists y Ly\beta}}{\exists x\exists y Lyx}$$

$$\cfrac{\forall x\exists y Lyx \qquad \cfrac{\overline{\exists y Ly\alpha}^{\,(2)} \qquad \cfrac{\overline{L\alpha\alpha}^{\,(1)}}{\bot} \qquad \cfrac{\overline{L\beta\alpha}^{\,(1)}}{\bot}}{\bot}{\scriptstyle(1)}}{\bot}{\scriptstyle(2)}$$

43. $\dfrac{\exists x\exists y Lyx}{\forall x\forall y Lyx}$

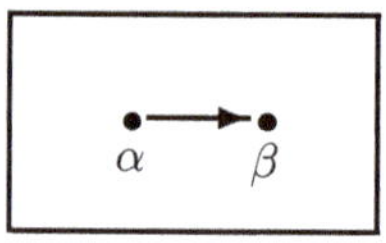

$$\cfrac{\cfrac{L\alpha\beta}{\exists y Ly\beta}}{\exists x\exists y Lyx}$$

$$\cfrac{\forall x\forall y Lyx \qquad \cfrac{\overline{\forall y Ly\alpha}^{\,(2)} \qquad \cfrac{\overline{L\alpha\alpha}^{\,(1)}}{\bot}}{\bot}{\scriptstyle(1)}}{\bot}{\scriptstyle(2)}$$

44. $\dfrac{\exists x\exists y Lyx}{\exists x Lxx}$

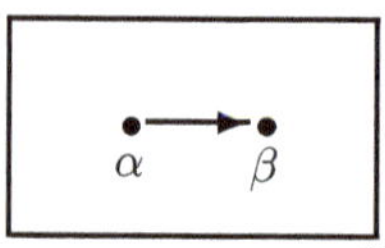

$$\cfrac{\cfrac{L\alpha\beta}{\exists y Ly\beta}}{\exists x\exists y Lyx}$$

$$\cfrac{\exists x Lxx \qquad \cfrac{\overline{L\alpha\alpha}^{\,(1)}}{\bot} \qquad \cfrac{\overline{L\beta\beta}^{\,(1)}}{\bot}}{\bot}{\scriptstyle(1)}$$

45. $\dfrac{\exists x\exists y Lyx}{\forall x Lxx}$

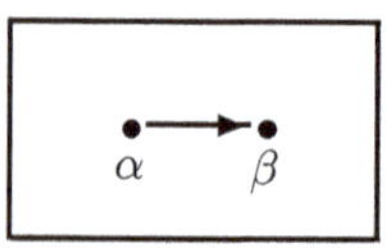

$$\dfrac{\dfrac{L\alpha\beta}{\exists y L y\beta}}{\exists x\exists y Lyx} \qquad \dfrac{\forall x Lxx \qquad \dfrac{\overline{L\alpha\alpha}^{(1)}}{\bot}}{\bot}{(1)}$$

46. $\dfrac{\exists x\forall y Lyx}{\exists x\exists y Lxy}$

$$\dfrac{\exists x\forall y Lyx \qquad \dfrac{\dfrac{\dfrac{\overline{\forall y Lya}^{(1)}}{Laa}}{\exists y Lay}}{\exists x\exists y Lxy}}{\exists x\exists y Lxy}{(1)}$$

47. $\dfrac{\exists x\forall y Lyx}{\exists x\forall y Lxy}$

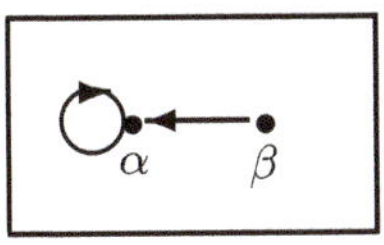

$$\dfrac{\dfrac{\overline{L\alpha\alpha} \qquad \overline{L\beta\alpha}}{\forall y Ly\alpha}}{\exists x\forall y Lyx} \qquad \dfrac{\exists x\forall y Lxy \qquad \dfrac{{}^{(3)}\overline{\forall y L\alpha y} \qquad \dfrac{\overline{L\alpha\beta}^{(1)}}{\bot}}{\bot}{(1)} \qquad \dfrac{{}^{(3)}\overline{\forall y L\beta y} \qquad \dfrac{\overline{L\beta\beta}^{(2)}}{\bot}}{\bot}{(2)}}{\bot}{(3)}$$

48. $\dfrac{\exists x\forall y Lyx}{\forall x\exists y Lxy}$

$$\dfrac{\exists x\forall y Lyx \qquad \dfrac{\dfrac{\dfrac{\overline{\forall y Lya}^{(1)}}{Lba}}{\exists y Lby}}{\forall x\exists y Lxy}}{\forall x\exists y Lxy}{(1)}$$

49. $\dfrac{\exists x\forall y Lyx}{\forall x\forall y Lxy}$

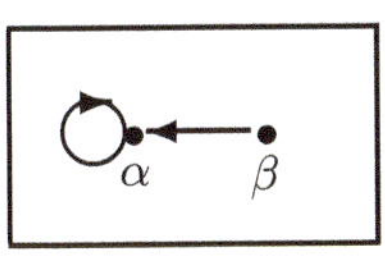

$$\dfrac{\dfrac{\overline{L\alpha\alpha} \qquad \overline{L\beta\alpha}}{\forall y Ly\alpha}}{\exists x\forall y Lyx} \qquad \dfrac{\forall x\forall y Lxy \qquad \dfrac{{}^{(2)}\overline{\forall y L\alpha y} \qquad \dfrac{\overline{L\alpha\beta}^{(1)}}{\bot}}{\bot}{(1)}}{\bot}{(2)}$$

50. $\dfrac{\exists x\forall y Lyx}{\exists x\exists y Lyx}$

$$\dfrac{\exists x\forall y Lyx \qquad \dfrac{\dfrac{\dfrac{\overline{\forall y Lya}^{(1)}}{Lba}}{\exists y Lya}}{\exists x\exists y Lyx}}{\exists x\exists y Lyx}{(1)}$$

51. $\dfrac{\exists x\forall y Lyx}{\forall x\exists y Lyx}$

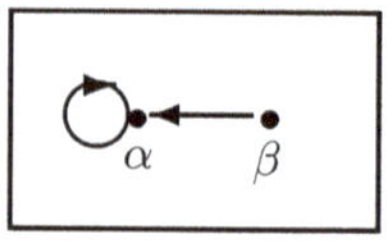

$$\dfrac{\dfrac{\overline{L\alpha\alpha} \qquad \overline{L\beta\alpha}}{\forall y Ly\alpha}}{\exists x\forall y Lyx} \qquad\qquad \dfrac{\forall x\exists y Lyx \qquad \dfrac{{}^{(2)}\overline{\exists y Ly\beta} \qquad \dfrac{\overline{L\alpha\beta}^{\,(1)}}{\bot} \qquad \dfrac{\overline{L\beta\beta}^{\,(1)}}{\bot}}{\bot}{}^{(1)}}{\bot}{}^{(2)}$$

52. $\dfrac{\exists x\forall y Lyx}{\forall x\forall y Lyx}$

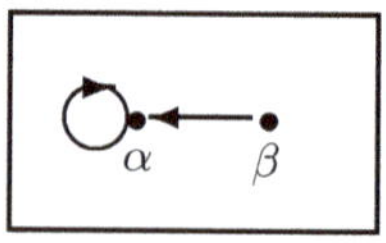

$$\dfrac{\dfrac{\overline{L\alpha\alpha} \qquad \overline{L\beta\alpha}}{\forall y Ly\alpha}}{\exists x\forall y Lyx} \qquad\qquad \dfrac{\forall x\forall y Lyx \qquad \dfrac{{}^{(2)}\overline{\forall y Ly\beta} \qquad \dfrac{\overline{L\alpha\beta}^{\,(1)}}{\bot}}{\bot}{}^{(1)}}{\bot}{}^{(2)}$$

53. $\dfrac{\exists x\forall y Lyx}{\exists x Lxx}$

$$\dfrac{\exists x\forall y Lyx \qquad \dfrac{\dfrac{\overline{\forall y Lya}^{\,(1)}}{Laa}}{\exists x Lxx}}{\exists x Lxx}{}^{(1)}$$

54. $\dfrac{\exists x\forall y Lyx}{\forall x Lxx}$

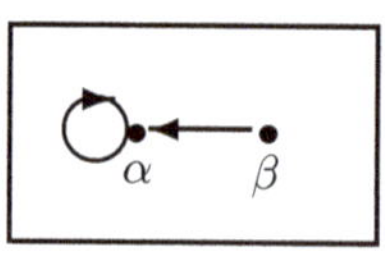

$$\dfrac{\dfrac{\overline{L\alpha\alpha} \qquad \overline{L\beta\alpha}}{\forall y Ly\alpha}}{\exists x\forall y Lyx} \qquad\qquad \dfrac{\forall x Lxx \qquad \dfrac{\overline{L\beta\beta}^{\,(1)}}{\bot}}{\bot}{}^{(1)}$$

55. $\dfrac{\forall x\exists y Lyx}{\exists x\exists y Lxy}$

$$\dfrac{\dfrac{\forall x\exists y Lyx}{\exists y Lya} \qquad \dfrac{\dfrac{\overline{Lba}^{\,(1)}}{\exists y Lby}}{\exists x\exists y Lxy}}{\exists x\exists y Lxy}{}^{(1)}$$

56. $\dfrac{\forall x\exists yLyx}{\exists x\forall yLxy}$

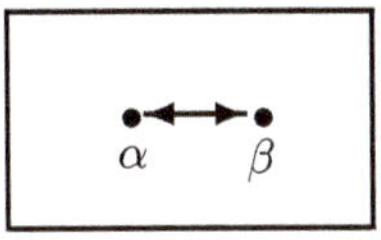

$$\dfrac{\dfrac{\overline{L\beta\alpha}}{\exists yLy\alpha}\qquad\dfrac{\overline{L\alpha\beta}}{\exists yLy\beta}}{\forall x\exists yLyx}$$

$$\dfrac{\exists x\forall yLxy\qquad\dfrac{{\scriptstyle(3)}\overline{\forall yL\alpha y}\qquad\dfrac{\overline{L\alpha\alpha}^{(1)}}{\bot}}{\bot}{\scriptstyle(1)}\qquad\dfrac{{\scriptstyle(3)}\overline{\forall yL\beta y}\qquad\dfrac{\overline{L\beta\beta}^{(2)}}{\bot}}{\bot}{\scriptstyle(2)}}{\bot}{\scriptstyle(3)}$$

57. $\dfrac{\forall x\exists yLyx}{\forall x\exists yLxy}$

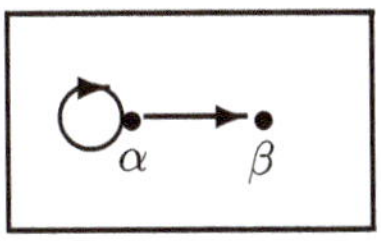

$$\dfrac{\dfrac{\overline{L\alpha\alpha}}{\exists yLy\alpha}\qquad\dfrac{\overline{L\alpha\beta}}{\exists yLy\beta}}{\forall x\exists yLyx}$$

$$\dfrac{\forall x\exists yLxy\qquad\dfrac{{\scriptstyle(2)}\overline{\exists yL\beta y}\qquad\dfrac{\overline{L\beta\alpha}^{(1)}}{\bot}\qquad\dfrac{\overline{L\beta\beta}^{(1)}}{\bot}}{\bot}{\scriptstyle(1)}}{\bot}{\scriptstyle(2)}$$

58. $\dfrac{\forall x\exists yLyx}{\forall x\forall yLxy}$

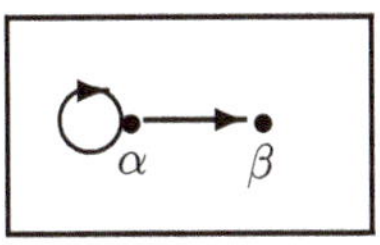

$$\dfrac{\dfrac{\overline{L\alpha\alpha}}{\exists yLy\alpha}\qquad\dfrac{\overline{L\alpha\beta}}{\exists yLy\beta}}{\forall x\exists yLyx}$$

$$\dfrac{\forall x\forall yLxy\qquad\dfrac{{\scriptstyle(2)}\overline{\forall yL\beta y}\qquad\dfrac{\overline{L\beta\alpha}^{(1)}}{\bot}}{\bot}{\scriptstyle(1)}}{\bot}{\scriptstyle(2)}$$

59. $\dfrac{\forall x\exists yLyx}{\exists x\exists yLyx}$

$$\dfrac{\dfrac{\forall x\exists yLyx}{\exists yLya}}{\exists x\exists yLyx}$$

60. $\dfrac{\forall x\exists yLyx}{\exists x\forall yLyx}$

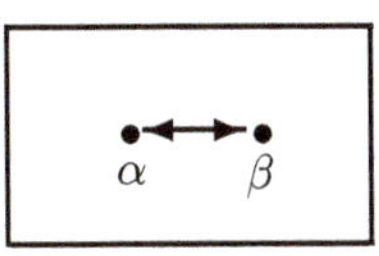

$$\dfrac{\dfrac{\overline{L\beta\alpha}}{\exists y L y\alpha} \quad \dfrac{\overline{L\alpha\beta}}{\exists y L y\beta}}{\forall x\exists y Lyx} \qquad \dfrac{\exists x\forall y Lyx \quad \dfrac{\overline{\forall y Ly\alpha}^{(3)} \quad \dfrac{\overline{L\alpha\alpha}^{(1)}}{\bot}}{\bot}{\scriptstyle (1)} \quad \dfrac{\overline{\forall y Ly\beta}^{(3)} \quad \dfrac{\overline{L\beta\beta}^{(2)}}{\bot}}{\bot}{\scriptstyle (2)}}{\bot}{\scriptstyle (3)}$$

61. $\dfrac{\forall x\exists y Lyx}{\forall x\forall y Lyx}$

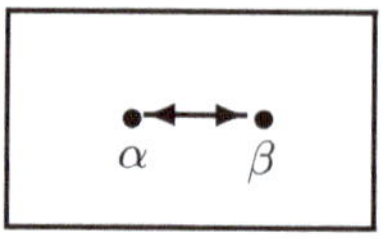

$$\dfrac{\dfrac{\overline{L\beta\alpha}}{\exists y L y\alpha} \quad \dfrac{\overline{L\alpha\beta}}{\exists y L y\beta}}{\forall x\exists y Lyx} \qquad \dfrac{\forall x\forall y Lyx \quad \dfrac{\overline{\forall y Ly\alpha}^{(2)} \quad \dfrac{\overline{L\alpha\alpha}^{(1)}}{\bot}}{\bot}{\scriptstyle (1)}}{\bot}{\scriptstyle (2)}$$

62. $\dfrac{\forall x\exists y Lyx}{\exists x Lxx}$

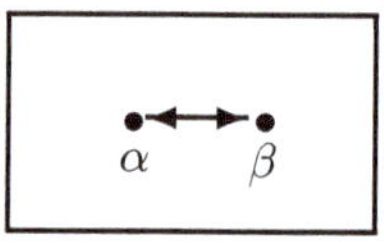

$$\dfrac{\dfrac{\overline{L\beta\alpha}}{\exists y L y\alpha} \quad \dfrac{\overline{L\alpha\beta}}{\exists y L y\beta}}{\forall x\exists y Lyx} \qquad \dfrac{\exists x Lxx \quad \dfrac{\overline{L\alpha\alpha}^{(1)}}{\bot} \quad \dfrac{\overline{L\beta\beta}^{(1)}}{\bot}}{\bot}{\scriptstyle (1)}$$

63. $\dfrac{\forall x\exists y Lyx}{\forall x Lxx}$

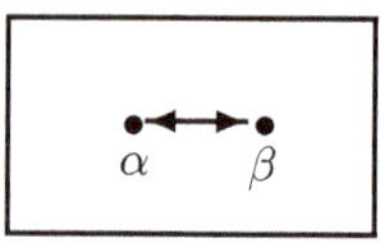

$$\dfrac{\dfrac{\overline{L\beta\alpha}}{\exists y L y\alpha} \quad \dfrac{\overline{L\alpha\beta}}{\exists y L y\beta}}{\forall x\exists y Lyx} \qquad \dfrac{\forall x Lxx \quad \dfrac{\overline{L\alpha\alpha}^{(1)}}{\bot}}{\bot}{\scriptstyle (1)}$$

64. $\dfrac{\forall x\forall y Lyx}{\exists x\exists y Lxy}$

$$\dfrac{\dfrac{\dfrac{\dfrac{\forall x\forall y Lyx}{\forall y Lya}}{Laa}}{\exists y Lay}}{\exists x\exists y Lxy}$$

65. $\dfrac{\forall x\forall yLyx}{\exists x\forall yLxy}$ $$\dfrac{\dfrac{\dfrac{\dfrac{\forall x\forall yLyx}{\forall yLya}}{Laa}}{\forall yLay}}{\exists x\forall yLxy}$$

66. $\dfrac{\forall x\forall yLyx}{\forall x\forall yLxy}$ $$\dfrac{\dfrac{\dfrac{\dfrac{\forall x\forall yLyx}{\forall yLya}}{Laa}}{\forall yLay}}{\forall x\forall yLxy}$$

67. $\dfrac{\forall x\forall yLyx}{\forall x\exists yLxy}$ $$\dfrac{\dfrac{\dfrac{\dfrac{\forall x\forall yLyx}{\forall yLya}}{Laa}}{\exists yLay}}{\forall x\exists yLxy}$$

68. $\dfrac{\forall x\forall yLyx}{\exists x\exists yLyx}$ $$\dfrac{\dfrac{\dfrac{\dfrac{\forall x\forall yLyx}{\forall yLya}}{Laa}}{\exists yLya}}{\exists x\exists yLyx}$$

69. $\dfrac{\forall x\forall yLyx}{\exists x\forall yLyx}$ $$\dfrac{\dfrac{\forall x\forall yLyx}{\forall yLya}}{\exists x\forall yLyx}$$

70. $\dfrac{\forall x\forall yLyx}{\forall x\exists yLyx}$ $$\dfrac{\dfrac{\dfrac{\dfrac{\forall x\forall yLyx}{\forall yLya}}{Laa}}{\exists yLya}}{\forall x\exists yLyx}$$

71. $\dfrac{\forall x\forall yLyx}{\exists xLxx}$ $$\dfrac{\dfrac{\dfrac{\forall x\forall yLyx}{\forall yLya}}{Laa}}{\exists xLxx}$$

72. $\dfrac{\forall x\forall yLyx}{\forall xLxx}$ $$\dfrac{\dfrac{\dfrac{\forall x\forall yLyx}{\forall yLya}}{Laa}}{\forall xLxx}$$

73. $\dfrac{\exists xLxx}{\exists x\exists yLxy}$ $$\dfrac{\exists xLxx \qquad \dfrac{\dfrac{\dfrac{}{Laa}{}^{(1)}}{\exists yLay}}{\exists x\exists yLxy}}{\exists x\exists yLxy}{}^{(1)}$$

74. $\dfrac{\exists x Lxx}{\exists x \forall y Lxy}$

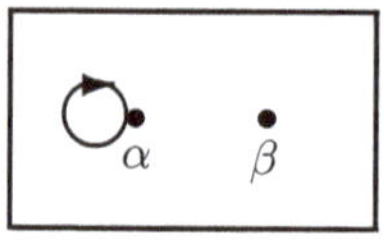

$$\dfrac{\overline{L\alpha\alpha}}{\exists x Lxx}$$

$$\dfrac{\exists x \forall y Lxy \quad \dfrac{{}^{(3)}\overline{\forall y L\alpha y} \quad \dfrac{\overline{L\alpha\beta}^{(1)}}{\bot}}{\bot}{\scriptstyle (1)} \quad \dfrac{{}^{(3)}\overline{\forall y L\beta y} \quad \dfrac{\overline{L\beta\beta}^{(2)}}{\bot}}{\bot}{\scriptstyle (2)}}{\bot}{\scriptstyle (3)}$$

75. $\dfrac{\exists x Lxx}{\forall x \exists y Lxy}$

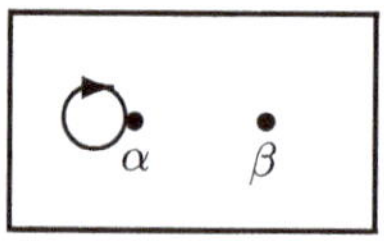

$$\dfrac{\overline{L\alpha\alpha}}{\exists x Lxx}$$

$$\dfrac{\forall x \exists y Lxy \quad \dfrac{{}^{(2)}\overline{\exists y L\beta y} \quad \dfrac{\overline{L\beta\alpha}^{(1)}}{\bot} \quad \dfrac{\overline{L\beta\beta}^{(1)}}{\bot}}{\bot}{\scriptstyle (1)}}{\bot}{\scriptstyle (2)}$$

76. $\dfrac{\exists x Lxx}{\forall x \forall y Lxy}$

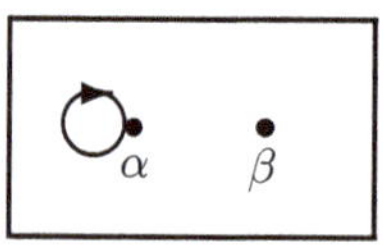

$$\dfrac{\overline{L\alpha\alpha}}{\exists x Lxx}$$

$$\dfrac{\forall x \forall y Lxy \quad \dfrac{{}^{(2)}\overline{\forall y L\alpha y} \quad \dfrac{\overline{L\alpha\beta}^{(1)}}{\bot}}{\bot}{\scriptstyle (1)}}{\bot}{\scriptstyle (2)}$$

77. $\dfrac{\exists x Lxx}{\exists x \exists y Lyx}$

$$\dfrac{\exists x Lxx \quad \dfrac{\dfrac{\overline{Laa}^{(1)}}{\exists y Lya}}{\exists x \exists y Lyx}}{\exists x \exists y Lyx}{\scriptstyle (1)}$$

78. $\dfrac{\exists x Lxx}{\exists x \forall y Lyx}$

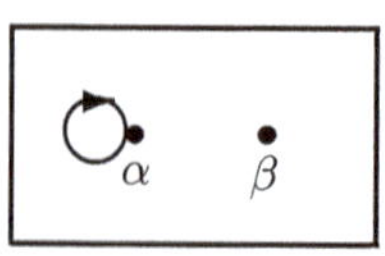

$$\dfrac{\overline{L\alpha\alpha}}{\exists x Lxx} \qquad \dfrac{\exists x\forall y Lyx \quad \dfrac{{}^{(3)}\overline{\forall y Ly\alpha} \quad \dfrac{\overline{L\beta\alpha}^{\,(1)}}{\bot}}{\bot}{\scriptstyle(1)} \quad \dfrac{{}^{(3)}\overline{\forall y Ly\beta} \quad \dfrac{\overline{L\beta\beta}^{\,(2)}}{\bot}}{\bot}{\scriptstyle(2)}}{\bot}{\scriptstyle(3)}$$

79. $\dfrac{\exists x Lxx}{\forall x\exists y Lyx}$

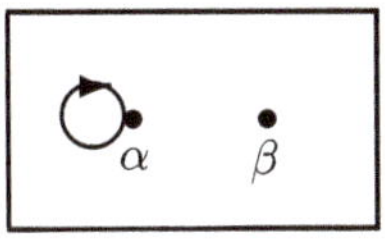

$$\dfrac{\overline{L\alpha\alpha}}{\exists x Lxx} \qquad \dfrac{\forall x\exists y Lyx \quad \dfrac{{}^{(2)}\overline{\exists y Ly\beta} \quad \dfrac{\overline{L\alpha\beta}^{\,(1)}}{\bot} \quad \dfrac{\overline{L\beta\beta}^{\,(1)}}{\bot}}{\bot}{\scriptstyle(1)}}{\bot}{\scriptstyle(2)}$$

80. $\dfrac{\exists x Lxx}{\forall x\forall y Lyx}$

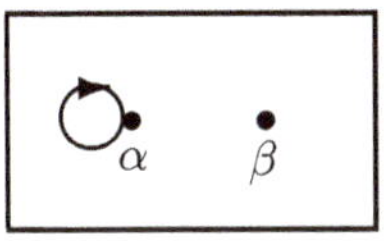

$$\dfrac{\overline{L\alpha\alpha}}{\exists x Lxx} \qquad \dfrac{\forall x\forall y Lyx \quad \dfrac{{}^{(2)}\overline{\forall y Ly\alpha} \quad \dfrac{\overline{L\beta\alpha}^{\,(1)}}{\bot}}{\bot}{\scriptstyle(1)}}{\bot}{\scriptstyle(2)}$$

81. $\dfrac{\exists x Lxx}{\forall x Lxx}$

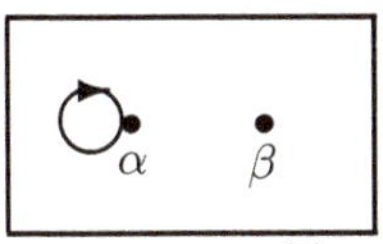

$$\dfrac{\overline{L\alpha\alpha}}{\exists x Lxx} \qquad \dfrac{\forall x Lxx \quad \dfrac{\overline{L\beta\beta}^{\,(1)}}{\bot}}{\bot}{\scriptstyle(1)}$$

82. $\dfrac{\forall x Lxx}{\exists x\exists y Lxy}$ $\qquad \dfrac{\dfrac{\dfrac{\forall x Lxx}{Laa}}{\exists y Lay}}{\exists x\exists y Lxy}$

83. $\dfrac{\forall x Lxx}{\exists x\forall y Lxy}$

$$\frac{\overline{L\alpha\alpha} \qquad \overline{L\beta\beta}}{\forall x Lxx}$$

$$\dfrac{\exists x\forall y Lxy \qquad \dfrac{{}^{(3)}\overline{\forall y L\alpha y} \qquad \dfrac{\overline{L\alpha\beta}^{\,(1)}}{\bot}}{\bot}{\scriptstyle (1)} \qquad \dfrac{{}^{(3)}\overline{\forall y L\beta y} \qquad \dfrac{\overline{L\beta\alpha}^{\,(2)}}{\bot}}{\bot}{\scriptstyle (2)}}{\bot}{\scriptstyle (3)}$$

84. $\dfrac{\forall x Lxx}{\forall x\exists y Lxy}$ $\qquad$ $\dfrac{\dfrac{\dfrac{\forall x Lxx}{Laa}}{\exists y Lay}}{\forall x\exists y Lxy}$

85. $\dfrac{\forall x Lxx}{\forall x\forall y Lxy}$

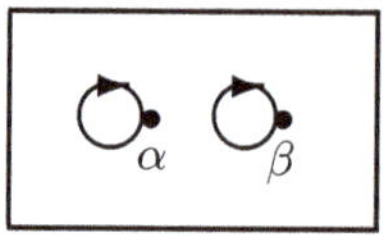

$$\frac{\overline{L\alpha\alpha} \qquad \overline{L\beta\beta}}{\forall x Lxx}$$

$$\dfrac{\forall x\forall y Lxy \qquad \dfrac{{}^{(2)}\overline{\forall y L\alpha y} \qquad \dfrac{\overline{L\alpha\beta}^{\,(1)}}{\bot}}{\bot}{\scriptstyle (1)}}{\bot}{\scriptstyle (2)}$$

86. $\dfrac{\forall x Lxx}{\exists x\exists y Lyx}$ $\qquad$ $\dfrac{\dfrac{\dfrac{\forall x Lxx}{Laa}}{\exists y Lay}}{\exists x\exists y Lxy}$

87. $\dfrac{\forall x Lxx}{\exists x\forall y Lyx}$

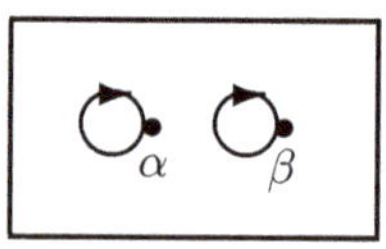

$$\frac{\overline{L\alpha\alpha} \qquad \overline{L\beta\beta}}{\forall x Lxx}$$

$$\dfrac{\exists x\forall y Lyx \qquad \dfrac{{}^{(3)}\overline{\forall y Ly\alpha} \qquad \dfrac{\overline{L\beta\alpha}^{\,(1)}}{\bot}}{\bot}{\scriptstyle (1)} \qquad \dfrac{{}^{(3)}\overline{\forall y Ly\beta} \qquad \dfrac{\overline{L\alpha\beta}^{\,(2)}}{\bot}}{\bot}{\scriptstyle (2)}}{\bot}{\scriptstyle (3)}$$

88. $\dfrac{\forall x Lxx}{\forall x\exists y Lyx}$ $\qquad$ $\dfrac{\dfrac{\dfrac{\forall x Lxx}{Laa}}{\exists y Lya}}{\forall x\exists y Lyx}$

89. $\dfrac{\forall x Lxx}{\forall x\forall y Lyx}$

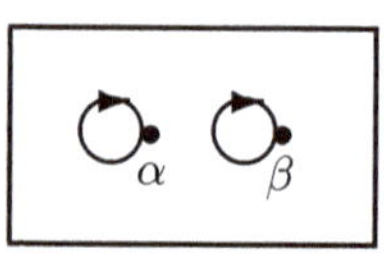

$$\dfrac{\overline{L\alpha\alpha} \quad \overline{L\beta\beta}}{\forall x Lxx} \qquad \dfrac{\forall x \forall y Lyx \qquad \dfrac{\overset{(2)}{\overline{\forall y L y\alpha}} \qquad \dfrac{\overset{(1)}{\overline{L\beta\alpha}}}{\bot}}{\bot}(1)}{\bot}(2)$$

90. $\dfrac{\forall x Lxx}{\exists x Lxx}$ $\qquad \dfrac{\dfrac{\forall x Lxx}{Laa}}{\exists x Lxx}$

We have examined the cases $n = 1$ and $n = 2$ in a sequence of languages $\mathbf{L}_n$, to be defined in general below. In the unary case ($n=1$), we got by with only one *individual variable*; and in the binary case ($n=2$) we got by with only two *individual variables*. (*Why* we could do so is obvious on brief reflection.) That said, we point out that alphabetical variation in choice of bound variables will not affect sentence identity. Thus, for example,

$$\forall x \exists y Lxy \quad \text{and} \quad \forall y \exists x Lyx$$

count as the same sentence. They each have the same 'pattern of linkage' established by their bound variables. This is why, when we described the language $\mathbf{L}^2$, we remarked that it had only ten sentences (disregarding different possible alphabetical choices of bound variables), and stuck to alphabetical ordering of variable-occurrences within quantifier prefixes:

$\exists x \exists y Lxy$, $\exists x \forall y Lxy$, $\forall x \exists y Lxy$, $\forall x \forall y Lxy$, $\exists x \exists y Lyx$,
$\exists x \forall y Lyx$, $\forall x \exists y Lyx$, $\forall x \forall y Lyx$, $\exists x Lxx$, $\forall x Lxx$.

12.3 The language $\mathbf{L}_3$

$\mathbf{L}_3$ has 74 distinct sentences. It therefore has $74 \times 73 = 5,402$ distinct single-premise arguments and $74 \times 73 \times 72 = 388,944$ distinct two-premise arguments.

Compare:

$\mathbf{L}_2$ has 10 distinct sentences; 90 distinct single-premises arguments; and 720 distinct two-premise arguments.

$\mathbf{L}_1$ has 2 distinct sentences; 2 distinct single-premises arguments; and no two-premise arguments.

It would appear that the numbers ramp up rather fast.

Exercise 35. Specify an algorithm that will determine, for any given single-premise argument in the language $\mathbf{L}_3$, whether it has a counterexample containing no more than 3 individuals in its domain.

12.4 Upping the arity

Definition 9. *We shall denote by* $\mathbf{L}_n$ *a first-order language satisfying the following conditions.*

1. *it does not have the identity predicate* $=$;
2. *it does not have any connectives;*
3. *it has exactly* n *individual variables;*
4. *it has exactly one* n*-ary predicate, which we shall call* L;
5. *it has the quantifiers* $\exists$ *and* $\forall$.

It is easy to see that no sentence of $\mathbf{L}_n$ is logically refutable—that is, there is no proof of $\perp$ from any sentence of $\mathbf{L}_n$. This is because $\perp$ may occur as a conclusion of a proof (i.e., of a disproof) *only when negations are involved*, allowing for the application, somewhere within the disproof, of the rule of Negation Elimination, whose conclusion is $\perp$. But $\mathbf{L}_n$, *ex hypothesi*, is negation-free.

Likewise, no sentence of $\mathbf{L}_n$ is a logical theorem (i.e., deducible from the empty set of assumptions). This is because the introduction rules for $\forall$ and $\exists$ do not effect discharge of assumptions, and their elimination rules have undischarged major premises. It follows that no proof made up only of applications of quantifier rules can have all its assumptions discharged.

Observation 1. *Every sentence of* $\mathbf{L}_n$ *is contingent—that is, there is a (finite) model that makes it true, and there is a (finite) model that makes it false.*

Exercise 36. How many sentences are in $\mathbf{L}_n$?

Exercise 37. Is it the case that for every sentence φ of $\mathbf{L}_n$

1. there is a finite model that makes φ true; and
2. there is a finite model that makes φ false?

Exercise 38. Suppose the answers to questions (1) and (2) in the previous exercise are affirmative. Is it the case that for every n there is a computable function $f(\,)$ such that for every sentence φ of $\mathbf{L}_n$

1. there is a model with at most $f(n)$ individuals that makes φ true; and
2. there is a model with at most $f(n)$ individuals that makes φ false?

If the answer to this last question is affirmative, then we shall have an effective procedure for deciding the validity of single-premise arguments in $\mathbf{L}_n$. We would need only to examine every possible model providing an extension for the n-ary predicate L in a domain containing at most $f(n)$ individuals, looking for one that will both make the premise true and make the premise false. If we find one, the argument is invalid; otherwise, it is valid.

Note that this method will be available only if we are possessed of the uniform bound $f(n)$ on the size of models to be inspected. It would be of no help to know only that for every sentence φ there is a finite model that makes φ true and there is a finite model that makes φ false. At any stage where our search for a finite countermodel (ordered, say, by size of domain) had turned out empty-handed, we would not know whether this was because there were no such (finite) countermodel to be had, or whether we simply had not yet searched far enough (along increasing sizes of domain). This holds even if we were to know (by whatever other means) that there could not be any *infinite* countermodel to the argument in hand.

Observation 2. *It would be wonderful if for all $n > 0$ we had $f(n) \leq n+1$. (Note that for $n = 2$ we had $f(n) = 2 \leq 3$)*

(We have to allow for $f(1) = 2$, since, as we have seen, the smallest countermodel to argument (1) in $\mathbf{L}_1$ has two individuals.)

Chapter 13

From Rules of Evaluation to Rules of Natural Deduction: Quantifiers

13.1 Morphing quantifiers' evaluation rules to their natural deduction rules

We remind the reader once again of the method expounded in Chapter 7 for the propositional case, dealing with just the connectives. The method continues to apply in the first-order case, but now with point (5) below as an extra tweak. We omit mention of anything relating to connectives in this regard, since they have already been dealt with. Our first three points are emphatic repetitions; the last two points are specific to the quantifiers.

1. Replace the sets Λ of *literals* in the evaluation rules with *arbitrary sets* Δ of sentences in the rules of natural deduction. This allows the resulting rules of deduction to be 'interpretation- (or model-) invariant'. They do not depend on any particular interpretation of the extra-logical vocabulary.

2. When morphing, **keep the boxes** appended to discharge strokes. This means that 'vacuous discharge' of assumptions is prohibited: one must have used the assumption(s) in question in the way in-

dicated, in order to be able to apply the rule that occasions the discharge. This ensures that the deductions maintain **relevance** between *premise-sets* and *conclusions*, by analogy with the way truth-value computations employ their *inputs* relevantly to produce their *outputs*.

3. When morphing, **keep major premises standing proud**, with no non-trivial proof-work above them. This ensures that all deductions will be in **normal form**.

4. In the $\mathcal{F}$-rules for $\forall$ and $\exists$, as we did for the $\mathcal{F}$-rules for $\wedge$, $\vee$, $\rightarrow$, allow for an arbitrary sentence θ as the conclusion, as an alternative to $\bot$. One can still, of course, have $\bot$ as a conclusion.

5. When morphing from $\forall$-$\mathcal{V}$ to $\forall$-I, and from $\exists$-$\mathcal{F}$ to $\exists$-E, replace the domain-many sub-evaluations for instances (verifications for $\forall$-$\mathcal{V}$, falsifications for $\exists$-$\mathcal{F}$) with single subproof templates involving a parameter a. This respects the fact that the aim of deduction is to preserve truth over all models, and to do so by means of finitary proofs.

Note that with the notation $\varphi(\boldsymbol{\alpha})$ below we are taking φ to be a unary formula whose (sole) free variable is x; and taking $\varphi(\boldsymbol{\alpha})$ to be the result of replacing x at each of its free occurrences with an occurrence of $\boldsymbol{\alpha}$.

Morphing $\forall$-$\mathcal{V}$ into $\forall$-I

$$(\forall\text{-}\mathcal{V})\quad \frac{\left\{\begin{matrix}\Lambda_{\boldsymbol{\alpha}}\\ V_{\boldsymbol{\alpha}}\\ \varphi(\boldsymbol{\alpha})\end{matrix}\right\}_{\boldsymbol{\alpha}\in M}}{\forall x\varphi(x)} \quad\rightsquigarrow\quad (\forall\text{-I})\quad \frac{\begin{matrix}\Delta^{\textcircled{a}}\\ \Pi\\ \varphi(a)\end{matrix}}{\forall x\varphi(x)}$$

Morphing $\exists$-$\mathcal{V}$ into $\exists$-I

$$(\exists\text{-}\mathcal{V})\quad \frac{\begin{matrix}\Lambda\\ V\\ \varphi^x_{\boldsymbol{\alpha}}\end{matrix}}{\exists x\varphi}\ \text{where } \boldsymbol{\alpha} \text{ is any individual in the domain} \quad\rightsquigarrow\quad (\exists\text{-I})\quad \frac{\begin{matrix}\Delta\\ \Pi\\ \varphi^x_t\end{matrix}}{\exists x\varphi}\ \text{where } t \text{ is any closed term}$$

Morphing $\forall$-$\mathcal{F}$ into $\forall$-E

$$(\forall\text{-}\mathcal{F})\quad \dfrac{\forall x\varphi(x) \qquad \begin{array}{c}\overset{\square\text{———}(i)}{\underbrace{\varphi(\boldsymbol{\alpha})\ ,\ \Lambda}}\\ F\\ \bot\end{array}}{\bot}{\scriptstyle(i)} \quad \text{where } \boldsymbol{\alpha} \text{ is any individual in the domain} \quad \rightsquigarrow \quad (\forall\text{-E})\quad \dfrac{\forall x\varphi(x) \qquad \begin{array}{c}\overset{\square\text{———}(i)}{\underbrace{\varphi(t)\ ,\ \Delta}}\\ \Pi\\ \theta\end{array}}{\theta}{\scriptstyle(i)}$$

where t is any closed term ;

hence, for increased economy:

$$\dfrac{\forall x\varphi(x) \qquad \begin{array}{c}{\scriptstyle(i)}\text{———}\cdots\square\cdots\text{———}{\scriptstyle(i)}\\ \underbrace{\varphi(t_1)\ ,\ \ldots\ ,\ \varphi(t_n)\ ,\ \Delta}\\ \Pi\\ \theta\end{array}}{\theta}{\scriptstyle(i)}$$

where $t_1, \ldots, t_n$ are closed terms

Morphing $\exists$-$\mathcal{F}$ into $\exists$-E

$$(\exists\text{-}\mathcal{F})\quad \dfrac{\exists x\varphi \quad \left(\begin{array}{c}\overset{\square\text{——}(i)}{\underbrace{\varphi^x_{\boldsymbol{\alpha}}\ ,\ \Lambda_{\boldsymbol{\alpha}}}}\\ F_{\boldsymbol{\alpha}}\\ \bot\end{array}\right)_{\boldsymbol{\alpha}\in M}}{\bot}{\scriptstyle(i)} \quad \rightsquigarrow \quad (\exists\text{-E})\quad \dfrac{\exists x\varphi^{@} \qquad \begin{array}{c}\overset{\square\text{——}(i)}{\underbrace{\varphi^x_a\ ,\ \Delta^{@}}}\\ \Pi\\ \theta^{@}\end{array}}{\theta^{@}}{\scriptstyle(i)}$$

13.2 Graphic statement of the rules of inference for classical first-order logic

We are now in a position to collect together all the *natural-deduction* rules of classical first-order logic that we have discovered by the method of morphing our 'proof-theoretically' *semantic* rules of verification and falsification. In the interests of thoroughness we shall be reprising later on in this section the natural-deduction rules for propositional logic that we have already stated in §7.4. We shall revert below to the usual way of stating such rules. We shall use vertical ellipses to gesture at possible proof-work implicitly there. We shall suppress explicit reminders of sets of undischarged side-assumptions. *But* we shall continue to be meticulous in annotating discharge rules with their boxes or diamonds, because this is so important for ensuring *relevance* of premises to conclusions of proofs.

Note that the following important features from the propositional case have now been preserved at first order, in §13.1: all elimination rules are in *parallelized* form, and their major premises *stand proud*; hence, all first-order proofs are in *normal form.*

13.2.1 Remarks about parallelized elimination rules

It is worth saying more at this juncture about our parallelized elimination rules. Precursors for our version of them in this work were first introduced into the logical literature by Schroeder-Heister [1984]. Schroeder-Heister, however, allowed for the possibility that major premises for his parallelized elimination rules could stand within licitly constructed proofs as conclusions of non-trivial proof-work above them. This possibility was conventional in the treatments of Gentzen and Prawitz, who also employed the *serial* forms of rules for eliminations of $\wedge$, $\rightarrow$, and $\forall$, that we saw in the two snippets from their *locis classicis* on p. 53 and p. 55. Our own add-on to Schroeder-Heister's treatment of parallelized elimination rules for *all* the logical operators was the further stipulation that their major premises should always *stand proud* within proofs, with *no* non-trivial proof-work above them.[1]

This add-on was motivated by two considerations. First, it afforded efficient constraints on proof search in automated deduction, without any loss at all of deductive completeness, and in general produced shorter proofs (as measured by the number of steps taken within them). Second, the resulting *natural deductions* would be in normal form, and therefore *isomorphic* to the corresponding *cut-free* sequent proofs of the results proved. That meant that algorithms for proof-search would be the same, whether the search would be for natural deductions or for sequent proofs. It also meant that the notion of *proof* itself could be under-

[1]See Tennant [1992] for the origin of this requirement.

stood more abstractly: proofs would be abstract objects underlying their *forms of representation*, be they natural deductions or the corresponding sequent proofs. Proofs would, macro-structurally, be identifiable as labeled *trees*. The main difference, then, between a natural-deduction tree and a sequent-proof tree, would be that in the former the tree's nodes would be labeled by *sentences*, whereas in the latter they would be labeled by *sequents*—that is, each node would be labeled by the sequent that had been established *by the stage at which one's downward descent reached the node in question.*

The beginning student is asked to take these theoretical assurances on trust. Their truth should provide strong motivation to memorize the rules that follow, and to gain fluency in applying them when constructing formal proofs of valid arguments. But it is not only beginners for whom these explanatory remarks about parallelized elimination rules are intended. They ought to be of help also to those instructors whose own past training in logic might have been confined to gaining fluency only in the more 'linearized' ways of setting out formal deductions—those more 'linear'-looking forms having *hidden* tree-structure within them notwithstanding. This remark is directed not only to Quine's intellectual progeny in logic in the USA, but also to those many Oxbridge-trained philosopher-logicians who have had only a squeeze of Lemmon. Fortunately the Gentzenian tradition is thriving in Continental Europe.

Graphic Deductive Rules for the Connectives

Introduction Rules

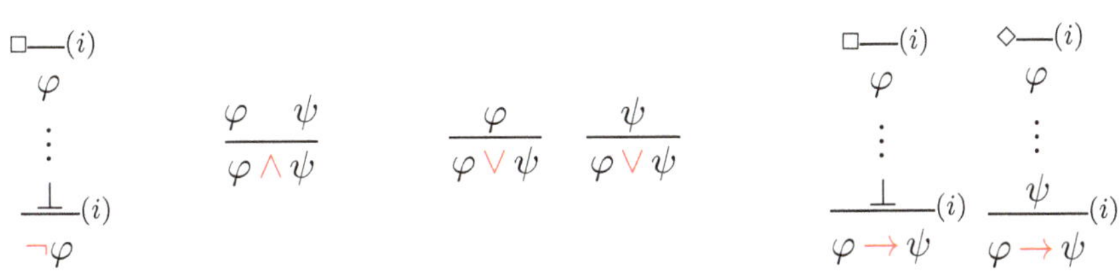

Elimination Rules (parallelized, MPEs standing proud)

$$\frac{\neg\varphi \quad \varphi}{\bot} \qquad \frac{\varphi\land\psi \quad \begin{array}{c}(i)\text{—}\square\text{—}(i)\\ \underbrace{\varphi\,,\psi}\\ \vdots\\ \theta\end{array}}{\theta}{\scriptstyle (i)} \qquad \frac{\varphi\lor\psi \quad \begin{array}{c}\square\text{—}(i)\\ \varphi\\ \vdots\\ \theta/\bot\end{array} \quad \begin{array}{c}\square\text{—}(i)\\ \psi\\ \vdots\\ \theta/\bot\end{array}}{\theta/\bot}{\scriptstyle (i)} \qquad \frac{\begin{array}{c}\\ \\ \vdots\\ \varphi\to\psi\end{array} \quad \begin{array}{c}\\ \\ \vdots\\ \varphi\end{array} \quad \begin{array}{c}\square\text{—}(i)\\ \psi\\ \vdots\\ \theta\end{array}}{\theta}{\scriptstyle (i)}$$

Graphic Deductive Rules for the Quantifiers

Introduction Rules

$$(\forall\text{-I}) \quad \frac{\begin{array}{c}ⓐ\\ \vdots\\ \varphi\end{array}}{\forall x\varphi^{a}_{x}} \qquad\qquad (\exists\text{-I}) \quad \frac{\varphi^{x}_{t}}{\exists x\varphi}$$

Elimination Rules (parallelized, MPEs standing proud)

$$(\forall\text{-E}) \quad \frac{\forall x\varphi(x) \quad \begin{array}{c}(i)\text{—}\cdots\ \square\ \cdots\text{—}(i)\\ \underbrace{\varphi^{x}_{t_1}\ ,\ \ldots\ ,\ \varphi^{x}_{t_n}}\\ \vdots\\ \theta\end{array}}{\theta}{\scriptstyle (i)}$$

where $t_1, \ldots, t_n$ are closed terms

$$(\exists\text{-E}) \quad \frac{\exists x\varphi^{ⓐ} \quad \begin{array}{c}\square\text{—}(i)\\ \underbrace{ⓐ\cdots\varphi^{x}_{a}\cdots ⓐ}\\ \vdots\\ \theta^{ⓐ}\end{array}}{\theta}{\scriptstyle (i)}$$

The Classical Rule of Dilemma

Dilemma

$$\frac{\begin{array}{c}\square\text{—}(i)\\ \varphi\\ \vdots\\ \psi\end{array} \quad \begin{array}{c}\square\text{—}(i)\\ \neg\varphi\\ \vdots\\ \psi\end{array}}{\psi}{\scriptstyle (i)} \qquad \frac{\begin{array}{c}\square\text{—}(i)\\ \varphi\\ \vdots\\ \psi\end{array} \quad \begin{array}{c}\square\text{—}(i)\\ \neg\varphi\\ \vdots\\ \bot\end{array}}{\psi}{\scriptstyle (i)}$$

13.3 The rules for identity

To be sure, we need also to state the rules that govern the identity predicate '='. But those rules are straightforward, even in the case of free logic. (See Chapter 16.) The Introduction rule is usually called *Reflexivity of Identity*; the Elimination rule is usually called *Substitutivity of Identicals.* In *unfree* logic, based on the background assumption that every singular term denotes, these rules are as follows:

(=-I) $$\frac{}{t = t}$$

(=-E) $$\frac{t = u \quad \varphi}{\psi} \qquad \text{where } \varphi^u_t = \psi^u_t \text{ and } \varphi \neq \psi$$

(=-E) allows for any number of interchanges of the terms t and u. An illustrative instance of its application is

$$\frac{t = u \quad L(t, u)}{L(u, t)} .$$

In (=-E) we can require, without loss of completeness, that the sentences φ and ψ be atomic. It is easy to derive, then, that (=-E) holds for non-atomic sentences φ and ψ.

Exercise 39. *Prove the following arguments:*
$t{=}u : u{=}t$ *(Symmetry of Identity);*
$t{=}u, u{=}v : t{=}v$ *(Transitivity of Identity).*

13.4 A recipe for proof construction

We are now at the point, forecast on p. 91, where the reader can enjoy their first taste of formal proofs to be constructed *on their own.* By this stage the rules of natural deduction that have been set out in §13.2 should be thoroughly memorized. This is crucial for the student to be able to apply the following strategic, if not fully algorithmic, advice.

Note that this recipe for proof search is iterative. In response to a deductive problem that has no immediate and obvious solution, it will generate new (but simpler) *sub*problems to solve. Eventually one will reach a stage at which, beginning with the 'innermost' subproblem-solutions, one can successively 'move outwards' (both 'upwards' and 'downwards') to create solutions to the earlier (sub)problems that generated the later (sub)problems. The end product will be a solution to the original problem itself (before one started generating any subproblems from it).

1. If the sought **conclusion** is complex, set out to derive it by means of the *introduction rule* for its dominant operator. This will generate a new (but simpler) *deductive subproblem*, to which this strategic advice will in turn apply.

2. If the sought **conclusion** is a negation, or a conjunction, or a conditional, or a universal, invoke the appropriate introduction rule in order to generate the appropriate new (but simpler) subproblem.

3. If the sought **conclusion** is a disjunction $\varphi \vee \psi$, you *might* succeed by proving just one of its disjuncts (a 'simpler' problem), and then applying $\vee$-I to finish your proof. But this is not the only way one might prove $\varphi \vee \psi$. It is possible that there is some other disjunction occurring positively in one of the premises; and that an eventual 'proof by cases' (application of $\vee$-E), using this *other* disjunction as its major premise, will furnish the wherewithal for you to derive your sought disjunctive conclusion $\varphi \vee \psi$. You might also try to use Classical Reductio as your final step in deriving a disjunction. To do so, assume (for the sake of argument) the negation $\neg(\varphi \vee \psi)$, and seek to derive absurdity ($\bot$). Likewise:

4. If the sought **conclusion** is an existential $\exists x\psi$, you *might* succeed by proving just one of its instances of the form ψ^x_t (a 'simpler' problem), for some strategically appropriate (or fortunate!) choice of a term t from among the terms in the offing. In that case you can apply $\exists$-I to finish your proof. But this is not the only way one might prove $\exists x\psi$. It is possible that there is some other existential occurring positively in one of the premises; and that an eventual 'existential elimination' (application of $\exists$-E), using this *other* existential as its major premise, will furnish the wherewithal for you

to derive your sought existential conclusion $\exists x\psi$. You might also try to use Classical Reductio as your final step in deriving an existential. To do so, assume (for the sake of argument) the negation $\neg\exists x\psi$, and seek to derive absurdity ($\bot$).

5. If there is a conjunction among the **premises**, replace it with its two conjuncts. Again, that will generate a new (but simpler) deductive subproblem. The aim would be to apply $\wedge$-E when the latter has eventually been solved.

6. If there is a disjunction among the **premises**, generate two new (but simpler) deductive subproblems: two prospective 'case proofs' for $\vee$-E, within which the respective disjuncts take the place of the disjunction.

7. If there is a conditional $\varphi \rightarrow \psi$ among the **premises**, generate two new (but simpler) deductive subproblems: seek the so-called *'minor'* proof of φ; and seek the so-called *'major'* proof of the conclusion you are after by using ψ as a premise in place of $\varphi \rightarrow \psi$.

8. If there is an existential $\exists x\psi$ among the **premises**, instantiate it with a fresh parameter a (say), and use the parametric assumption ψ^x_a along with the other permitted premises to derive the conclusion sought. This new subproblem is simpler.

9. If there is a universal $\forall x\psi$ among the **premises**, consider all the strategically sensible instantiations of the form ψ^x_t that you can obtain by using the various *terms t* that are in view. Importantly: *these can include parameters.* If you succeed in this, apply $\forall$-E. There are at most finitely many instantiations of the universal to be used as premises in place of the universal itself; and they are all simpler than it. So the new subproblem is itself simpler than the problem from which it has been generated.

10. If among the **premises** there are both existentials and universals, try to apply existential eliminations first, to introduce as many 'new' parameters as you can—and then 'hit them on the head' with prospective universal eliminations, so as to generate (or 'unpack') as much 'new' information as you can.

11. You will find yourself 'toggling' back and forth between 'constructing upwards from below' and 'building downwards from above' as you follow these pieces of advice. In 'building downwards' you are extracting information from the premises by means of eliminations. In 'building upwards' you are re-assembling such information, or 're-packaging' it, so as to reach the conclusion by means of introductions. The aim is to 'glue together the bits in the middle' when enough logical operators have been dealt with to expose the potential 'glueing points'—the bottommost bits from 'building downwards' and the uppermost bits from 'constructing upwards'.

12. If it is a sequent proof you are after, read 'Right rule' and 'Left rule' for 'introduction rule' and 'elimination rule' respectively. The foregoing advice about proof search remains the same, because in our system natural deductions (with parallelized elimination rules whose major premises stand proud) are structurally isomorphic to the corresponding sequent proofs.

Exercise 40. Construct natural deductions for the following valid arguments in propositional logic.

1. $\dfrac{}{P\to P}$
2. $\dfrac{P}{\neg\neg P}$
3. $\dfrac{}{P\to\neg\neg P}$
4. $\dfrac{}{\neg(P\wedge\neg P)}$
5. $\dfrac{}{\neg\neg\neg P\to\neg P}$
6. $\dfrac{}{\neg\neg(P\vee\neg P)}$
7. $\dfrac{P}{P\wedge P}$
8. $\dfrac{P\vee P}{P}$
9. $\dfrac{}{P\to(Q\to P)}$
10. $\dfrac{}{P\to((P\to Q)\to Q)}$
11. $\dfrac{P\to(Q\to R)}{Q\to(P\to R)}$
12. $\dfrac{P\wedge Q}{P\vee Q}$
13. $\dfrac{P\wedge Q}{Q\vee R}$
14. $\dfrac{P\wedge Q}{Q\wedge P}$
15. $\dfrac{P\vee Q}{Q\vee P}$
16. $\dfrac{P\wedge(Q\wedge R)}{(P\wedge Q)\wedge R}$
17. $\dfrac{P\vee(Q\vee R)}{(P\vee Q)\vee R}$
18. $\dfrac{P\wedge Q}{R\to P}$
19. $\dfrac{\neg P}{P\to Q}$
20. $\dfrac{P\to Q}{\neg Q\to\neg P}$
21. $\dfrac{P\to\neg Q}{Q\to\neg P}$
22. $\dfrac{P\to(Q\wedge R)}{(P\to Q)\wedge(P\to R)}$
23. $\dfrac{(P\to R)\vee(Q\to R)}{(P\wedge Q)\to R}$
24. $\dfrac{(P\to Q)\wedge(P\to R)}{P\to(Q\wedge R)}$
25. $\dfrac{(P\vee Q)\to R}{(P\to R)\wedge(Q\to R)}$
26. $\dfrac{P\wedge(Q\vee R)}{(P\wedge Q)\vee R}$
27. $\dfrac{\begin{array}{c}P\to Q\\ Q\to R\end{array}}{P\to R}$

28. $\dfrac{(P\land Q)\to R}{P\to(Q\to R)}$

29. $\dfrac{P\to(Q\to R)}{(P\land Q)\to R}$

30. $\dfrac{P\to(P\to Q)}{P\to Q}$

31. $\dfrac{P\to Q}{(Q\to R)\to(P\to R)}$

32. $\dfrac{\neg(P\lor Q)}{\neg P}$

33. $\dfrac{\neg(P\lor Q)}{\neg Q}$

34. $\dfrac{\neg(P\to Q)}{\neg Q}$

35. $\dfrac{\neg P}{\neg(P\land Q)}$

36. $\dfrac{\neg Q}{\neg(P\land Q)}$

37. $\dfrac{\begin{array}{l}\neg P\\ \neg Q\end{array}}{\neg(P\lor Q)}$

38. $\dfrac{\begin{array}{l}P\\ \neg Q\end{array}}{\neg(P\to Q)}$

39. $\dfrac{}{(P\land\neg P)\to Q}$

40. $\dfrac{}{\neg(P\to P)\to Q}$

41. $\dfrac{\begin{array}{l}\neg P\\ P\lor Q\end{array}}{Q}$

42. $\dfrac{\neg P\lor\neg Q}{\neg(P\land Q)}$

43. $\dfrac{\neg P\land\neg Q}{\neg(P\lor Q)}$

44. $\dfrac{\neg(P\lor Q)}{\neg P\land\neg Q}$

45. $\dfrac{P\land(Q\lor R)}{(P\land Q)\lor(P\land R)}$

46. $\dfrac{P\lor(Q\land R)}{(P\lor Q)\land(P\lor R)}$

47. $\dfrac{\neg P\lor Q}{P\to Q}$

48. $\dfrac{P\to Q}{\neg(P\land\neg Q)}$

49. $\dfrac{P\to(Q\to R)}{(P\to\neg R)\to(P\to\neg Q)}$

50. $\dfrac{P\to(Q\land R)}{\neg Q\to\neg P}$

51. $\dfrac{P\to\neg(Q\lor R)}{Q\to\neg P}$

52. $\dfrac{P\to(Q\lor R)}{\neg R\to(\neg Q\to\neg P)}$

53. $\dfrac{P\to(Q\lor R)}{\neg R\to(P\to Q)}$

54. (classical) $\dfrac{}{\neg\neg P\to P}$

55. (classical) $\dfrac{}{P\lor\neg P}$

56. (classical) $\dfrac{\neg(P\land Q)}{\neg P\lor\neg Q}$

57. (classical) $\dfrac{P\to(Q\lor R)}{\neg Q\to(R\lor\neg P)}$

58. (classical) $\dfrac{P\to(Q\lor R)}{(P\to Q)\lor(P\to R)}$

59. (classical) $\dfrac{}{(P\to Q)\lor(Q\to P)}$

Exercise 41. *For each of the following arguments provide either a proof or a counterexample. In the right column are the converses of the arguments in the left column.*

1. $\dfrac{(A\to C)\land(B\to C)}{(A\land B)\to C}$

1c. $\dfrac{(A\land B)\to C}{(A\to C)\land(B\to C)}$

2. $\dfrac{(A\to C)\land(B\to C)}{(A\lor B)\to C}$

2c. $\dfrac{(A\lor B)\to C}{(A\to C)\land(B\to C)}$

3. $\dfrac{(A\to C)\lor(B\to C)}{(A\land B)\to C}$

3c. $\dfrac{(A\land B)\to C}{(A\to C)\lor(B\to C)}$

4. $\dfrac{(A\to C)\lor(B\to C)}{(A\lor B)\to C}$

4c. $\dfrac{(A\lor B)\to C}{(A\to C)\lor(B\to C)}$

5. $\dfrac{(A \rightarrow B) \wedge (A \rightarrow C)}{A \rightarrow (B \wedge C)}$ 5c. $\dfrac{A \rightarrow (B \wedge C)}{(A \rightarrow B) \wedge (A \rightarrow C)}$

6. $\dfrac{(A \rightarrow B) \wedge (A \rightarrow C)}{A \rightarrow (B \vee C)}$ 6c. $\dfrac{A \rightarrow (B \vee C)}{(A \rightarrow B) \wedge (A \rightarrow C)}$

7. $\dfrac{(A \rightarrow B) \vee (A \rightarrow C)}{A \rightarrow (B \wedge C)}$ 7c. $\dfrac{A \rightarrow (B \wedge C)}{(A \rightarrow B) \vee (A \rightarrow C)}$

8. $\dfrac{(A \rightarrow B) \vee (A \rightarrow C)}{A \rightarrow (B \vee C)}$ 8c. $\dfrac{A \rightarrow (B \vee C)}{(A \rightarrow B) \vee (A \rightarrow C)}$

Exercise 42. Recall the Sheffer connective stroke from §9.5. Write down its verification and falsification rules. Morph them into its introduction and elimination rules. Then write down its Right and Left sequent rules. For each system of proof (natural deduction and the sequent calculus) write down an elegant strictly classical rule that will make the system complete for classical propositional logic based on just the Sheffer stroke.

13.5 The challenge ahead

In §13.2 we furnished a methodologically adequate set of rules of inference by means of which one can fully formalize all of the deductive reasoning in mathematics and science. That one can make do with them in mathematics and science will be argued in Chapter 24. There we shall set out the *adequacy conditions* for a logic that is to be successful in this regard; and we shall inform the reader of the various metatheorems that establish the adequacy of our proof systems—the system for constructive reasoning, and the one for classical reasoning. The metatheorems themselves will be covered in our planned more advanced sequel to this primer. It is important for the reader of this primer to know that it will usher them to those important foundational results. They are important for mathematicians, scientists, philosophers, computer scientists, and logicians—especially those working in automated deduction.

Exercise 43. Construct natural deductions for the following valid one-premise arguments. If you use the well-known serial elimination rules for

$\wedge$, $\rightarrow$, and $\forall$, you should then re-cast your proofs using the *parallelized* elimination rules for those operators, ensuring that major premises for eliminations stand proud, with no non-trivial proof-work above them.

We shall illustrate with a model answer to such an exercise relating to the following argument:

$$\frac{\forall x(Fx \rightarrow Gx)}{\forall xFx \rightarrow \forall xGx}\ .$$

Model Answer

Here is a natural deduction that uses the serial forms of $\forall$-E and $\rightarrow$-E:

$$\dfrac{\dfrac{\dfrac{\dfrac{\forall x(Fx \rightarrow Gx)}{Fa \rightarrow Ga} \qquad \dfrac{\overline{\forall xFx}^{\,(1)}}{Fa}}{Ga}}{\forall xGx}}{\forall xFx \rightarrow \forall xGx}\,(1)$$

Here is a natural deduction that uses the parallelized forms of $\forall$-E and $\rightarrow$-E:

$$\dfrac{\dfrac{{}^{(4)}\overline{\forall xFx} \qquad \dfrac{\forall x(Fx \rightarrow Gx) \qquad \dfrac{{}^{(2)}\overline{Fa \rightarrow Ga} \qquad \overline{Fa}^{\,(3)} \qquad \overline{Ga}^{\,(1)}}{Ga}\,(1)}{Ga}\,(2)}{\dfrac{Ga}{\forall xGx}}\,(3)}{\forall xFx \rightarrow \forall xGx}\,(4)$$

The reader is now tasked to undertake similar exercises with the following deductive problems.

1. $\dfrac{\exists xFx \vee \exists xGx}{\exists x(Fx \vee Gx)}$

2. $\dfrac{\forall xFx \vee \forall xGx}{\forall x(Fx \vee Gx)}$

3. $\dfrac{\forall x(Fx \wedge Gx)}{\forall xFx \wedge \forall xGx}$

4. $\dfrac{\forall xFx \wedge \forall xGx}{\forall x(Fx \wedge Gx)}$

5. $$\frac{\exists x Fx \quad \forall x(Fx \to Gx)}{\exists x Gx}$$

6. $$\frac{\forall x(Fx \to Gx)}{\exists x Fx \to \exists x Gx}$$

7. $$\frac{\forall x(Fx \to G)}{\exists x Fx \to G}$$ (where x does not occur free in G)

8. $$\frac{\exists x Fx \to G}{\forall x(Fx \to G)}$$ (where x does not occur free in G)

9. $$\frac{\forall x Fx \vee G}{\forall x(Fx \vee G)}$$

(where x does not occur free in G)

10. $$\frac{\neg \exists x Fx}{\forall x \neg Fx}$$

11. $$\frac{\forall x \neg Fx}{\neg \exists x Fx}$$

12. $$\frac{\exists x \neg Fx}{\neg \forall x Fx}$$

13. $$\frac{\exists y \forall x Lxy}{\forall x \exists y Lxy}$$

14. $$\frac{\forall x \forall y(Lxy \to \neg Lyx)}{\forall x \neg Lxx}$$

15. $$\frac{\forall x \forall y Lxy}{\forall y \forall x Lxy}$$

16. $$\frac{\exists x \exists y Lxy}{\exists y \exists x Lxy}$$

17. $$\frac{\exists x(Fx \wedge Gx)}{\exists x Fx}$$

18. $$\frac{\exists x(Fx \wedge Gx)}{\exists x Gx}$$

19. $$\frac{\exists x(Fx \wedge Gx)}{\exists x Fx \wedge \exists x Gx}$$

20. $$\frac{\exists x(Fx \vee Gx)}{\exists x Fx \vee \exists x Gx}$$

21. (classical) $$\frac{\neg \forall x Fx}{\exists x \neg Fx}$$

22. (classical) $$\frac{\forall x(Fx \vee G)}{\forall x Fx \vee G}$$

(where x does not occur free in G)

23. (classical) $$\frac{\forall x(Fx \vee Gx)}{\exists x Fx \vee \forall x Gx}$$

24. (classical) $$\frac{}{\exists x Fx \vee \forall x \neg Fx}$$

25. (classical) $$\frac{}{\exists x \neg Fx \vee \forall x Fx}$$

Exercise 44. Construct natural deductions for the following valid two-premise arguments. Be on the lookout for possible applications of the Maxim of Shallow Analysis, in cases where one does not need to use all of the logical structure that has been exposed in the premises and/or the conclusion.

1.
$$\frac{\begin{array}{l}\forall x \exists y Lxy \\ \forall x \forall y (Lxy \rightarrow Lyx)\end{array}}{\forall x \exists y Lyx}$$

2.
$$\frac{\begin{array}{l}\forall x \exists y Lxy \\ \forall z (\exists y Lyz \rightarrow Fz)\end{array}}{\exists w Fw}$$

3.
$$\frac{\begin{array}{l}\forall x \exists y Lxy \\ \forall z (\exists x \exists y Lxy \rightarrow Lzz)\end{array}}{\forall w Lww}$$

4.
$$\frac{\begin{array}{l}\forall x \exists y Lxy \\ \forall z (\exists y Lzy \rightarrow Fz)\end{array}}{\forall w Fw}$$

Proving a more challenging valid argument

Everyone loves someone or is loved by someone
All who love someone love themselves
All who are loved by someone love themselves
―――――――――――――――――――――――――
Everyone loves and is loved by someone

Here is a detailed formalization of the premises and conclusion of this argument:

$$\frac{\begin{array}{l}\forall x (\exists y Lxy \vee \exists y Lyx) \\ \forall x (\exists y Lxy \rightarrow Lxx) \\ \forall x (\exists y Lyx \rightarrow Lxx)\end{array}}{\forall x \exists y (Lxy \wedge Lyx)}$$

The argument's validity, however, can be established by means of a proof that uses only the amount of logical structure displayed below. We therefore apply the aforementioned Maxim of Shallow Analysis.

$$\frac{\begin{array}{l}\forall x (Gx \vee Hx) \\ \forall x (Gx \rightarrow Lxx) \\ \forall x (Hx \rightarrow Lxx)\end{array}}{\forall x \exists y (Lxy \wedge Lyx)}$$

Here is the proof that uses serial elimination rules for $\rightarrow$ and $\forall$. You should check each step in it, to make sure you understand what natural-deduction rule of inference is being applied when that step is taken.

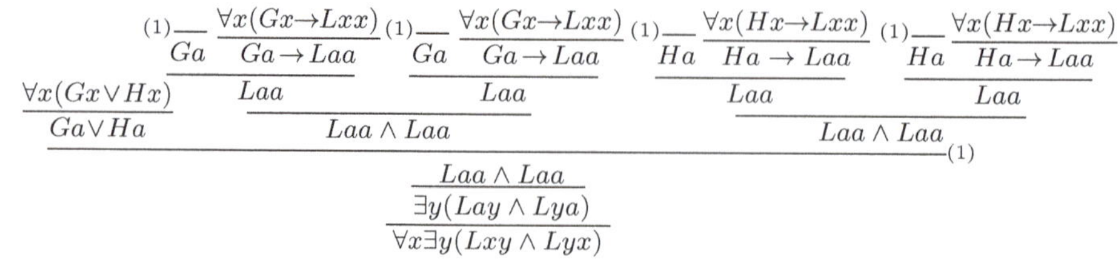

This proof has 14 steps, because of its serial ∀- and →-Eliminations. We can reduce the number of steps to 10, by *parallelizing* these eliminations:

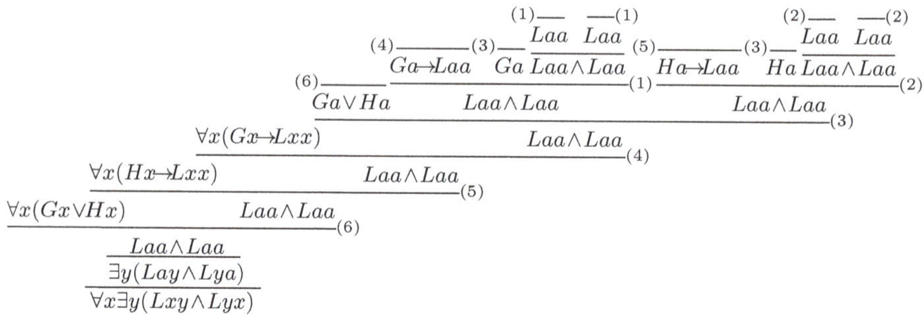

Note that the steps labeled (4), (5), and (6), are applications of *parallelized* ∀-Elimination, respectively discharging the *assumptions* higher up that are labeled (4), (5), and (6). The step labeled (3) is a proof by cases (an application of ∨-E), discharging the case-assumptions Ga and Ha. The steps labeled (1) and (2) are applications of parallelized →-E, respectively discharging the consequents-as-assumptions Laa. The two conditional major premises differ only in their antecedents. Note how every major premise for an elimination in this proof stands proud, with no non-trivial proof-work above it.

Another way to prove this last result (in two stages), is to use an *interpolant*. With serial eliminations, using the particular interpolant $\forall x Lxx$ reduces the total number of steps from 14 to 12:

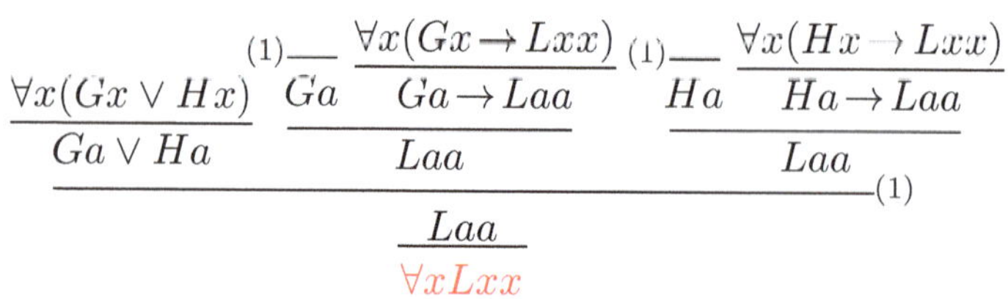

$$
\dfrac{\dfrac{\dfrac{\dfrac{\forall x Lxx}{Laa} \quad \dfrac{\forall x Lxx}{Laa}}{Laa \wedge Laa}}{\exists y(Lay \wedge Lya)}}{\forall x \exists y(Lxy \wedge Lyx)}
$$

In this example, the interpolant $\forall x Lxx$ implies, but is not implied by, the final conclusion $\forall x \exists y(Lxy \wedge Lyx)$. So the interpolant is a logically stronger result, from the given premises, than is the final conclusion. This is not in general the case in mathematics.

With parallelized eliminations (exercise left to reader), using the foregoing interpolant reduces the total number of steps to from 12 to11:

	Without interpolant	With interpolant
Serialized	14	12
Parallelized	10	11

In general, parallelizing the eliminations shortens the overall length n of proof; and using a well-chosen interpolant can do so dramatically:

$$
\begin{array}{lccc}
 & \text{Without interpolant} & & \text{With interpolant} \\
\text{Serialized} & n_s & \gg & n_{si} \\
 & \vee\!\!\vee & & \vee\!\!\vee \\
\text{Parallelized} & n_p & \gg & n_{pi}
\end{array}
$$

This is why mathematicians, starting from their axioms, prove mathematical *lemmas* on their way to proving mathematical theorems. Lemmas are interpolants; they can help to reduce (dramatically!—indeed, hyper-logarithmically) the overall length of proof of a theorem, as measured by the total number of primitive steps involved.

Chapter 14

From Evaluation Rules to Natural Deduction Rules to Sequent Rules

As already explained in Chapter 10, but worth reminding the reader here, a *sequent* is composed of an *antecedent* and a *succedent*, in that order, separated by a colon. The antecedent is a finite, possibly empty set of sentences. The succedent is either the empty set $\emptyset$ or a singleton $\{\theta\}$ (where θ is a sentence).

We shall often write $\varphi_1, \ldots, \varphi_n$ instead of $\{\varphi_1, \ldots, \varphi_n\}$.

This chapter aims to bring together the three main kinds of formal rules that we have treated in this work: the rules of evaluation; the rules of natural deduction; and the rules of the sequent calculus. We summarize how both the natural deduction rules and the sequent rules, which are *deductive*, i.e., *model-invariant*, have their origin (via morphing) in the *model-relative* evaluation rules (i.e., rules of verification and of falsification). The *verification* rules give rise to the *introduction rules* of natural deduction, and to the corresponding *Right* rules of the sequent calculus. The *falsification* rules give rise to the *elimination rules* of natural deduction, and to the corresponding *Left* rules of the sequent calculus. The rules of natural deduction and the rules of the sequent calculus are simply different ways of representing the fundamentally licit, primitive, deductive inferences that one can make with the logical operators concerned. The major difference in their respective upshots concerns how to identify the premises and the conclusion of the overall argument that the deduction establishes. With a natural deduction, to identify the *premises* of the argument that it proves, one has to gather up those sentences at leaf nodes that have not been discharged by rule-applications lower down; while the *conclusion* of the argument is the sentence at the root-node of the proof-tree. With a sequent proof, by contrast, one simply looks at the sequent occupying the root node of the proof-tree. The *premises* of the argument established are to the left of the colon in that sequent; while the *conclusion* of the argument is to the right of the colon.

In the following pairs of trios of rules for each logical operator @, we state, first, its verification rule (@-$\mathcal{V}$); then its natural-deduction introduction rule (@-I); and finally its Right-rule (@-R) in the sequent calculus. Second, we state its falsification rule (@-$\mathcal{F}$); then its natural-deduction elimination rule (@-E); and finally its Left-rule (@-L) in the sequent calculus.

14.1 Rules for ¬

(¬-$\mathcal{V}$)
$$\dfrac{\begin{array}{c}\overset{\square\!\!-\!\!-(i)}{\underbrace{\varphi\ ,\ \Lambda}}\\ \vdots \\ \bot\end{array}}{\neg\varphi}\,(i)$$

(¬-I)
$$\dfrac{\begin{array}{c}\overset{\square\!\!-\!\!-(i)}{\underbrace{\varphi\ ,\ \Delta}}\\ \vdots \\ \bot\end{array}}{\neg\varphi}\,(i)$$

(¬-R)
$$\dfrac{\varphi, \Delta : \bot}{\Delta : \neg\varphi}$$

(¬-$\mathcal{F}$)
$$\dfrac{\neg\varphi \qquad \begin{array}{c}\Lambda\\ \vdots \\ \varphi\end{array}}{\bot}$$

(¬-E)
$$\dfrac{\neg\varphi \qquad \begin{array}{c}\Delta\\ \vdots \\ \varphi\end{array}}{\bot}$$

(¬-L)
$$\dfrac{\Delta : \varphi}{\Delta, \neg\varphi : \bot}$$

14.2 Rules for $\wedge$

$$(\wedge\text{-}\mathcal{V})\quad \dfrac{\begin{matrix}\Lambda_1 & & \Lambda_2\\ \vdots & & \vdots\\ \varphi & & \psi\end{matrix}}{\varphi\wedge\psi} \qquad (\wedge\text{-I})\quad \dfrac{\begin{matrix}\Delta_1 & & \Delta_2\\ \vdots & & \vdots\\ \varphi & & \psi\end{matrix}}{\varphi\wedge\psi} \qquad (\wedge\text{-R})\quad \dfrac{\Delta_1:\varphi \quad \Delta_2:\psi}{\Delta,\Delta_2:\varphi\wedge\psi}$$

$$(\wedge\text{-}\mathcal{F})\ \left\{\begin{matrix}\dfrac{\varphi\wedge\psi \quad \begin{matrix}\overset{\square\text{---}(i)}{\underbrace{\varphi\ ,\ \Lambda}}\\ \vdots\\ \bot\end{matrix}}{\bot}{\scriptstyle(i)}\\[2em] \dfrac{\varphi\wedge\psi \quad \begin{matrix}\overset{\square\text{---}(i)}{\underbrace{\psi\ ,\ \Lambda}}\\ \vdots\\ \bot\end{matrix}}{\bot}{\scriptstyle(i)}\end{matrix}\right. \qquad (\wedge\text{-E})\ \dfrac{\varphi\wedge\psi \quad \begin{matrix}(i)\text{---}\square\text{---}(i)\\ \underbrace{\varphi,\psi\ ,\ \Delta}\\ \vdots\\ \theta\end{matrix}}{\theta}{\scriptstyle(i)} \qquad (\wedge\text{-L})\ \left\{\begin{matrix}\dfrac{\Delta,\varphi,\psi:\theta}{\Delta,\varphi\wedge\psi:\theta}\\[1em] \dfrac{\Delta,\varphi:\theta}{\Delta,\varphi\wedge\psi:\theta}\\[1em] \dfrac{\Delta,\psi:\theta}{\Delta,\varphi\wedge\psi:\theta}\end{matrix}\right.$$

14.3 Rules for $\vee$

$$(\vee\text{-}\mathcal{V})\quad \dfrac{\begin{matrix}\Lambda\\ \vdots\\ \varphi\end{matrix}}{\varphi\vee\psi} \quad \dfrac{\begin{matrix}\Lambda\\ \vdots\\ \psi\end{matrix}}{\varphi\vee\psi} \qquad (\vee\text{-I})\quad \dfrac{\begin{matrix}\Delta\\ \vdots\\ \varphi\end{matrix}}{\varphi\vee\psi} \quad \dfrac{\begin{matrix}\Delta\\ \vdots\\ \psi\end{matrix}}{\varphi\vee\psi} \qquad (\vee\text{-R})\quad \begin{matrix}\dfrac{\Delta:\varphi}{\Delta:\varphi\vee\psi}\\[1em] \dfrac{\Delta:\psi}{\Delta:\varphi\vee\psi}\end{matrix}$$

$$(\vee\text{-}\mathcal{F})\ \dfrac{\varphi\vee\psi \quad \begin{matrix}\overset{\square\text{---}(i)}{\underbrace{\varphi\ ,\Lambda_1}}\\ \vdots\\ \bot\end{matrix} \quad \begin{matrix}\overset{\square\text{---}(i)}{\underbrace{\psi\ ,\Lambda_2}}\\ \vdots\\ \bot\end{matrix}}{\bot}{\scriptstyle(i)} \qquad (\vee\text{-E})\ \dfrac{\varphi\vee\psi \quad \begin{matrix}\overset{\square\text{---}(i)}{\underbrace{\varphi\ ,\Delta_1}}\\ \vdots\\ \theta/\bot\end{matrix} \quad \begin{matrix}\overset{\square\text{---}(i)}{\underbrace{\psi\ ,\Delta_2}}\\ \vdots\\ \theta/\bot\end{matrix}}{\theta/\bot}{\scriptstyle(i)} \qquad (\vee\text{-L})\ \dfrac{\Delta_1,\varphi:\theta/\bot \quad \Delta_2,\psi:\theta/\bot}{\Delta_1,\Delta_2,\varphi\vee\psi:\theta/\bot}$$

14.4 Rules for →

$$(\to\text{-}\mathcal{V})\quad \frac{\begin{array}{c}\underbrace{\overset{\square\text{—}(i)}{\varphi\,,\,\Lambda}}\\ \vdots\\ \bot\end{array}}{\varphi\to\psi}(i)\qquad \frac{\begin{array}{c}\Lambda\\ \vdots\\ \psi\end{array}}{\varphi\to\psi}$$

$$(\to\text{-I})\quad \frac{\begin{array}{c}\underbrace{\overset{\square\text{—}(i)}{\varphi\,,\,\Delta}}\\ \vdots\\ \bot\end{array}}{\varphi\to\psi}(i)\qquad \frac{\begin{array}{c}\underbrace{\overset{\diamond\text{—}(i)}{\varphi\,,\,\Delta}}\\ \vdots\\ \psi\end{array}}{\varphi\to\psi}(i)$$

$$(\to\text{-R})\quad \frac{\Delta,\varphi:\bot}{\Delta:\varphi\to\psi}\qquad \frac{\Delta:\psi}{\Delta\backslash\{\varphi\}:\varphi\to\psi}$$

$$(\to\text{-}\mathcal{F})\quad \frac{\varphi\to\psi\quad \begin{array}{c}\Lambda_1\\ \vdots\\ \varphi\end{array}\quad \begin{array}{c}\underbrace{\overset{\square\text{—}(i)}{\psi\,,\,\Lambda_2}}\\ \vdots\\ \bot\end{array}}{\bot}(i)$$

$$(\to\text{-E})\quad \frac{\varphi\to\psi\quad \begin{array}{c}\Delta_1\\ \vdots\\ \varphi\end{array}\quad \begin{array}{c}\underbrace{\overset{\square\text{—}(i)}{\psi\,,\,\Delta_2}}\\ \vdots\\ \theta\end{array}}{\theta}(i)$$

$$(\to\text{-L})\quad \frac{\Delta_1:\varphi\qquad \Delta_2,\psi:\theta}{\Delta_1,\Delta_2,\varphi\to\psi:\theta}$$

14.5 Rules for ∃

$$(\exists\text{-}\mathcal{V})\quad \frac{\begin{array}{c}\Lambda\\ \vdots\\ \varphi^x_{\boldsymbol{\alpha}}\end{array}}{\exists x\varphi}$$

where $\boldsymbol{\alpha}\in \boldsymbol{M}$

$$(\exists\text{-I})\quad \frac{\begin{array}{c}\Delta\\ \vdots\\ \varphi^x_t\end{array}}{\exists x\varphi}$$

$$(\exists\text{-R})\quad \frac{\Delta:\varphi^x_t}{\Delta:\exists x\varphi}$$

$$(\exists\text{-}\mathcal{F})\quad \frac{\exists x\varphi\quad \left\{\begin{array}{c}\underbrace{\overset{\square\text{—}(i)}{\varphi^x_{\boldsymbol{\alpha}}\,,\,\Lambda_{\boldsymbol{\alpha}}}}\\ \vdots\\ \bot\end{array}\right\}^{\boldsymbol{\alpha}\in M}}{\bot}(i)$$

$$(\exists\text{-E})\quad \frac{\exists x\varphi^{@}\quad \begin{array}{c}\underbrace{\overset{\square\text{—}(i)}{\varphi^x_a\,,\,\Delta^{@}}}\\ \vdots\\ \theta^{@}\end{array}}{\theta^{@}}(i)$$

$$(\exists\text{-L})\quad \frac{\Delta,\varphi^x_a:\theta}{\Delta,\exists x\varphi:\theta}$$

where a does not occur in the bottom sequent

14.6 Rules for $\forall$

$$(\forall\text{-}\mathcal{V})\quad \frac{\left\{\begin{array}{c}\Lambda_{\boldsymbol{\alpha}}\\ \vdots\\ \psi(\boldsymbol{\alpha})\end{array}\right\}_{\boldsymbol{\alpha}\in M}}{\forall x\psi(x)}$$

$$(\forall\text{-I})\quad \frac{\begin{array}{c}\Delta^{@}\\ \vdots\\ \psi(a)\end{array}}{\forall x\psi(x)}$$

$$(\forall\text{-R})\quad \frac{\Delta : \psi(a)}{\Delta : \forall x\psi(x)}$$

where a does not occur in the bottom sequent

$$(\forall\text{-}\mathcal{F})\quad \frac{\forall x\psi \qquad \begin{array}{c}\Box\!-\!\!-^{(i)}\\ \underbrace{\psi^x_{\boldsymbol{\alpha}}\,,\Lambda}\\ \vdots\\ \bot\end{array}}{\bot}\,(i)$$

where $\boldsymbol{\alpha}\in M$

$$(\forall\text{-E})\quad \frac{\forall x\psi \qquad \begin{array}{c}{}^{(i)}\!-\!\!-\cdots\Box\cdots-\!\!-^{(i)}\\ \underbrace{\psi^x_{t_1},\ldots,\psi^x_{t_n}\,,\Delta}\\ \vdots\\ \theta\end{array}}{\theta}\,(i)$$

where $t_1,\ldots,t_n$ are closed terms

$$(\forall\text{-L})\quad \frac{\Delta,\psi^x_{t_1},\ldots,\psi^x_{t_n} : \theta}{\Delta,\forall x\psi : \theta}$$

14.7 Pausing to reflect

We have now generated the *sequent-calculus* rules for all the logical operators, including the *quantifiers*. The reader will recall that we had generated the sequent rules for the *connectives* earlier, in §10.4 (pp. 204 *ff*), and had included with them the sequent version of the classical rule of Dilemma.

Now that we have in hand all the rules of the sequent calculus, we can re-pose as exercises in the *sequent calculus* all the earlier exercises that called for natural deductions as solutions. The reader should be advised that if they intend to use their natural deductions as starting points for conversion into sequent proofs, they should ensure that they are using natural deductions in which all eliminations are in *parallelized* form. They can let themselves be guided in this conversion process by the remarkable isomorphism, described in §10.5, between *sequent proofs* and *natural deductions with parallelized eliminations* (in which all MPEs stand proud).

Exercise 45. For each of the following arguments in sequent form: if it is valid, give a *sequent* proof; if not, give a counterexample.

1. $Fa : Fa$
2. $Fa : \bot$
3. $: Fa$
4. $Fa : \exists x Fx$
5. $\exists x Fx : Fa$
6. $Fa : \forall x Fx$
7. $\forall x Fx : Fa$
8. $\forall x Fx : \exists x Fx$
9. $\exists x Fx : \forall x Fx$
10. $\forall x \neg Fx : \neg \exists x Fx$
11. $\neg \exists x Fx : \forall x \neg Fx$
12. $\exists x \neg Fx : \neg \forall x Fx$
13. $\exists x (Fx \wedge Gx) : \exists x Fx$
14. $\forall x (Fx \wedge Gx) : \forall x Fx$
15. $\exists x (Fx \vee Gx) : \exists x Fx$
16. $\forall x (Fx \vee Gx) : \forall x Fx$
17. $\exists x (Fx \to Gx) : \exists x Fx$
18. $\forall x (Fx \to Gx) : \forall x Fx$
19. $\exists x (Fx \to Gx) : \exists x Gx$
20. $\forall x (Fx \to Gx) : \forall x Gx$
21. $\exists x Fx : \exists x (Fx \wedge Gx)$
22. $\forall x Fx : \forall x (Fx \wedge Gx)$
23. $\exists x Fx : \exists x (Fx \vee Gx)$
24. $\forall x Fx : \forall x (Fx \vee Gx)$
25. $\exists x Fx : \exists x (Fx \to Gx)$
26. $\forall x Fx : \forall x (Fx \to Gx)$
27. $\exists x Gx : \exists x (Fx \to Gx)$
28. $\forall x Gx : \forall x (Fx \to Gx)$
29. $\exists x \neg Fx : \exists x (Fx \to Gx)$
30. $\forall x \neg Fx : \forall x (Fx \to Gx)$
31. $\exists x \neg Gx : \exists x (Fx \to Gx)$
32. $\forall x \neg Gx : \forall x (Fx \to Gx)$
33. $\exists x Fx : \neg \exists x (Fx \to Gx)$
34. $\forall x Fx : \neg \forall x (Fx \to Gx)$
35. $\exists x Gx : \neg \exists x (Fx \to Gx)$
36. $\forall x Gx : \neg \forall x (Fx \to Gx)$

Exercise 46. *Provide sequent-calculus proofs of the valid arguments among the ninety two-quantifier deductive problems of Exercise 34 on p. 234.*

Exercise 47. *Provide sequent-calculus proofs for all the deductive problems posed in Exercise 40 (p. 266).*

Exercise 48. *Provide sequent-calculus proofs for all the deductive problems posed in Exercise 41 (p. 267).*

Exercise 49. *Provide sequent-calculus proofs for all the deductive problems posed in Exercise 43 (p. 269).*

Exercise 50. *Provide sequent-calculus proofs for all the deductive problems posed in Exercise 44 (p. 270).*

Chapter 15

Inductive Definition of Proof in Natural Deduction

15.1 Proofs as generated by rules of inference

We have seen how terms and formulae (among the latter, sentences as special cases) are co-inductively defined. We have also seen how model-relative verifications and falsifications (which collectively we call evaluations) are co-inductively defined. Because of these inductive definitions, meta-results about all formulae (sentences) and about all evaluations (verifications/falsifications) can be established 'by induction' on the complexity of those items.

What is seldom appreciated by, let alone explicitly conveyed to learners of logic, is that the same general approach applies (one might say: with a vengeance) to *proofs*—in this study, *natural deductions* and/or *sequent proofs*. Proofs, too, are complex abstract objects that are inductively defined.[1] Indeed, if one partitions them into proofs (of sentences, not of $\bot$) and disproofs (i.e., proofs whose conclusions *are* $\bot$), then these items too admit of co-inductive definition.[2] Upon such partition, the

[1]Such a definition of natural deductions was given in Tennant [1978]. See also Tennant [2021] for an argument that rules of inference should be construed as inductive (formation) clauses in one's inductive definition of proofs.

[2]See Tennant [1999] for a co-inductive definition of proofs and disproofs.

analogy will be as tight as possible with the case of co-inductively defined terms and formulae (when the language contains variable-binding term-forming operators) and with the case of co-inductively defined model-relative verifications and falsifications.

General results about items in an inductively defined class are established in a way that is now familiar. There is a *basis step* that shows the desired generalizable property is possessed by the very simplest items of the class in question. Then one makes the *inductive hypothesis* that the property is possessed by all items less complex than an 'arbitrary' one that is itself complex. Thereafter one proceeds to take the *inductive step*, usually 'by cases', to show that the property will be transmitted from ones to which the inductive hypothesis applies, to newer, more complex ones that the rules of formation permit one to form out of the former. After investigating each of the permitted ways of forming more complex items of the class out of items already in it, and showing that this property-transmission will indeed take place, the reasoner concludes that *all* items in the class do indeed possess the property in question.

One therefore has to understand that the 'ways of forming more complex items out of simpler ones' are covered by the *inductive clauses* in the inductive definition of the class in question. It is because of the *closure clause* of the inductive definition that one knows that one's work is done after investigating each of the explicitly permitted ways of forming new items in the class out of simpler ones. The method of proof by induction is therefore the inferentialist's way of capturing the gist of the closure clause.

The simplest possible illustration of the general idea here is to be had from the natural (i.e., the counting) numbers. Few students realize that the natural numbers form an inductively defined class—indeed, they form the *paradigm example* of an inductively defined class! For here is the inductive definition of what—please excuse the pun—counts as a natural number.

Definition 10.
Basis clause: 0 is a natural number.
Inductive clause: The successor of any natural number is a natural number.
Closure clause: Every natural number is obtained in this way—i.e., by taking successors, after beginning with 0.

A more formal, 'inferentialist' way of framing this definition is the following. The predicate $N(x)$ is to be read as 'x is a natural number'.

Definition 11.

Basis: $\overline{N(0)}$

Inductive clause: $\dfrac{N(t)}{N(s(t))}$

Closure clause:

$$\dfrac{N(t) \quad \Phi(0) \quad \begin{array}{c} \underbrace{\overline{N(a)}^{(i)} \, , \, \overline{\Phi(a)}^{(i)}} \\ \vdots \\ \Phi(s(a)) \end{array}}{\Phi(t)}{\scriptstyle (i)}$$

where the parameter a occcurs only where indicated, and t is a closed term

One can easily see that the inferentialist's rendering of the closure clause in the inductive definition of the natural numbers reveals that it is none other than the *Principle of Mathematical Induction* for the natural numbers. In order to show that any natural number t has the property Φ, show, first, that 0 has the property Φ (this is the so-called 'Basis step'); show, second, that if an arbitrary natural number a has the property Φ, then so too does its successor $s(a)$ (this is the so-called 'Inductive step'); and be done! Note that t in the conclusion $\Phi(t)$ can be any closed term. So it could be a fresh parameter b, say. The rule $\forall$-I would then be applicable (on 'arbitrary b'), to deliver the explicitly universal generalization as follows:

$$\dfrac{\dfrac{\dfrac{{}^{(1)}\overline{N(b)} \quad \Phi(0) \quad \begin{array}{c} \underbrace{\overline{N(a)}^{(i)} \, , \, \overline{\Phi(a)}^{(i)}} \\ \vdots \\ \Phi(s(a)) \end{array}}{\Phi(b)}{\scriptstyle (i)}}{N(b) \rightarrow \Phi(b)}{\scriptstyle (1)}}{\forall n(N(n) \rightarrow \Phi(n))}$$

where the parameter a occcurs only where indicated

The basic insight here is that the closure clause of an inductive definition is what affords the definer the right to use the method of *proof by induction* when establishing general properties of the kind of items being defined. Once one appreciates this in the (degenerate) case of the natural numbers, it enables one to see how the method of proof by induction will apply also in the case of inductively defined objects of more intricate kinds. The greater intricacy would come from the various ways in which one is permitted to form new items of that kind from existing items. In the case of the natural numbers, one simply takes a (indeed, *the*) successor. In the case of more intricate kinds—such as formulae, or evaluations—one does more than this, as we have seen. One can apply expression-forming operators, like connectives and quantifiers, in the case of formulae. One can apply verification and falsification rules, in the case of model-relative evaluations. And finally, in the case of *proofs*: one can apply *rules of inference* to form new, more complex proofs out of simpler ones. That is why rules of inference are to be understood as *inductive clauses in the inductive definition* of the notion of proof. Then, if we wish, as meta-reasoners, to establish results about proofs in general, we will be able to proceed 'by induction', showing how the general property in question will be transmitted by each rule of inference that can be employed to form more complex proofs from simpler ones.

15.2 Notational preliminaries

'$\mathcal{P}(\Pi, \varphi, \Delta)$' is short for '$\Pi$ is a proof of the conclusion φ from the set Δ of undischarged assumptions'.

'$[\varphi, \Pi_1, \ldots, \Pi_n]$' is a singular term, and is short for 'the tree of sentence-occurrences with root-node φ, and immediate sub-trees $\Pi_1, \ldots, \Pi_n$ (in that order)'.

We can assume we already know exactly what the latter expression means. The following inductive definition now pins down exactly what the expression '$\mathcal{P}(\Pi, \varphi, \Delta)$' means.

15.3 The inductive definition of the notion of proof

Basis Clause

$\overline{\mathcal{P}(\varphi, \varphi, \{\varphi\})}$. Graphically: φ .

Every occurrence of a sentence φ is a proof of φ from the singleton-set $\{\varphi\}$ of undischarged assumptions. Or: φ is a proof of itself from itself.

Inductive Clause (¬-I)

Here is the inferential metalinguistic form of this clause:

$$\frac{\mathcal{P}(\Pi, \bot, \Delta) \qquad \varphi \in \Delta}{\mathcal{P}([\neg\varphi, \Pi], \neg\varphi, \Delta \setminus \{\varphi\})}$$

Note:
(1) the assumption φ must have been *used* in deriving $\bot$;
(2) the assumption φ is *discharged* upon application of the rule.

Graphically: 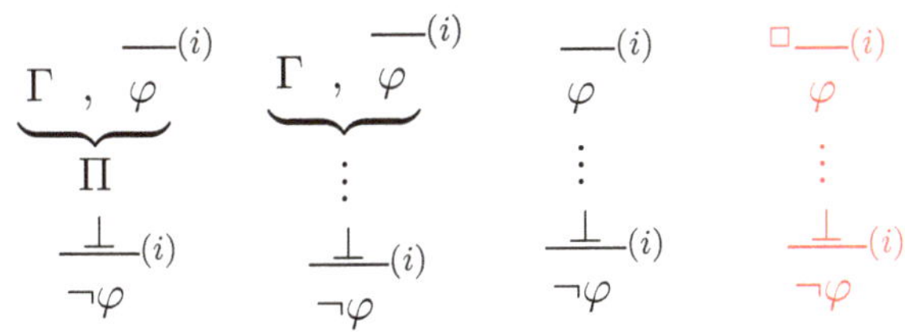

Note the progressive *simplifications through suppression of details* as we move from left to right. The leftmost graphic form contains all the details one could ask for. In the next graphic form we drop the placeholder Π for the immediate subproof (which is actually a *dis*proof in this case), and insert a vertical ellipsis to indicate deductive descent. The form after that drops the placeholder Γ for the side-assumptions which, one must bear in mind, will in general be there. The rightmost graphic form appends a box to the discharge stroke over φ, to register explicitly the requirement that the assumption φ must have been *used* as a premise in the subproof.

This 'graphic attenuation' will characterize all the left-to-right alterations of graphic forms that will be presented below. By memorizing the *rightmost* forms, and understanding their provenance in the *leftmost* forms, the beginner in Logic will rapidly gain mastery of the method of formal proof.

Inductive Clause ($\wedge$-I)

$$\frac{\mathcal{P}(\Pi,\varphi,\Delta)\qquad\qquad \mathcal{P}(\Sigma,\psi,\Gamma)}{\mathcal{P}([\varphi\wedge\psi,\Pi,\Sigma],\varphi\wedge\psi,\Delta\cup\Gamma)}$$

Graphically: $$\dfrac{\begin{matrix}\Delta & \Gamma\\ \Pi & \Sigma\\ \varphi & \psi\end{matrix}}{\varphi\wedge\psi} \qquad \dfrac{\begin{matrix}\Delta&\Gamma\\ \vdots&\vdots\\ \varphi&\psi\end{matrix}}{\varphi\wedge\psi}\qquad \dfrac{\begin{matrix}\vdots&\vdots\\ \varphi&\psi\end{matrix}}{\varphi\wedge\psi}\qquad \dfrac{\varphi\quad\psi}{\varphi\wedge\psi}$$

Inductive Clause ($\vee$-I)

$$\frac{\mathcal{P}(\Pi,\varphi,\Delta)}{\mathcal{P}([\varphi\vee\psi,\Pi],\varphi\vee\psi,\Delta)}\qquad \frac{\mathcal{P}(\Pi,\psi,\Delta)}{\mathcal{P}([\varphi\vee\psi,\Pi],\varphi\vee\psi,\Delta)}$$

Graphically:

$$\dfrac{\begin{matrix}\Delta\\ \Pi\\ \varphi\end{matrix}}{\varphi\vee\psi}\quad \dfrac{\begin{matrix}\Delta\\ \vdots\\ \varphi\end{matrix}}{\varphi\vee\psi}\quad \dfrac{\begin{matrix}\vdots\\ \varphi\end{matrix}}{\varphi\vee\psi}\quad \dfrac{\varphi}{\varphi\vee\psi}\ ;\quad \dfrac{\begin{matrix}\Delta\\ \Pi\\ \psi\end{matrix}}{\varphi\vee\psi}\quad \dfrac{\begin{matrix}\Delta\\ \vdots\\ \psi\end{matrix}}{\varphi\vee\psi}\quad \dfrac{\begin{matrix}\vdots\\ \psi\end{matrix}}{\varphi\vee\psi}\quad \dfrac{\psi}{\varphi\vee\psi}$$

Inductive Clause ($\rightarrow$-I)

$$\frac{\mathcal{P}(\Pi,\bot,\Delta)\qquad\qquad \varphi\in\Delta}{\mathcal{P}([\varphi\to\psi,\Pi],\varphi\to\psi,\Delta\setminus\{\varphi\})}\qquad \frac{\mathcal{P}(\Pi,\psi,\Delta)}{\mathcal{P}([\varphi\to\psi,\Pi],\varphi\to\psi,\Delta\setminus\{\varphi\})}$$

Graphically:

$$\dfrac{\begin{matrix}\overset{\square\overline{}(i)}{\underbrace{\Gamma\ ,\ \varphi}}\\ \Pi\\ \bot\end{matrix}}{\varphi\to\psi}(i)\quad \dfrac{\begin{matrix}\overset{\square\overline{}(i)}{\underbrace{\Gamma\ ,\ \varphi}}\\ \vdots\\ \bot\end{matrix}}{\varphi\to\psi}(i)\quad \dfrac{\begin{matrix}\overset{\square\overline{}(i)}{\varphi}\\ \vdots\\ \bot\end{matrix}}{\varphi\to\psi}(i)\ ;\quad \dfrac{\begin{matrix}\overset{\diamond\overline{}(i)}{\underbrace{\Gamma\ ,\ \varphi}}\\ \Pi\\ \psi\end{matrix}}{\varphi\to\psi}(i)\quad \dfrac{\begin{matrix}\overset{\diamond\overline{}(i)}{\underbrace{\Gamma\ ,\ \varphi}}\\ \vdots\\ \psi\end{matrix}}{\varphi\to\psi}(i)\quad \dfrac{\begin{matrix}\overset{\diamond\overline{}(i)}{\varphi}\\ \vdots\\ \psi\end{matrix}}{\varphi\to\psi}(i)$$

Assumption φ must have been used, and is discharged.

Assumption φ is discharged if it has been used.

Inductive Clause ($\exists$-I)

$$\frac{\mathcal{P}(\Pi, \varphi^x_t, \Delta)}{\mathcal{P}([\exists x\varphi, \Pi], \exists x\varphi, \Delta)}$$

Graphically:

$$\begin{array}{c}\Delta\\ \Pi\\ \dfrac{\varphi^x_t}{\exists x\varphi}\end{array} \qquad \begin{array}{c}\Delta\\ \vdots\\ \dfrac{\varphi^x_t}{\exists x\varphi}\end{array} \qquad \begin{array}{c}\vdots\\ \dfrac{\varphi^x_t}{\exists x\varphi}\end{array} \qquad \dfrac{\varphi^x_t}{\exists x\varphi}$$

Note that in passing downwards to the main conclusion, the term t gets replaced by x at some, but not necessarily all, of its occurrences in the subordinate conclusion.

Inductive Clause ($\forall$-I)

$$\frac{\mathcal{P}(\Pi, \varphi, \Delta) \qquad a \text{ occurs in } \varphi \qquad a \text{ does not occur in } \Delta}{\mathcal{P}([\forall x\varphi^a_x, \Pi], \forall x\varphi, \Delta)}$$

Graphically:

$$\begin{array}{c}\Delta^{@}\\ \Pi\\ \dfrac{\varphi}{\forall x\varphi^a_x}\end{array} \qquad \begin{array}{c}\Delta^{@}\\ \vdots\\ \dfrac{\varphi}{\forall x\varphi^a_x}\end{array} \qquad \begin{array}{c}@\\ \vdots\\ \dfrac{\varphi}{\forall x\varphi^a_x}\end{array}$$

Note that in passing downwards to the main conclusion, the parameter a gets replaced by x at *all* its occurrences in the subordinate conclusion.

Inductive Clause ($\neg$-E)

$$\frac{\mathcal{P}(\Pi, \varphi, \Delta)}{\mathcal{P}([\bot, \neg\varphi, \Pi], \bot, \{\neg\varphi\} \cup \Delta)}$$

Graphically:

$$\begin{array}{c}\qquad\Delta\\ \qquad\Pi\\ \dfrac{\neg\varphi \quad \varphi}{\bot}\end{array} \qquad \begin{array}{c}\qquad\Delta\\ \qquad\vdots\\ \dfrac{\neg\varphi \quad \varphi}{\bot}\end{array} \qquad \begin{array}{c}\qquad\vdots\\ \dfrac{\neg\varphi \quad \varphi}{\bot}\end{array} \qquad \dfrac{\neg\varphi \quad \varphi}{\bot}$$

Note: the major premise $\neg\varphi$ *stands proud*, with no proof-work above it.

Inductive Clause ($\wedge$-E)

$$\frac{\mathcal{P}(\Sigma, \theta, \Gamma) \qquad \varphi \in \Gamma \text{ or } \psi \in \Gamma}{\mathcal{P}([\theta, \varphi \wedge \psi, \Sigma], \theta, \{\varphi \wedge \psi\} \cup (\Gamma \setminus \{\varphi, \psi\}))}$$

Note:
(1) either the assumption φ or the assumption ψ (or both) must have been *used* in deriving θ;
(2) whichever of these assumptions *was* used gets *discharged* upon application of the rule;
(3) θ might be $\bot$.

Graphically:

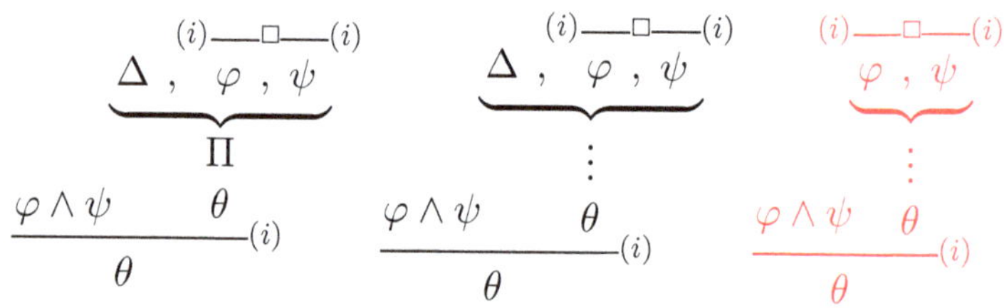

Note: the major premise $\varphi \wedge \psi$ *stands proud*, with no proof-work above it.

Inductive Clause ($\vee$-E)

$$\frac{\mathcal{P}(\Pi, \theta, \Delta) \qquad \mathcal{P}(\Sigma, \bot, \Gamma) \qquad \varphi \in \Delta \qquad \psi \in \Gamma}{\mathcal{P}([\theta, \varphi \vee \psi, \Pi, \Sigma], \theta, \{\varphi \vee \psi\} \cup (\Delta \setminus \{\varphi\}) \cup (\Gamma \setminus \{\psi\}))}$$

$$\frac{\mathcal{P}(\Pi, \bot, \Delta) \qquad \mathcal{P}(\Sigma, \theta, \Gamma) \qquad \varphi \in \Delta \qquad \psi \in \Gamma}{\mathcal{P}([\theta, \varphi \vee \psi, \Pi, \Sigma], \theta, \{\varphi \vee \psi\} \cup (\Delta \setminus \{\varphi\}) \cup (\Gamma \setminus \{\psi\}))}$$

$$\frac{\mathcal{P}(\Pi, \theta, \Delta) \qquad \mathcal{P}(\Sigma, \theta, \Gamma) \qquad \varphi \in \Delta \qquad \psi \in \Gamma}{\mathcal{P}([\theta, \varphi \vee \psi, \Pi, \Sigma], \theta, \{\varphi \vee \psi\} \cup (\Delta \setminus \{\varphi\}) \cup (\Gamma \setminus \{\psi\}))}$$

Graphically:

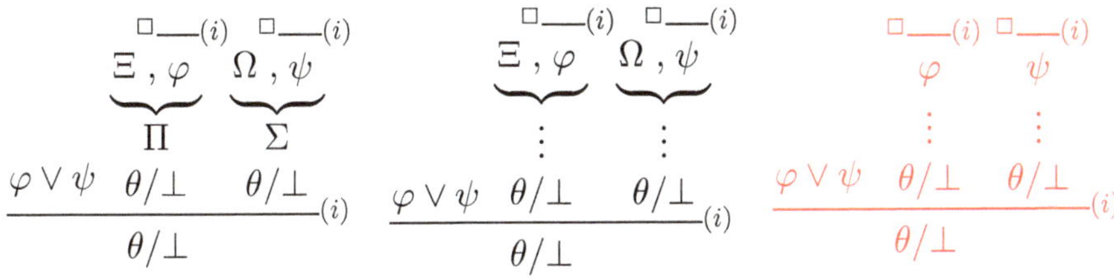

Note: the major premise $\varphi \vee \psi$ *stands proud*, with no proof-work above it. The two case proofs must use the respective case-assumptions, which get *discharged* upon application of the rule.

Inductive Clause ($\rightarrow$-E)

$$\frac{\mathcal{P}(\Pi, \varphi, \Delta) \qquad\qquad \mathcal{P}(\Sigma, \theta, \Gamma) \qquad\qquad \psi \in \Gamma}{\mathcal{P}([\theta, \varphi \rightarrow \psi, \Pi, \Sigma], \theta, \{\varphi \rightarrow \psi\} \cup \Delta \cup (\Gamma \setminus \{\psi\}))}$$

Graphically:

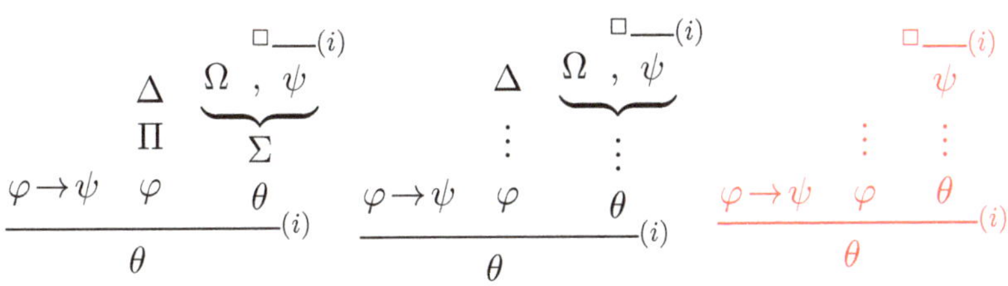

Note:
1. the major premise $\varphi \rightarrow \psi$ *stands proud*, with no proof-work above it;
2. the major subproof (of θ, above the line) must use the assumption ψ, which gets discharged by the applications of the rule;
3. θ might be $\bot$.

Inductive Clause ($\exists$-E)

$$\frac{\mathcal{P}(\Pi, \psi, \Delta) \qquad \varphi^x_a \in \Delta \qquad a \text{ does not occur in } \exists x\varphi, \psi \text{ or } \Delta \setminus \{\varphi^x_a\}}{\mathcal{P}([\psi, \exists x\varphi, \Pi], \psi, \{\exists x\varphi\} \cup (\Delta \setminus \{\varphi^x_a\}))}$$

Graphically:

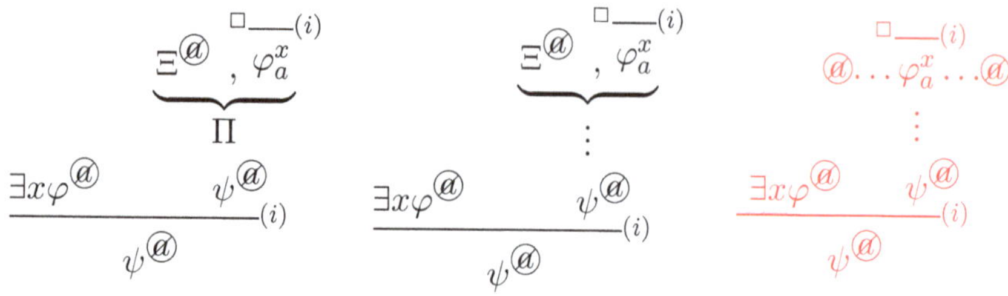

Note:
1. the major premise $\exists x\varphi$ *stands proud*, with no proof-work above it;
2. the parametric assumption φ^x_a must be used in the subordinate proof of ψ above the line, and gets *discharged* by the application of the rule;
3. ψ might be $\perp$.

Inductive Clause (∀-E)

$$\frac{\mathcal{P}(\Pi, \psi, \Delta \cup \{\varphi^x_{t_1}, \ldots, \varphi^x_{t_n}\}) \qquad \{\varphi^x_{t_1}, \ldots, \varphi^x_{t_n}\} \cap \Delta = \emptyset}{\mathcal{P}([\psi, \forall x\varphi, \Pi], \psi, \{\forall x\varphi\} \cup \Delta)}$$

Graphically:

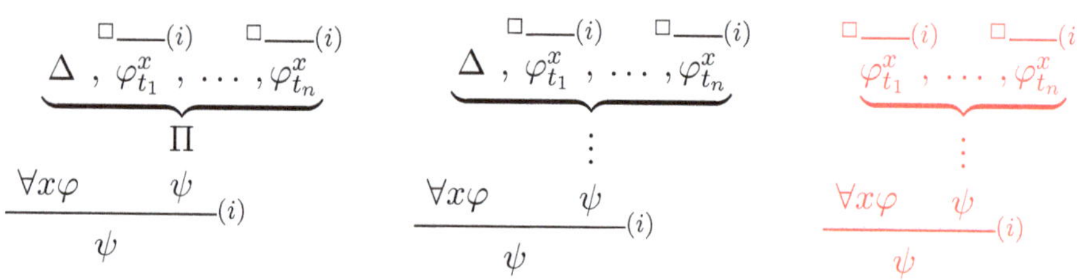

Note:
1. the major premise $\forall x\varphi$ *stands proud*, with no proof-work above it.
2. in the subordinate proof of ψ above the line, at least one instance of the quantified predicate φ must have been used as an assumption; and all such instances get discharged by the application of the rule;
3. ψ might be $\perp$.

Closure Clause

If $\mathcal{P}(\Pi, \varphi, \Delta)$, then this can be shown by appeal to the foregoing clauses.

Exercise 51. *State a similarly rigorous and detailed inductive definition of the notion* 'Π *is a* sequent proof *of* $\Delta : \varphi$'.

Chapter 16

Free Logic

16.1 The tradition of unfree logic

As explained in Chapter 2, Frege, in his *Begriffsschrift*, invented the first formal language that could accommodate both singular reference and existential and universal quantification. He subsequently applied it in the (inconsistent) theory of classes that he developed later in his *Grundgesetze der Arithmetik*.[1] In both these works he made the simplifying assumption that he was dealing with a *logically perfect language*, in which *every well-formed singular term denotes*.[2]

Logics that are *free* of such an assumption are called *free logics*. A free logic is furthermore *universally* free when in addition it is no longer based on the assumption (in standard logic) that the universe is non-empty. It allows, that is, for the 'empty universe'.[3] Presumably the definite article is in order here; it is hard to imagine there being *distinct* empty universes.

There are many good reasons for adopting a free logic, rather than a Fregean one. We shall now reprise some of the points made in §4.10, when we first made mention of free logic.

---

[1] Frege [1893; reprinted 1962] and Frege [1903; reprinted 1962].

[2] In the interests of complete rigor and clarity, one should add the condition that the terms in question do not contain any parameters.

[3] The most influential early work developing free logic is Lambert [1967]. He worked, however, in an axiomatic system. A universally free logic, as a system of natural deduction, was presented in Tennant [1978].

First, many a 'name' lacks a bearer. Such is the case in myth, legend, and fiction, for example—there is no such individual as Pegasus, the mythical winged flying horse, nor is there any such individual as Santa Claus. It is also the case in mistaken science. There was (and still is) no planet (to be called) Vulcan, when Urbain Le Verrier mistakenly conjectured, in 1859, that there was such a planet between Mercury and the Sun, and that its presence would account for the precession of Mercury's perihelion.[4]

One wishes, in realistic vein, to be able to assert 'There is no such individual as Pegasus [Santa Claus; Vulcan]', or to assert 'Pegasus [Santa Claus; Vulcan] does not exist'. For such a *negative existential*

$$\neg\exists x\, x = t$$

to be true, however, the singular term t within it must fail to denote. Yet in a so-called 'logically perfect' language the singular term t *would* denote—albeit some artificially chosen entity, which is chosen only in order to ensure logical 'perfection'.[5] Moreover, when one and the same artificially chosen entity is taken as the denotation of *all* such problematic singular terms, one has the added embarrassment of having identity claims such as 'Pegasus is Santa Claus' coming out *true*. All these embarrassments are avoided by taking the hard-nosed line that even *names* can fail to denote.

Second, many a functional term lacks a bearer. Equivalently: many functions are only *partial*, not total. In a language for real number theory in which division is represented by a primitive function sign $\div$, the careful mathematician will say that the value of any expression of the form $t \div 0$ is 'undefined'. That is another way of saying that expressions of that form do not denote: the division function is partial. Likewise, in real number theory (but not in complex number theory) any expression of the form $\sqrt{-\mid t \mid}$ (where t does not denote 0) fails to denote. The square-

[4] See Airy [1846].

[5] For Frege, there were other requirements, in addition, for logical perfection, such as the requirement that all concepts should have sharp boundaries (that is, that predicates should not be vague).

root function is partial, not being defined on negative real numbers.[6]

Thirdly, some first-order languages contain abstraction terms, such as definite descriptive terms $\iota x \varphi x$ and set-abstraction terms $\{x \mid \varphi x\}$. Whether such terms denote depends on the formula φ. In Russell's famous example (1905), 'the present King of France' fails to denote, but only contingently.[7] The descriptive term $\iota x \neg(x = x)$, by contrast, *cannot* denote. And the lesson to be extracted with hindsight from Russell's Paradox (1903) is that the set-term $\{x \mid \neg(x \in x)\}$ likewise cannot denote.[8] So, if we wish to have these abstraction operators as primitives in the language, we have to accept that the language will contain terms that do not merely *happen* not to denote; they *cannot* denote.

As soon as one allows for non-denoting terms—whether names, functional terms or abstraction terms—one's logic has to be modified. No longer, for example, would the simple inference forms ($\forall$-E) and ($\exists$-I):

$$\frac{\forall x F(x)}{F(t)} \qquad\qquad \frac{F(t)}{\exists x F(x)}$$

be valid. Instead, their conclusions would be correctly drawn only on the extra assumption that the term t involved denotes:

$$\frac{\forall x F(x) \quad \exists! t}{F(t)} \qquad\qquad \frac{F(t) \quad \exists! t}{\exists x F(x)}$$

The claim that the term t denotes is expressed by the sentence $\exists x\, x = t$, which we abbreviate here to $\exists! t$.

Free v. unfree logic

In unfree logic we have, for every closed term t,

$$\vdash \exists! t.$$

[6]The non-mathematician might also appreciate being informed that the 'absolute value' of the real-valued function $\mid t \mid$ of t is always non-negative. Thus $- \mid t \mid$ is negative; hence there is no such *real* number as $\sqrt{- \mid t \mid}$. This is because the square of *any* real number is non-negative.

[7]Russell [1905].

[8]Russell [1903].

We also have

$$\vdash \exists x\, x = x.$$

So unfree logic is committed to a denotation for every closed term, and to the existence of at least one thing.

In free logic, by contrast, we do not assume that every closed term denotes. But we *do* insist that no non-denoting term can occur in a true atomic relational statement. Put another way: if an atomic claim $A(\ldots t \ldots)$ is true, then so too is $\exists! t$. So in free logic we need to state the following rule, called the Rule of Atomic Denotation:

$$\frac{A(\ldots t \ldots)}{\exists! t} \quad \text{where } A \text{ is atomic}$$

Atomic predicates include, of course, the identity predicate =; and, in set theory, the membership predicate $\in$.

In free logic we also need to modify the Introduction and Elimination rules for the quantifiers, and the rule of Reflexivity of Identity—which, in *un*free logic, is simply

$$\frac{}{t = t}\ .$$

The modifications of these quantifier rules make explicit various existential presuppositions:

Universal Introduction ($\forall$-I)

$$\begin{array}{c} \overline{\exists! a}^{\,(i)} \\ \vdots \\ \dfrac{\varphi}{\forall x \varphi^a_x}{\scriptstyle (i)} \end{array} \quad \text{where } a \text{ is } \forall\text{-parametric in the subproof}$$

Universal Elimination ($\forall$-E)

$$\frac{\exists! t \quad \forall x \varphi}{\varphi^x_t}$$

Existential Introduction ($\exists$-I)

$$\frac{\exists! t \quad \varphi^x_t}{\exists x \varphi}$$

Existential Elimination (∃-E)

$$\dfrac{\exists x\varphi \quad \begin{matrix} \overset{(i)}{\overline{\exists! a}} \, , \, \overline{\varphi^x_a}^{(i)} \\ \underbrace{\qquad\qquad} \\ \vdots \\ \psi \end{matrix}}{\psi}\,(i)$$

where a is ∃-parametric in the subproof on the right

Reflexivity of Identity

$$\frac{\exists! t}{t = t}$$

This modified rule of Reflexivity of Identity in free logic can be *derived* if one is permitted to use the following 'introduction rule' for the identity predicate:

(=-I) $\dfrac{A(t)}{t = t}$ where $A(t)$ is atomic

Here is the derivation:

$$\dfrac{\begin{matrix}\exists! t, \text{ i.e.,} \\ \exists x\, x = t\end{matrix} \qquad \dfrac{\overline{a = t}^{\,(1)}}{t = t}\,{=}\text{-I}}{t = t}\,(1)\ \exists\text{-E}$$

This introduction rule (=-I) just stated tells one that identity is *the minimal reflexive (binary) relation.* Note that this rule's premise and conclusion are both atomic.

Now one *should* be able, in principle, to specify meaning-constituting introduction *and* elimination rules for the identity predicate that do *not* rely on the presence, in the language, of any logical operators—not even ∃. For, surely, there is room for a grasp of identity even in a language without quantifiers, but with atomic sentences capable of expressing that things have certain properties, and stand to one another in certain relations. So we shall take *Substitutivity of Identicals* as the appropriate elimination rule to accompany our new introduction rule for identity:

$$(=\text{-I})\quad \frac{A(t)}{t=t} \qquad\qquad (=\text{-E})\quad \frac{t=u \quad \varphi}{\varphi(t|u)}$$

In this statement of the rule (=-E) it is to be understood that the minor premise φ contains at least one occurrence of t or of u; and that the conclusion $\varphi(t|u)$ can be any result of substituting t for at least one occurrence of u in φ, and/or substituting u for at least one occurrence of t in φ. Thus the following three instances of (=-E) involving the binary predicate L would all be licit:

$$\frac{t=u \quad L(t,u)}{L(t,t)} \quad \frac{t=u \quad L(t,u)}{L(u,t)} \quad \frac{t=u \quad L(t,u)}{L(u,u)}$$

Exercise 52. *Prove the following theorem of free logic with identity:* $\forall x \forall y(\forall z(z{=}x \leftrightarrow z{=}y) \rightarrow x{=}y)$.
(Recall the definition of $\leftrightarrow$ in §9.4.1, on p. 197.)

Some reflections on universal elimination

Note that the modified elimination rule that we stated above for $\forall$ in free logic was in serial form:

$$\text{Serial } (\forall\text{-E}) \qquad \frac{\exists! t \quad \forall x \varphi}{\varphi^x_t}$$

Its parallelized form would be

$$\text{Parallelized } (\forall\text{-E}) \qquad \frac{\forall x\varphi \quad \exists! t \quad \begin{array}{c}\overline{\varphi^x_t}^{\,(i)} \\ \vdots \\ \theta\end{array}}{\theta}{\scriptstyle (i)}$$

This, however, countenances appeal to only *one* instance of the universal to propel one's reasoning further. It would be more efficient, in general, to allow for more than one such instance. Multiple applications of this single-instance parallelized rule would justify the use of the following *multiple-instances* parallelized form:

Parallelized ($\forall$-E)

$$\dfrac{\forall x\varphi \qquad \begin{array}{c}\vdots\\ \exists! t_1\end{array} \quad \dots \quad \begin{array}{c}\vdots\\ \exists! t_n\end{array} \qquad \begin{array}{c}\underbrace{\overline{\varphi^x_{t_1}}^{\,(i)} \ \dots\ \overline{\varphi^x_{t_n}}^{\,(i)}}\\ \vdots \\ \theta\end{array}}{\theta}\,(i)$$

In construing this graphic form as corresponding to an inductive clause in the definition of proof, the clause in question must be carefully formulated to capture the following condition:

> Finitely many instances $\varphi^x_{t_1} \dots, \varphi^x_{t_n}$ $(n \geq 1)$ of φ may feature as premises in the subordinate proof of θ that is indicated by the vertically descending dots; for each such instance $\varphi^x_{t_i}$ there must be a minor subproof with conclusion $\exists! t_i$; and upon applying the rule, all the assumptions $\varphi^x_{t_1} \dots, \varphi^x_{t_n}$ in the subordinate proof must be discharged.

Let us now investigate how proofs in free logic might have to be a little more complicated than proofs (of the same argument) in unfree logic.

Consider the very simple argument that we gave early on:

$$\dfrac{\begin{array}{l}\text{Socrates is a man}\\ \text{All men are mortal}\end{array}}{\text{Socrates is mortal}}$$

In symbolic translation, using our earlier translation manual on p. 58, it became

$$\dfrac{\begin{array}{l}Ms\\ \forall x(Mx \rightarrow Dx)\end{array}}{Ds}$$

Our earlier proof, which took only 2 steps, was given in *un*free logic, using serial forms of elimination:

$$\dfrac{\dfrac{\forall x(Mx \rightarrow Dx)}{Ms \rightarrow Ds} \qquad Ms}{Ds}$$

Here now is a proof of this argument, in *free* logic, and also in *serial* form. Because it is in *free* logic, it takes 3 steps, beause we need to winkle out Socrates's existence by an application of RAD, in order then to be able to apply the free-logical version of ∀-E:

$$\cfrac{\cfrac{\cfrac{Ms}{\exists! s} \qquad \forall x(Mx \to Dx)}{Ms \to Ds} \qquad Ms}{Ds}$$

And here now is the argument's 3-step proof, in free logic, in *parallelized* form:

$$\cfrac{\forall x(Mx \to Dx) \qquad \cfrac{Ms}{\exists! s} \qquad \cfrac{\overline{Ms \to Ds}^{\,(2)} \qquad Ms \qquad \overline{Ds}^{\,(1)}}{Ds}{\scriptstyle (1)}}{Ds}{\scriptstyle (2)}$$

It would appear that parallelization (in free logic) does not increase the number of steps in a proof—at least, in a proof of an argument as simple as this example. But that *might* be an over-hasty judgment.

Let us make the existence of Socrates—the needed existential presupposition for the application of ∀-E—a little more difficult to 'pry out' from the premises. Consider the argument

$$\cfrac{\text{Socrates loves someone} \qquad \text{Everyone who loves someone loves themselves}}{\text{Socrates loves himself}}$$

In fully detailed symbolic translation:

$$\cfrac{\exists y Lsy \qquad \forall x(\exists y Lxy \to Lxx)}{Lss}$$

Applying the Maxim of Shallow Analysis we can re-write this formal argument as

$$\cfrac{\exists y Lsy \qquad \forall x(\exists y Lxy \to \Phi x)}{\Phi s}$$

The proof of this re-written rgument in *un*free logic, using serial forms of elimination, takes only 2 steps:

$$\dfrac{\exists yLsy \qquad \dfrac{\forall x(\exists yLxy \to \Phi x)}{\exists yLsy \to \Phi s}}{\Phi s}$$

But how would the argument be proved in *free* logic? The coarsening of logical structure (using Φx for Lxx) does not make it any easier to pry out the existence of Socrates. Here is a *4-step* proof of the coarsened argument, in free logic, in *serial* form:

$$\dfrac{\exists yLsy \qquad \dfrac{\exists yLsy \qquad \dfrac{\dfrac{\overline{Lsa}^{\,(1)}}{\exists! s} \quad \forall x(\exists yLxy \to \Phi x)}{\exists yLsy \to \Phi s}}{\Phi s}}{\Phi s}{}^{(1)}$$

Note the unavoidable *double* use of the premise $\exists yLsy$. And here is 'the same' 4-step proof, in free logic, in *parallelized* form:

$$\dfrac{\exists yLsy \qquad \dfrac{\forall x(\exists yLxy \to \Phi x) \quad \dfrac{\overline{Lsa}^{\,(3)}}{\exists! s} \quad \dfrac{\overline{\exists yLsy \to \Phi s}^{\,(2)} \quad \exists yLsy \quad \overline{\Phi s}^{\,(1)}}{\Phi s}{}^{(1)}}{\Phi s}{}^{(2)}}{\Phi s}{}^{(3)}$$

Exercise 53. *Re-write every quantificational proof that you have provided in any earlier exercise in this text as a proof in* free *logic.*

Exercise 54. *In Chapter 15 we saw how to frame inductive clauses in the inductive definition of proof, with each such clause corresponding to a rule of inference in the proof system. Now that we have the modified rules of inference for free logic, the reader should revise those inductive clauses for proof-formation, to take the free-logical modifications into account.*

16.2 Systems of (universally) free logic, with identity

In order to make comparisons among systems of free logic easier to draw, we assume that the language contains no function signs. The language may, if we wish, contain variable-binding term-forming operators, such as the definite-description operator or the set-abstraction operator. Our sortal placeholder t can then be understood as place-holding for definite descriptive terms, and/or set-abstraction terms, in addition to names and parameters. Provided only that the language contains some closed terms other than parameters, all relevant contrasts among systems can be made. Note that the language *is* assumed to contain the *identity* predicate. It can be useful (but it is not obligatory) to employ a 'thinghood' predicate in order to give expression to the existential presuppositions called for in the rules of free logic.

Four systems are generated by the presence or absence of each of two features:

($\mathcal{U}$) allowing for the empty universe; and

(**T**) using a 'thinghood' predicate.

We shall accordingly obtain four systems of free logic (with identity) that we shall call $\mathcal{F}$, $\mathcal{FU}$, $\mathcal{F}\mathbf{T}$, and $\mathcal{FU}\mathbf{T}$. All four systems have the common aim of accommodating non-denoting singular terms. The four systems are very intimately related.

$\mathcal{FU}$ is an amended version of the system (call it $\mathcal{FU}_{NL}$) of universally free logic whose classical extension was proved sound and complete in Tennant [1978]. The systems $\mathcal{FU}$ and $\mathcal{FU}_{NL}$ have the same *identity-free* rules if one interprets '$\exists!$' as primitive and not defined.

All four free-logical systems become classical upon addition of any of the strictly classical rules of inference —the law of excluded middle, classical *reductio ad absurdum*, double-negation elimination, or dilemma. The discussion of (different kinds of) freedom in a logic is orthogonal to the debate over whether constructive logic is to be preferred to classical logic, or *vice versa*.

The interested reader can find further discussion of free-logical systems in Tennant [2007].

Chapter 17

Logic for 'The' and 'Set of'

17.1 Definite descriptions

We introduced the definite description operator ι ('iota') in Chapter 4, as one of three interesting term-forming variable-binding operators. The other two were the number abstraction operator # and the set abstraction operator $\{\ |\ldots\}$. The iota appeared again in Chapter 16, in our discussion of the need for a free logic to deal with terms that can fail to denote—not just contingently, but also necessarily, depending on their logical structure. An example of such a term is $\iota x \neg x = x$ ('the thing that is not identical to itself').

We have yet, however, to inform the reader about the more detailed logical behavior of iota—how and why it really is the best regimentation of the simple word 'the', both in ordinary English and in mathematical discourse using English in combination with formal symbols (so-called 'logician's English'). The present chapter remedies that omission. It supplies the needed rules of inference governing iota. To get to this juncture, however, we needed to set out the basics of free logic, which has now been done in Chapter 16.

The rule of introduction (in free logic) for the variable-binding abstraction operator ι that forms definite descriptive terms from predicates is

$$\iota\text{-I}\qquad \dfrac{\begin{array}{c}\underbrace{\overset{(i)}{\overline{\varphi(a)}}\ ,\ \overset{(i)}{\overline{\exists!a}}}\\ \vdots\\ a=t\end{array}\qquad \exists!t \qquad \begin{array}{c}\overset{(i)}{\overline{a=t}}\\ \vdots\\ \varphi(a)\end{array}}{t=\iota x\varphi(x)}\,{}^{(i)}\quad ,$$

where a is parametric.[1]

Note how the canonical conclusion

$$t = \iota x \varphi(x)$$

of ι-I has t on its left-hand side, as a placeholder for *any singular term whatsoever*, including the *parameters* (conventionally a, b, c, ...) that can be used for reasoning involving existentials and universals. On the right-hand side of the identity is a definite descriptive term, formed by means of a *dominant* occurrence of the variable-binding abstraction operator ι. This operator may be applied to a formula φ to form the definite-descriptive term $\iota x \varphi$ if, but only if, the variable x has a free occurrence in φ.

I call such a rule for the introduction of a variable-binding term-forming operator a *single-barreled* rule. The rule concerns a *single* occurrence of the operator in 'as dominant a position as possible' within the conclusion, which is an identity statement.

The elimination rules corresponding to the introduction rule stated above for ι are the following three, each one employing the canonical identity statement

$$t = \iota x \varphi(x)$$

as its major premise (to the left, immediately above the inference stroke). The minor premises (or subproofs) of the first and third rules correspond, respectively, to the first and third immediate subproofs of the introduction rule. This is a convincing sign that the elimination rules are in harmony with the introduction rule that begets them. Bear in mind

[1]Note that since $=$ is an *atomic* binary predicate, the assumption $a = t$ in the rightmost subordinate proof implies $\exists!a$ (by the Rule of Atomic Denotation). So it is not necessary to have $\exists!a$ as a further dischargeable assumption in that subordinate proof.

that the major premise $t = \iota x\varphi(x)$ stands proud in any application of an elimination rule for ι.

ι-E$_1$ $\dfrac{t=\iota x\varphi \quad \exists!u \quad \varphi^x_u}{u=t}$; or, in parallelized form:

$$\dfrac{t=\iota x\varphi \quad \exists!u \quad \varphi^x_u \quad \begin{array}{c}\square\overline{\;u=t\;}^{(i)}\\ \vdots \\ \psi\end{array}}{\psi}{\scriptstyle (i)}$$

ι-E$_2$ $\dfrac{t=\iota x\varphi}{\exists!t}$; or, in parallelized form:

$$\dfrac{t=\iota x\varphi \quad \begin{array}{c}\square\overline{\;\exists!t\;}^{(i)}\\ \vdots \\ \psi\end{array}}{\psi}{\scriptstyle (i)}$$

ι-E$_3$ $\dfrac{t=\iota x\varphi \quad u=t}{\varphi^x_u}$; or, in parallelized form:

$$\dfrac{t=\iota x\varphi \quad u=t \quad \begin{array}{c}\square\overline{\;\varphi^x_u\;}^{(i)}\\ \vdots \\ \psi\end{array}}{\psi}{\scriptstyle (i)}$$

Exercise 55. *Using only ι-I, prove the argument $\exists!t : t=\iota x(x=t)$.*

17.2 Set abstraction

The rule of introduction (in free logic) for the variable-binding abstraction operator that forms set-terms from predicates is

$$\{\,\}\text{-I} \qquad \dfrac{\begin{array}{c}\underbrace{{}^{(i)}\overline{\varphi(a)}\;,\;\overline{\exists!a}^{(i)}}\\ \vdots \\ a\in t\end{array} \qquad \exists!t \qquad \begin{array}{c}\overline{\;a\in t\;}^{(i)}\\ \vdots \\ \varphi(a)\end{array}}{t=\{x|\varphi(x)\}}{\scriptstyle (i)} \quad , \quad \text{where } a \text{ is parametric.}^{2}$$

---

[2]Note that since $\in$ is an *atomic* binary predicate, the assumption $a \in t$ in the rightmost subordinate proof implies $\exists!a$ (by the Rule of Atomic Denotation). So it is not necessary to have $\exists!a$ as a further dischargeable assumption in that subordinate proof.

Note how the canonical conclusion

$$t = \{x|\Phi\}$$

of { }-I has t on its left-hand side, as a placeholder for *any singular term whatsoever*, including the *parameters* (conventionally a, b, c, ...) that can be used for reasoning involving existentials and universals. On the right-hand side of the identity is a set-abstraction term, formed by means of a *dominant* occurrence of the variable-binding abstraction operator $\{x|\ldots x\ldots\}$. This operator may be applied to a formula φ to form the set-abstraction term $\{x|\varphi\}$ if, but only if, the variable x has a free occurrence in φ.

The reader will have noted that the rule for the introduction of the variable-binding term-forming operator of set abstraction is, like that for iota, a *single-barreled* rule. The rule concerns a *single* occurrence of the operator in 'as dominant a position as possible' within the conclusion, which is an identity statement.

The elimination rules corresponding to the introduction rule stated above for { } are the following three, each one employing the canonical identity statement

$$t = \{x|\Phi\}$$

as its major premise (to the left, immediately above the inference stroke). The minor premises (or subproofs) of the first and third elimination rules correspond, respectively, to the first and third immediate subproofs of the introduction rule. This is a convincing sign that the elimination rules are in harmony with the introduction rule that begets them. Bear in mind that the major premise $t = \{x|\Phi\}$ stands proud in any application of an elimination rule for { }.

{ }-E$_1$ $\dfrac{t=\{x|\varphi\} \quad \exists!u \quad \varphi^x_u}{u \in t}$; or, in parallelized form: $\dfrac{t=\{x|\varphi\} \quad \exists!u \quad \varphi^x_u \quad \begin{matrix}\overline{u\in t}^{(i)}\\ \vdots \\ \psi\end{matrix}}{\psi}{}_{(i)}$

{ }-E$_2$ $\dfrac{t=\{x|\varphi\}}{\exists!t}$; or, in parallelized form: $\dfrac{t=\{x|\varphi\} \quad \begin{matrix}\overline{\exists!t}^{(i)}\\ \vdots \\ \psi\end{matrix}}{\psi}{}_{(i)}$

{ }-E$_3$ $\dfrac{t=\{x|\varphi\} \quad u \in t}{\varphi^x_u}$; or, in parallelized form: $\dfrac{t=\{x|\varphi\} \quad u\in t \quad \begin{matrix}\overline{\varphi^x_u}^{(i)}\\ \vdots \\ \psi\end{matrix}}{\psi}{}_{(i)}$

The rule { }-E$_2$ is a special case of the Rule of Atomic Denotation.

The rule { }-E$_3$ needs to be in the parallelized form shown, in order to avoid having non-trivial proof-work above φ^x_u should it happen to stand as the major premise of an elimination within the immediate subproof indicated.[3]

The introduction rule and the elimination rules just stated for { } are, as just intimated, in harmony. Harmony requires that there at least be reduction procedures that will eliminate from proofs 'maximal sentence occurrences'—conclusions of introductions that are also major premises of eliminations. (Whether harmony requires *more* than this is a more complicated issue that we shall not broach here.)

In the intended sequel to this primer we shall develop the basics of set theory more thoroughly by using the set-abstraction operator { }

[3]This risk would be incurred if { }-E$_3$ were to be used in the serial form

$$\frac{t=\{x|\varphi\} \quad u\in t}{\varphi^x_u}\ .$$

as a grammatical primitive, and by framing rules of introduction and elimination for a great variety of familiar set-theoretical operators and predicates.[4]

Exercise 56. *Using only* { }-I, *prove the argument* $\exists !t : t=\{x|x \in t\}$.

Exercise 57. *Prove* $\forall x \forall y(\forall z(z \in x \leftrightarrow z \in y) \rightarrow x = y)$ *as a theorem of the foregoing logic of sets.*

(Hint: use your answer to Exercise 56 along with { }-I.)

[4]See Tennant [2024].

Chapter 18

Constructive v. Strictly Classical Reasoning

18.1 Kinds of validity

The student of logic needs to acquire the ability to recognize, when presented with a reasonably *simple* argument, into which class it would fall in the three-way 'classification' on the following page. In the case of *propositional logic*, there is in principle an effective method for making the correct determination, because in both constructive and classical propositional logic there is a mechanical procedure for determining, of any given argument, whether it is valid or invalid in the logic in question. In the case of *first-order* logic, however (with at least one binary predicate) there is *no* mechanical procedure for making any such determination, whether of classical or of constructive validity.[1] That is why, in the first-order case, we use the qualification 'reasonably *simple*' in our description of the logical abilities that the student ought to be able to develop.

The three-way classification is very straightforward. It consists of the following pairwise disjoint, and collectively exhaustive, classes of

[1]This is a corollary to the undecidability of first-order logical theoremhood (Church [1936]) and the strong completeness of first-order logic (Gödel [1930]).

arguments with finitely many premises:

constructively valid
classically valid but not constructively valid
classically invalid

It is impossible for any argument to be constructively valid but classically invalid. This is clear from the following classificatory diagram, in which points within the boxes represent arguments.

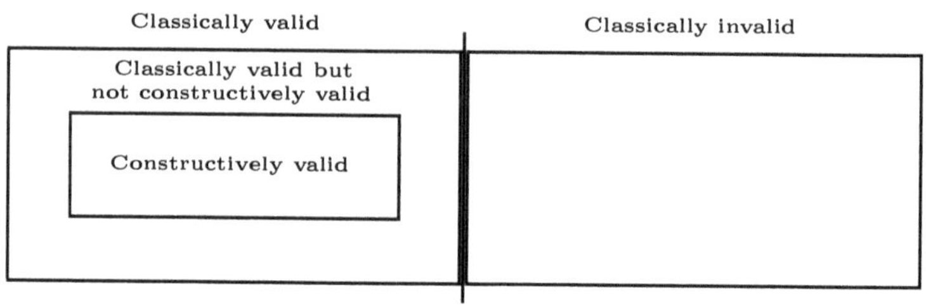

In §18.2 we shall illustrate this classification with a plethora of examples. Constructively valid arguments will be provided with constructive proofs, which of course are free of any applications of strictly classical rules. Arguments that are classically but not constructively valid will be provided with proofs in which, of course, an application of a strictly classical rule is required. The reader will notice that *which* strictly classical rule finds application in general depends on the circumstances. It is always a pragmatic choice, dictated by considerations of economy and convenience. Finally, invalid arguments are given counterexamples, in the form of simple models, or diagrams. It will be left to the reader to construct, relative to the model provided, verifications of the argument's premises and a falsification of its conclusion. All the models given here are not only finite (in having finite domains of individuals) but are also *small*.

Note that it is *not* the case that every classically invalid first-order argument has a finite countermodel.[2] There are some classically invalid arguments whose (finitely many) premises can be made true only in *infinite* models. (In these cases, the conjunction of the premises is called

[2]Warning to Linguistics students: innocence of this fact is a well-known defect of Johnson-Laird [1983].

an ‘axiom of infinity’.) For these arguments, one’s description of a countermodel would have to take some form other than a finite diagram like the ones employed below. No such arguments, however, will appear in the list of examples below.

After studying these thirty examples, for the first three of which we provide complete solutions, the reader should begin to acquire a ‘feel’ for the difference between constructive and strictly classical validity. At this stage the reader is not yet theoretically equipped to be absolutely certain that the ‘classically but not constructively’ valid arguments furnished here *really are not constructively valid.* All we can do at this stage is convince the reader of such arguments’ classical validity, by providing a classical proof. The reader should rest assured, however, that there is indeed a failsafe method of constructing counterexamples to *constructively invalid* arguments that establishes their *constructive* invalidity even when they are *classically* valid. (The method in question will be covered in the sequel to this primer.) Moreover, in the propositional case (as already implied by what has been said above) this method is effective, or algorithmic.

18.2 An instructive list of examples

We shall permit ourselves, in proofs, to use the *serial* forms for elimination of $\wedge$, $\rightarrow$, and $\forall$:

$$\frac{\varphi \wedge \psi}{\varphi} \quad \frac{\varphi \wedge \psi}{\psi} \qquad \frac{\varphi \rightarrow \psi \quad \varphi}{\psi} \qquad \frac{\forall x\psi}{\psi^x_t}$$

These serial forms are interderivable with the parallelized forms that have been obtained by our morphing of the rules of falsification in Chapter 7 (for the connectives) and in Chapter 13 (for the quantifiers). We have demonstrated this interderivability for $\wedge$ and $\rightarrow$ in §7.5. For the quantifier $\forall$, the interderivability of its serial and parallelized forms of elimination is similarly demonstrable as follows:

Derivation of parallelized form using serial form

$$\begin{array}{c}\underbrace{\dfrac{\forall x\psi}{\psi^x_{t_1}} \;\cdots\; \dfrac{\forall x\psi}{\psi^x_{t_n}}} \\ \vdots \\ \theta\end{array}$$

Derivation of serial form using parallelized form

$$\dfrac{\forall x\psi \quad \overline{\psi^x_t}^{(1)}}{\psi^x_t}{\scriptstyle (1)}$$

The three worked examples to follow illustrate how to give complete answers to the questions *Is this argument valid?*, and *If valid, is it constructively so?* The arguments are listed in Exercise 58 starting on p. 313. An invalid argument calls for a counterexample; a constructively valid argument calls for a constructive proof; and a classically valid argument that is not constructively valid calls for a proof using a strictly classical rule of inference.

Example #1

$$\frac{(P \to Q) \wedge (P \to R)}{P \to (Q \wedge R)} \qquad \text{Constructively valid}$$

$$\dfrac{\dfrac{\dfrac{{}^{(1)}\overline{P} \qquad \dfrac{(P \to Q) \wedge (P \to R)}{P \to Q}}{Q} \qquad \dfrac{{}^{(1)}\overline{P} \qquad \dfrac{(P \to Q) \wedge (P \to R)}{P \to R}}{R}}{Q \wedge R}}{P \to (Q \wedge R)}{\scriptstyle (1)}$$

Example #2

$$\frac{\exists x(Fx \vee Gx)}{\exists x Fx \wedge \exists x Gx} \qquad \text{Classically invalid}$$

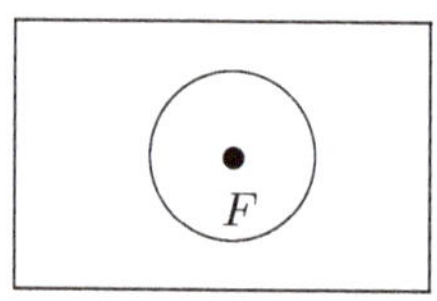

Example #3

$$\frac{\forall x(Fx \vee Gx)}{\exists x Fx \vee \forall x Gx} \qquad \text{Classically but not constructively valid}$$

$$\dfrac{\dfrac{\dfrac{{}^{(2)}\overline{\exists x Fx}}{\exists x Fx \vee \forall x Gx} \qquad \dfrac{\dfrac{\dfrac{\dfrac{\forall x(Fx \vee Gx)}{Fa \vee Ga} \qquad \dfrac{\dfrac{\overline{Fa}^{(1)}}{\exists x Fx} \qquad \overline{\neg \exists x Fx}^{(2)}}{\bot} \qquad \overline{Ga}^{(1)}}{Ga}{\scriptstyle (1)}}{\forall x Gx}}{\exists x Fx \vee \forall x Gx}}{\exists x Fx \vee \forall x Gx}{\scriptstyle (2)}}{}$$

Our remaining twenty eight examples are now left to the reader as the following exercise.

Exercise 58. Classify each of the following arguments as either
(i) constructively valid, or
(ii) classically but not constructively valid, or
(iii) classically invalid.
Amplify your answers by providing
(i) for each constructively valid argument a constructive proof;
(ii) for each classically but not constructively valid argument a strictly classical proof; and
(iii) for each classically invalid argument a counterexample.

1. $\dfrac{P}{\neg\neg P}$

2. $\dfrac{\exists x \neg Fx}{\neg \exists x Fx}$

3. $\dfrac{}{\neg(P \wedge \neg P)}$

4. $\dfrac{\neg \forall x Fx}{\forall x \neg Fx}$

5. $\dfrac{}{\exists x Fx \vee \forall x \neg Fx}$

6. $\dfrac{}{\neg\neg\neg P \rightarrow \neg P}$

7. $\dfrac{}{\neg\neg(P \vee \neg P)}$

8. $\dfrac{\exists x \neg Fx}{\forall x \neg Fx}$

9. $\dfrac{}{(P \rightarrow Q) \vee (Q \rightarrow P)}$

10. $\dfrac{\neg P}{P \rightarrow Q}$

11. $\dfrac{P \rightarrow \neg Q}{Q \rightarrow \neg P}$

12. $\dfrac{(P \vee Q) \rightarrow R}{(P \rightarrow R) \wedge (Q \rightarrow R)}$

13. $\dfrac{\forall x Fx \rightarrow \forall x Gx}{\forall x(Fx \rightarrow Gx)}$

14. $\dfrac{\neg \forall x Fx}{\exists x \neg Fx}$

15. $\dfrac{\neg \forall x Fx}{\neg \exists x Fx}$

16. $\dfrac{\neg P \vee Q}{P \rightarrow Q}$

17. $\dfrac{P \rightarrow (Q \vee R)}{(P \rightarrow Q) \vee (P \rightarrow R)}$

18. $\dfrac{P \rightarrow Q}{\neg(P \wedge \neg Q)}$

19. $\dfrac{\exists x Fx \wedge \exists x Gx}{\exists x(Fx \wedge Gx)}$

20. $\dfrac{\forall y \exists x Lxy}{\exists x \forall y Lxy}$

21. $\dfrac{\forall x \neg Fx}{\neg \exists x Fx}$

22. $\dfrac{\exists y \forall x Lxy}{\forall x \exists y Lxy}$

23. $\dfrac{\exists x(Fx \vee Gx)}{\exists x Fx \vee \exists x Gx}$

24. $\dfrac{\forall x Lxx}{\forall x \forall y Lxy}$

25. $\dfrac{}{\neg\neg P \rightarrow P}$

26. $\dfrac{\exists x(Fx \wedge Gx)}{\forall x(Fx \vee Gx)}$

27. $\dfrac{}{P \vee \neg P}$

28. $\dfrac{\neg(P \wedge Q)}{\neg P \vee \neg Q}$

Exercise 59.
The following arguments are all constructively valid. Prove them using only the rules of constructive logic.[3]

1. $A \to A$
2. $A \to B, B \to C : A \to C$
3. $A \to (B \to C) : B \to (A \to C)$
4. $A \to (B \to C) : (A \wedge B) \to C$
5. $(A \wedge B) \to C : A \to (B \to C)$
6. $A \to B : (B \to C) \to (A \to C)$
7. $A \to B : (C \to A) \to (C \to B)$
8. $A \to B : (A \wedge C) \to (B \wedge C)$
9. $A \to B : (C \wedge A) \to (C \wedge B)$
10. $A \to B : (A \vee C) \to (B \vee C)$
11. $A \to B : (C \vee A) \to (C \vee B)$
12. $\neg A : A \to B$
13. $A : \neg A \to B$
14. $B : A \to B$
15. $A \to B : \neg B \to \neg A$
16. $A \to \neg B : B \to \neg A$
17. $A \to \neg B : \neg\neg B \to \neg A$
18. $(A \to B) \wedge (B \to A)), A : B$
19. $(A \to B) \wedge (B \to A), B : A$
20. $(A \to B) \wedge (B \to A), (B \to C) \wedge (C \to B) : (A \to C) \wedge (C \to A)$
21. $A \to (B \to C)), \neg\neg A, \neg\neg B : \neg\neg C$
22. $\neg\neg(A \to B) : \neg\neg A \to \neg\neg B$
23. $\neg\neg(A \to B), \neg\neg(B \to C) : \neg\neg(A \to C)$
24. $: (\neg\neg(A \wedge B) \to (\neg\neg A \wedge \neg\neg B)) \wedge ((\neg\neg A \wedge \neg\neg B) \to \neg\neg(A \wedge B))$
25. $: (\neg\neg((A \to B) \wedge (B \to A)) \to (\neg\neg(A \to B) \wedge \neg\neg(B \to A)))$
    $\wedge$
    $(\neg\neg(A \to B) \wedge \neg\neg(B \to A)) \to \neg\neg((A \to B) \wedge (B \to A)))$
26. $(A \to B) \wedge (B \to A) : ((A \to C) \to (B \to C)) \wedge ((B \to C) \to (A \to C))$

---

[3]These are constructively valid examples drawn from the lists of well-known deducibilities in Kleene [1952], pp. 113–115 and pp. 118–119. (Kleene included in his lists a few deducibilities that are strictly classical and not constructive.) He used & for $\wedge$, and $\supset$ for $\to$.

27. $(A \to B) \wedge (B \to A) : ((A \wedge C) \to (B \wedge C)) \wedge ((B \wedge C) \to (A \wedge C))$
28. $(A \to B) \wedge (B \to A) : ((C \to A) \to (C \to B)) \wedge ((C \wedge B) \to (C \wedge A))$
29. $(A \to B) \wedge (B \to A) : ((C \wedge A) \to (C \wedge B)) \wedge ((C \wedge B) \to (C \wedge A))$
30. $(A \to B) \wedge (B \to A) : (\neg A \to \neg B) \wedge (\neg B \to \neg A)$
31. $: (((A \wedge B) \wedge C) \to (A \wedge (B \wedge C))) \wedge ((A \wedge (B \wedge C)) \to ((A \wedge B) \wedge C))$
32. $: ((A \wedge B) \to (B \wedge A)) \wedge ((B \wedge A) \to (A \wedge B))$
33. $: (A \to (A \wedge A)) \wedge ((A \wedge A) \to A)$
34. $A : ((A \to B) \to B) \wedge (B \to (A \to B))$
35. $B : ((A \to B) \to B) \wedge (B \to (A \to B))$
36. $\neg A : ((A \to B) \to \neg A) \wedge (\neg A \to (A \to B))$
37. $\neg B : ((A \to B) \to \neg A) \wedge (\neg A \to (A \to B))$
38. $B : ((A \wedge B) \to A) \wedge (A \to (A \wedge B))$
39. $\neg B : ((A \wedge B) \to B) \wedge (B \to (A \wedge B))$
40. $: A \to \neg\neg A$
41. $: (\neg\neg\neg A \to \neg A) \wedge (\neg A \to \neg\neg\neg A)$
42. $: \neg(A \wedge \neg A)$
43. $: \neg((A \to \neg A) \wedge (\neg A \to A)))$
44. $: \neg\neg(\neg\neg A \to A)$
45. $: ((A \wedge (B \wedge \neg B)) \to (B \wedge \neg B)) \wedge ((B \wedge \neg B) \to (A \wedge (B \wedge \neg B)))$
46. $: (A \to B) \to \neg(A \wedge \neg B)$
47. $: ((A \to \neg B) \to \neg(A \wedge B)) \wedge (\neg(A \wedge B) \to (A \to \neg B))$
48. $: (\neg(A \wedge B) \to (\neg\neg A \to \neg B)) \wedge ((\neg\neg A \to \neg B) \to \neg(A \wedge B))$
49. $\neg\neg B \to B : ((\neg\neg A \to B) \to (A \to B)) \wedge ((A \to B) \to (\neg\neg A \to B))$
50. $\neg\neg B \to B : ((A \to B) \to \neg(A \wedge \neg B)) \wedge (\neg(A \wedge \neg B)) \to (A \to B))$
51. $: (\neg\neg A \to B) \to \neg(A \wedge \neg B)$
52. $: (A \wedge B) \to \neg(A \to \neg B)$
53. $: (A \wedge \neg B) \to \neg(A \to B)$
54. $: (\neg\neg A \wedge B) \to \neg(A \to \neg B)$
55. $: ((\neg\neg A \wedge \neg B) \to \neg(A \to B)) \wedge (\neg(A \to B)) \to (\neg\neg A \wedge \neg B))$
56. $: (\neg(A \to B) \to \neg\neg(A \wedge \neg B)) \wedge (\neg\neg(A \wedge \neg B) \to \neg(A \to B))$
57. $: (\neg\neg(A \to B) \to \neg(A \wedge \neg B)) \wedge (\neg(A \wedge \neg B) \to \neg\neg(A \to B))$
58. $: (\neg(A \wedge \neg B)) \to (A \to \neg\neg B)) \wedge ((A \to \neg\neg B) \to \neg(A \wedge \neg B))$
59. $: ((A \to \neg\neg B) \to (\neg\neg A \to \neg\neg B)) \wedge ((\neg\neg A \to \neg\neg B) \to (A \to \neg\neg B))$

Chapter 19

Describing Finite Models Categorically

We shall now describe a method to give an absolutely exact (or so-called *categorical*) description of a given *finite* situation (or model) M. A model-description is categorical just in case one who receives the description can think of only one model that it describes—where 'only one' means 'only one, up to isomorphism'.

The *basic facts* in a model M for a first-order language L are given by the extensions that M provides for the *primitive items* of extra-logical vocabulary in L. These include

- which individual is denoted by a particular name;
- which individuals are in the extension of a particular one-place predicate;
- which ordered pairs of individuals are in the extension of a particular two-place predicate; and so on.

Suppose M and N are models for the same extra-logical expressions of a first-order language. (We say that the models M and N have the 'same signature'.) By 'model N is *isomorphic* to model M' we mean the following:

> There is a 1-1 correspondence (that is, a 1-1 mapping from the domain of N onto the domain of M) under which the basic facts in N correspond to the basic facts in M.

By 'sentence φ *categorically* (i.e., *exactly*) describes model M' we mean the following:

1. φ is true in M; **and**

2. any model in which φ is true is isomorphic to M.

Some model-theoretic concepts

Let M be a model of L with domain D.

Let ϕ be a one-one mapping with domain D. Set $D' = \phi[D]$, the range of ϕ. (Note that D' can be distinct from D.) The model $\phi[M]$ (call it M' for short) with domain D' is defined as the 'image' of the model M under the mapping ϕ. That is, for any primitive non-logical expression ε of L, we have

$$M'[\varepsilon] = \phi[M[\varepsilon]].$$

This identity says that the extension in M' of the non-logical expression ε is the image, under the mapping ϕ, of its extension in the model M. When this holds for all such expressions ε (names, function signs, predicates) the models M and M' are said to be *isomorphic*. The mapping ϕ is an *isomorphism* between them.

Metatheorem 1. *Suppose M, D, ϕ, and M' are as described above. Let ψ be any formula of L. Let t be any term of L. Let f assign individuals in D to the free variables in the formula ψ and in the term t. Let $\phi$of be the composition of f and ϕ—the result of first applying f and then applying ϕ. It follows that*

$$\phi(M_f[t]) = M'_{\phi o f}[t]$$

and

$$M_f \Vdash \psi \Leftrightarrow M'_{\phi o f} \Vdash \psi.$$

Proof. Easy by induction on the complexity of terms and formulae, with the Basis Step taken care of by the main supposition. □

Corollary 1. *Isomorphic models of L make the same L-sentences true.*

Definition 12. *Let M be a model, and let Δ be a set of sentences in the language that is interpreted in M. We say that Δ* categorically describes *M just in case $M \Vdash \Delta$ and any model of Δ is isomorphic to M. We also say that Δ* describes *M* 'up to isomorphism'.

19.1 Describing a finite model categorically

We shall demonstrate the 'how' by means of a simple example, using a first-order language L of very simple signature, to keep matters manageable. The *extra-logical* vocabulary of the language L for the purposes of this example consists of the following (very few) *primitive items*:

exactly one name a;
exactly one one-place predicate F; and
exactly one two-place predicate L.

The *logical* vocabulary of L consists of

the negation sign $\neg$;
the signs $\wedge$, $\vee$, $\rightarrow$ for conjunction, disjunction and implication;
the existential and universal quantifiers $\exists$, $\forall$; and
the two-place relation $=$ of identity.

Now we shall simplify even further, by working with very small models.

Consider all the models for our chosen language L that can be constructed on the basis of *exactly two* individuals, one of which will be named a (the other one remaining nameless). The models must therefore interpret the name a, the one-place predicate F, and the two-place predicate L.

For each such model M, provide the following:

1. A **diagram** depicting M, consisting of two dots for the individuals (one of which will be labeled a); a closed curve indicating the extension of F in M; and a selection of arrows representing the extension of L in M.

2. A **sentence** of L that exactly describes M.

Our task, then, in metalogical parlance, is to categorically describe, 'up to isomorphism', every two-element model for L. How many such models are there?

In a 2-element domain $\{\alpha, \beta\}$, there are $2^2 = 4$ possible extensions for the one-place predicate F:

$$\emptyset; \quad \{\alpha\}; \quad \{\beta\}; \quad \{\alpha, \beta\}.$$

In a 2-element domain $\{\alpha, \beta\}$, there are $2^4 = 16$ possible extensions for the two-place predicate L, ranging from the fullest possible extension

$$\{\langle\alpha, \alpha\rangle, \langle\alpha, \beta\rangle, \langle\beta, \alpha\rangle, \langle\beta, \beta\rangle, \}$$

to the empty one. It follows that there are $4 \times 16 = 64$ possible 2-element models for the language L.

Note that the availability of the name a (which we can let denote the individual α—or β—throughout) breaks any symmetries that might otherwise have reduced the number of isomorphism classes of models.

19.2 A fully worked-out example

We shall work out fully one example (from the 64 possible ones), to get the reader going on the remaining 63. After working through this example, we shall describe rigorously the general method employed to find the required solution. The general method will work with any language with finite signature, and any finite model of that signature. But we are keeping matters manageable here, by working with the chosen language L above, and the two-element model M that we are about to describe below.

After illustrating the method with our chosen example, the reader can employ the method to address the remaining 63 problems concerning models with exactly two individuals, for the language whose extra-logical primitives are the name a, the one-place predicate F, and the two-place predicate L.

In the diagram for the particular model of this kind that we have chosen—let us call it M—the arrows represent the extension

$$\{\langle\alpha, \alpha\rangle, \langle\alpha, \beta\rangle\}$$

in M of the two-place predicate L. The one-place predicate F has extension $\{\beta\}$ in M. The name a denotes individual α in M.

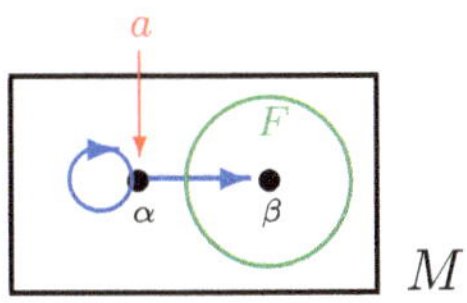

The basic *positive* facts, using the name a in place of our metalinguistic designation α, are:

$$Laa, La\beta, F\beta.$$

The basic *negative* facts are

$$\neg L\beta a, \neg L\beta\beta, \neg F\alpha, \neg\, \alpha = \beta.$$

A sentence that categorically describes this model is

$$\exists x(\neg\, a = x \wedge \forall y(y = a \vee y = x) \wedge Laa \wedge Lax \wedge Fx \wedge \neg Lxa \wedge \neg Lxx \wedge \neg F\alpha.)$$

What was the general method that we employed in order to find this sentence?

19.3 The general method just illustrated

Metatheorem 2. *If M is a finite model, then M is categorically describable.*

Proof. Since M is finite, so too is the set of primitive extralogical expressions of the language L of M. For simplicity's sake, assume that the denotation mapping on names is one-one. There is at least one sentence φ of L that describes M categorically. One can construct such a sentence φ as follows.

Assign variables uniquely, as needed, to provide 'temporary names' for any unnamed individuals in the domain of M. Let the variables used for this purpose be $x_1, \ldots, x_n$. For the k named individuals in M, let the names respectively be $a_1, \ldots, a_k$. So there are exactly $(k + n)$ individuals in the domain of M; for there are exactly k named ones, and exactly n unnamed ones. Consider all non-identities of the form

$\neg t = u$, where t and u are drawn from the list $a_1, \ldots, a_k, x_1, \ldots, x_k$, with t preceding u. Let the conjunction of these non-identities be ξ.

Form the conjunction ψ of all the (positive or negative) literals true in M that can be formed from the non-logical vocabulary of L (which includes the names $a_1, \ldots, a_k$) along with $x_1, \ldots, x_n$.

The sought sentence φ is

$$\exists x_1 \ldots \exists x_n(\xi \wedge \forall x(x = x_1 \vee \ldots \vee x = x_n \vee x = a_1 \vee \ldots \vee x = a_k) \wedge \psi).$$

□

Note that the exposition here of isomorphism of models applies to *all* models, finite or infinite. We need to fix the idea of isomorphism in order to show that all *finite* models can be described categorically. *Infinite* models, however, are a different kettle of fish. Even the set of all first-order sentences in the language of arithmetic that are true in the 'standard model' $\mathbb{N}$ has models with *countably* infinite domains that are not isomorphic to $\mathbb{N}$. Likewise, the set of all first-order sentences in the language of set theory that are true in the (countably infinite) 'standard model' V_ω of what are known as the 'hereditarily finite pure sets' has models with *countably* infinite domains that are not isomorphic to V_ω. Our arithmetical theorizing and our set-theoretical theorizing at first order are hopelessly 'non-categorical', even when one is seeking to attain the much more provincial notion of categoricity 'in the countably infinite cardinality $\aleph_0$'—let alone outright categoricity, regardless of the cardinality of the domain.

We shall see, in the intended sequel to this work, an isolated but comforting exception to this rather demoralizing limitation of expressive power at first order (i.e. power to describe a particular model 'up to isomorphism' among models of the same cardinality). The exception in question is the theory of the ordering of the rational numbers. Cantor famously proved, using his 'back and forth' method, that any two countable models of this theory are isomorphic. (The theory has no finite models, so any countable model of it is countably *infinite*, like the rational numbers.)

Chapter 20

Proofs of Aristotelian Syllogisms

20.1 Notation

As we mentioned in §2.2, Aristotle employed the following symbolic 'logical forms':

aAB	Every A is B
eAB	No A is B
iAB	Some A is B
oAB	Some A is not B

Note that Aristotle did not employ variables with his quantifiers. *That* innovation, by Frege, was to come only two millenia later.

In what follows, each q_i is one of Aristotle's binary quantifiers: a, e, i or o. Our aim here is eventually to give *natural deductions* to establish the validity of Aristotle's syllogisms, upon translating them into the formal language of first-order logic. But first a word will be in order about how Aristotle classified his syllogisms, using the terms *Figure, Mood, Major*, and *Minor*.

20.2 Aristotle's syllogistic terminology

The *major term* of an Aristotelian syllogistic argument is the rightmost one in its conclusion. It is rendered as P in the following. The major term occurs in exactly one of the argument's two premises, which is accordingly called the *major premise!Aristotelian sense.* The other premise is called the *minor premise!Aristotelian sense.* The major premise is always listed first when an Aristotelian argument is stated:

Major premise
Minor premise
———
Conclusion

We shall ambiguously denote the quantifier expression of the major premise by q_1; that of the minor premise by q_2; and that of the conclusion by q_3. Remember, there are four possible values for these q_j: a, i, e and o.

One of the three terms of the argument, called the *middle term,* occurs in *both* of the premises, but *not* in the conclusion. It is rendered as M in the following. Since the major premise contains both the major term P and the middle term M, it (the major premise) is of the form q_1MP or q_1PM.

The third term of the trio is called the *minor term.* It is rendered as S in the following. The conclusion contains the minor term S leftmost. It follows from what we have specified that the minor premise contains the middle term M and the minor term S. So it (the minor premise) is of the form q_2SM or q_2MS.

It also follows from the foregoing stipulations that the conclusion of an Aristotelian syllogistic argument must have the form q_3SP. Thus arguments can vary in their patterns of term-occurrences only in respect of their major and minor premises, and the way the latter combine the major term P, the middle term M and the minor term S. The major premise contains P and M (in either order). The minor premise contains M and S (in either order). So, bearing in mind that the major premise is always listed first, the only possible combinations of terms *within the two premises* are the following four, called *figures*:

	First Figure	Second Figure	Third Figure	Fourth Figure
Major premise	q_1MP	q_1PM	q_1MP	q_1PM
Minor premise	q_2SM	q_2SM	q_2MS	q_2MS

In all four figures the conclusion will be of the form q_3SP.

The *mood* of an argument is defined as its ordered triple

$$\langle q_1, q_2, q_3\rangle.$$

Clearly, there are $4^3 = 64$ possible moods within any one figure. Matters can be simplified, though. It is obvious that both e and i are 'symmetric' quantifier expressions. That is, eFG is equivalent to eGF, and iFG is equivalent to iGF. Thus there are really only 16 moods within any one figure that deserve serious independent attention.

It is the combination of figure and mood, for two-premise arguments involving sentences of the above forms, under the constraints of major-, minor-, and middle-term occurrences within premises and conclusion, that determine the *logical form* of the argument in question. An argument of mood $\langle q_1, q_2, q_3\rangle$ and of figure k will be said here to have the form $\langle q_1, q_2, q_3\rangle$-$k$. Aristotle gave each syllogistic form a name—see below. The systematic notation $\langle q_1, q_2, q_3\rangle$-$k$, however, is completely specific and is a helpful mnemonic—and is therefore preferable. In deference to tradition, however, we also give the medieval scholastics' names below.[1] Note how the vowels in the names codify the moods of the syllogisms.

Remember: a syllogism is a *valid* argument. Some of Aristotle's syllogisms, it turns out, need to be supplemented with certain existential premises in order to make them valid in the modern sense. These supplementations are supplied below, with asterisks, but without further comment.

The *mood* of an argument is defined as its ordered triple $\langle q_1, q_2, q_3\rangle$.

We now display the forms of the syllogisms of the four different figures.

[1]The mnemonic poem

Barbara, Celarent, Darii, Ferio que prioris;
Cesare, Camestres, Festino, Baroko secundae;
Tertia, Darapti, Disamis, Datisi, Felapton, Bokardo, Ferison, habet;
Quarta in super addit Bramantip, Camenes, Dimaris, Fesapo, Fresison

is of uncertain provenance, but is thought to be at least as old as William of Sherwood (1190–1249).

Syllogisms of the *First Figure* have the form
$$\frac{\begin{array}{l} q_1\mathrm{MP} \\ q_2\mathrm{SM} \end{array}}{q_3\mathrm{SP}}\ .$$

Syllogisms of the *Second Figure* have the form
$$\frac{\begin{array}{l} q_1\mathrm{PM} \\ q_2\mathrm{SM} \end{array}}{q_3\mathrm{SP}}\ .$$

Syllogisms of the *Third Figure* have the form
$$\frac{\begin{array}{l} q_1\mathrm{MP} \\ q_2\mathrm{MS} \end{array}}{q_3\mathrm{SP}}\ .$$

Syllogisms of the *Fourth Figure* have the form
$$\frac{\begin{array}{l} q_1\mathrm{PM} \\ q_2\mathrm{MS} \end{array}}{q_3\mathrm{SP}}\ .$$

Historical note

Aristotle himself, in his *Prior Analytics*, investigated only syllogisms in the first three Figures. His pupil Theophrastus added the Fourth Figure, which has also been attributed to Galen. It is strange that Aristotle omitted the Fourth Figure, given his usual systematic thoroughness. There has been debate over many centuries as to whether Aristotle ought to have recognized the Fourth Figure, or whether its arguments can really be assigned, 'indirectly', to the First. The Fourth Figure was recognized by Peter of Mantua in 1483, and was argued for by Peter Tartaret (ca. 1480), by Richard Crackenthorpe in 1622, and by Antoine Arnauld in 1662. For a scrupulously scholarly account of these matters, see Rescher [1966]. Another useful source is Henle [1949]. Here we follow the post-*Port Royal*, English tradition of giving the Fourth its due. Smiley [1994] §III offers an intriguing explanation for Aristotle's exclusion of the Fourth Figure, by suggesting 'a connection between it and the role of Platonic division in shaping syllogistic logic'.

20.3 Syllogisms of the *First Figure*

$\langle a, a, a\rangle$**-1**

(Barbara)

$$\begin{array}{l} \text{Every M is P} \\ \underline{\text{Every S is M}} \\ \text{Every S is P} \end{array} \qquad \begin{array}{l} \forall x(Mx\to Px) \\ \underline{\forall x(Sx\to Mx)} \\ \forall x(Sx\to Px) \end{array} \qquad \dfrac{\dfrac{\dfrac{\dfrac{\dfrac{}{Sa}{\scriptstyle(1)} \quad \dfrac{\forall x(Sx\to Mx)}{Sa\to Ma}}{Ma} \quad \dfrac{\forall x(Mx\to Px)}{Ma\to Pa}}{Pa}}{Sa\to Pa}{\scriptstyle(1)}}{\forall x(Sx\to Px)}$$

$\langle e, a, e\rangle$**-1 (Celarent)**

$$\begin{array}{l} \text{No M is P} \\ \underline{\text{Every S is M}} \\ \text{No S is P} \end{array} \qquad \begin{array}{l} \neg\exists x(Mx\wedge Px) \\ \underline{\forall x(Sx\to Mx)} \\ \neg\exists x(Sx\wedge Px) \end{array} \qquad \dfrac{\dfrac{\dfrac{}{\exists x(Sx\wedge Px)}{\scriptstyle(1)} \quad \dfrac{\dfrac{\dfrac{\dfrac{\dfrac{\dfrac{}{Sa\wedge Pa}{\scriptstyle(2)}}{Sa} \quad \dfrac{\forall x(Sx\to Mx)}{Sa\to Ma}}{Ma} \quad \dfrac{\dfrac{}{Sa\wedge Pa}{\scriptstyle(2)}}{Pa}}{Ma\wedge Pa}}{\exists x(Mx\wedge Px)} \quad \neg\exists x(Mx\wedge Px)}{\bot}}{\bot}{\scriptstyle(2)}}{\neg\exists x(Sx\wedge Px)}{\scriptstyle(1)}$$

$\langle a, i, i\rangle$**-1 (Darii)**

$$\begin{array}{l} \text{Every M is P} \\ \underline{\text{Some S is M}} \\ \text{Some S is P} \end{array} \qquad \begin{array}{l} \forall x(Mx\to Px) \\ \underline{\exists x(Sx\wedge Mx)} \\ \exists x(Sx\wedge Px) \end{array} \qquad \dfrac{\exists x(Sx\wedge Mx) \quad \dfrac{\dfrac{\dfrac{\dfrac{}{Sa\wedge Ma}{\scriptstyle(1)}}{Sa} \quad \dfrac{\dfrac{\dfrac{}{Sa\wedge Ma}{\scriptstyle(1)}}{Ma} \quad \dfrac{\forall x(Mx\to Px)}{Ma\to Pa}}{Pa}}{Sa\wedge Pa}}{\exists x(Sx\wedge Px)}}{\exists x(Sx\wedge Px)}{\scriptstyle(1)}$$

$\langle e, i, o \rangle$-1 (Ferio)

No M is P	$\neg\exists x(Mx \wedge Px)$
Some S is M	$\exists x(Sx \wedge Mx)$
Some S is not P	$\exists x(Sx \wedge \neg Px)$

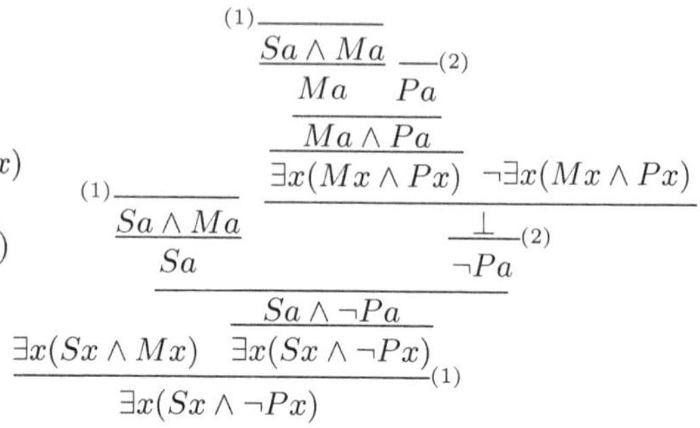

$\langle a, a, i \rangle$-1 (Barbari)

*There is some S	$\exists x Sx$
Every M is P	$\forall x(Mx \rightarrow Px)$
Every S is M	$\forall x(Sx \rightarrow Mx)$
Some S is P	$\exists x(Sx \wedge Px)$

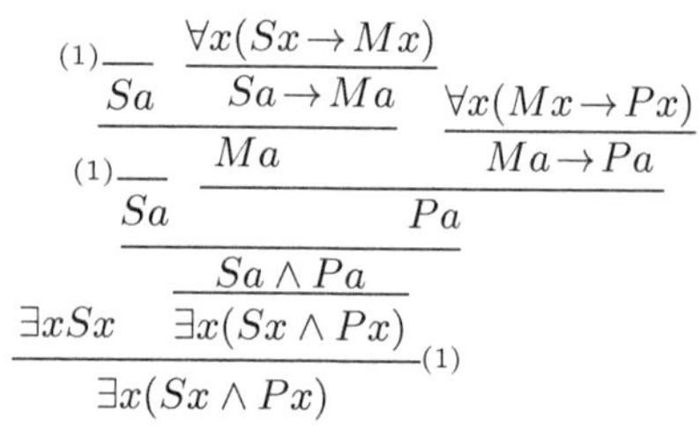

$\langle e, a, o \rangle$-1 (Celaront)

*There is some S	$\exists x Sx$
No M is P	$\neg\exists x(Mx \wedge Px)$
Every S is M	$\forall x(Sx \rightarrow Mx)$
Some S is not P	$\exists x(Sx \wedge \neg Px)$

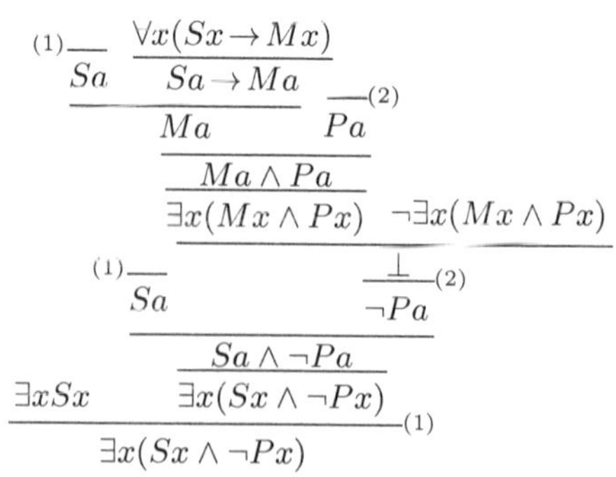

20.4 Syllogisms of the *Second Figure*

$\langle e, a, e \rangle$-2 (Cesare)

No P is M	$\neg\exists x(Px \wedge Mx)$
Every S is M	$\forall x(Sx \rightarrow Mx)$
No S is P	$\neg\exists x(Sx \wedge Px)$

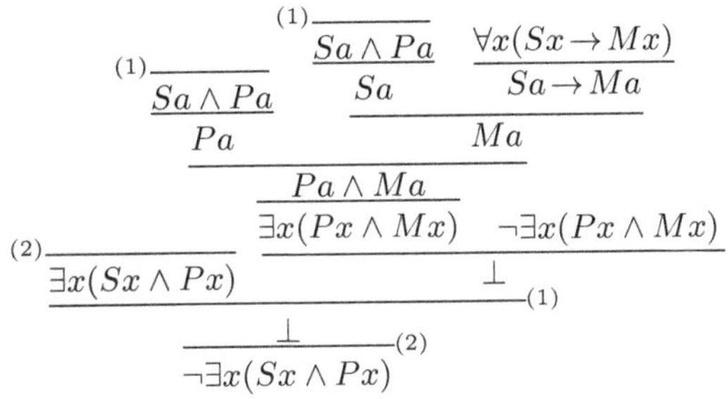

$\langle a, e, e \rangle$-2 (Camestres)

Every P is M	$\forall x(Px \rightarrow Mx)$
No S is M	$\neg\exists x(Sx \wedge Mx)$
No S is P	$\neg\exists x(Sx \wedge Px)$

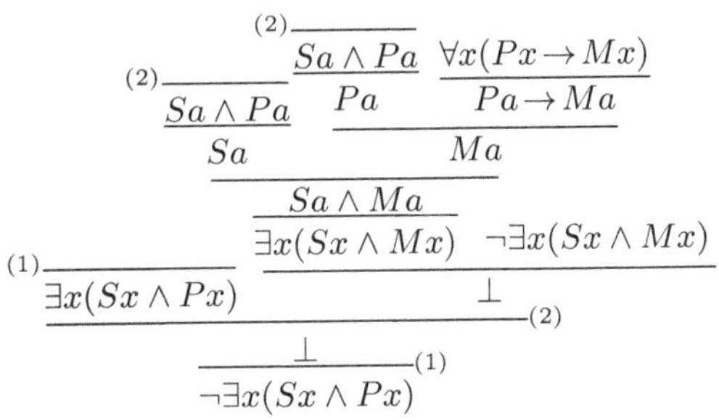

$\langle e, i, o \rangle$-2 (Festino)

No P is M	$\neg\exists x(Px \wedge Mx)$
Some S is M	$\exists x(Sx \wedge Mx)$
Some S is not P	$\exists x(Sx \wedge \neg Px)$

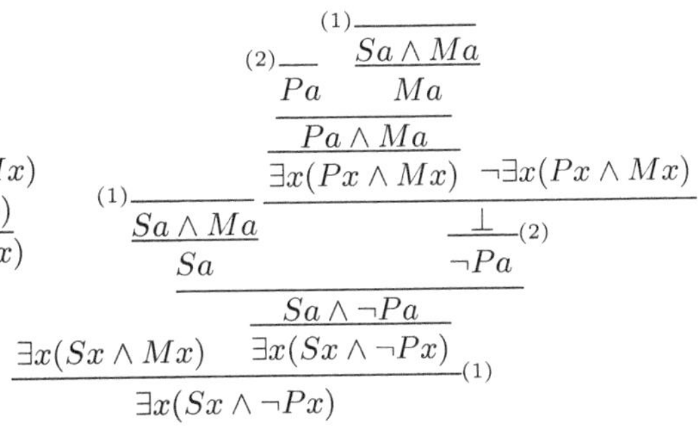

$\langle a, o, o\rangle$**-2 (Baroco)**

Every P is M	$\forall x(Px \to Mx)$
Some S is not M	$\exists x(Sx \wedge \neg Mx)$
Some S is not P	$\exists x(Sx \wedge \neg Px)$

$$\dfrac{\exists x(Sx \wedge \neg Mx) \qquad \dfrac{\dfrac{\dfrac{{}^{(1)}\dfrac{}{Sa \wedge \neg Ma}}{Sa} \qquad \dfrac{\dfrac{\dfrac{{}^{(2)}\dfrac{}{Pa} \qquad \dfrac{\forall x(Px \to Mx)}{Pa \to Ma}}{Ma} \qquad \dfrac{\dfrac{}{Sa \wedge \neg Ma}{}^{(1)}}{\neg Ma}}{\bot}}{\neg Pa}{}^{(2)}}{Sa \wedge \neg Pa}}{\exists x(Sx \wedge \neg Px)}}{\exists x(Sx \wedge \neg Px)}{}^{(1)}$$

$\langle e, a, o\rangle$**-2 (Cesaro)**

*There is some S	$\exists x Sx$
No P is M	$\neg\exists x(Px \wedge Mx)$
Every S is M	$\forall x(Sx \to Mx)$
Some S is not P	$\exists x(Sx \wedge \neg Px)$

$$\dfrac{\exists x Sx \qquad \dfrac{\dfrac{{}^{(1)}\dfrac{}{Sa} \qquad \dfrac{\dfrac{\dfrac{\dfrac{{}^{(2)}\dfrac{}{Pa} \qquad \dfrac{{}^{(1)}\dfrac{}{Sa} \qquad \dfrac{\forall x(Sx \to Mx)}{Sa \to Ma}}{Ma}}{Pa \wedge Ma}}{\exists x(Px \wedge Mx)} \qquad \neg\exists x(Px \wedge Mx)}{\bot}}{\neg Pa}{}^{(2)}}{Sa \wedge \neg Pa}}{\exists x(Sx \wedge \neg Px)}}{\exists x(Sx \wedge \neg Px)}{}^{(1)}$$

$\langle a, e, o\rangle$**-2 (Camestros)**

*There is some S	$\exists x Sx$
Every P is M	$\forall x(Px \to Mx)$
No S is M	$\neg\exists x(Sx \wedge Mx)$
Some S is not P	$\exists x(Sx \wedge \neg Px)$

$$\dfrac{\exists x Sx \qquad \dfrac{\dfrac{{}^{(1)}\dfrac{}{Sa} \qquad \dfrac{\dfrac{\dfrac{\dfrac{{}^{(1)}\dfrac{}{Sa} \qquad \dfrac{{}^{(2)}\dfrac{}{Pa} \qquad \dfrac{\forall x(Px \to Mx)}{Pa \to Ma}}{Ma}}{Sa \wedge Ma}}{\exists x(Sx \wedge Mx)} \qquad \neg\exists x(Sx \wedge Mx)}{\bot}}{\neg Pa}{}^{(2)}}{Sa \wedge \neg Pa}}{\exists x(Sx \wedge \neg Px)}}{\exists x(Sx \wedge \neg Px)}{}^{(1)}$$

20.5 Syllogisms of the *Third Figure*

$\langle a, a, i\rangle$-3 (Darapti)

*There is some M	$\exists x Mx$
Every M is P	$\forall x(Mx \to Px)$
Every M is S	$\forall x(Mx \to Sx)$
Some S is P	$\exists x(Sx \wedge Px)$

$$\dfrac{\exists x Mx \qquad \dfrac{\dfrac{\dfrac{\dfrac{}{Ma}{\scriptstyle(1)} \quad \dfrac{\forall x(Mx \to Sx)}{Ma \to Sa}}{Sa} \qquad \dfrac{\dfrac{}{Ma}{\scriptstyle(1)} \quad \dfrac{\forall x(Mx \to Px)}{Ma \to Pa}}{Pa}}{Sa \wedge Pa}}{\exists x(Sx \wedge Px)}}{\exists x(Sx \wedge Px)}{\scriptstyle(1)}$$

$\langle e, a, o\rangle$-3 (Felapton)

*There is some M	$\exists x Mx$
No M is P	$\neg\exists x(Mx \wedge Px)$
Every M is S	$\forall x(Mx \to Sx)$
Some S is not P	$\exists x(Sx \wedge \neg Px)$

$$\dfrac{\exists x Mx \qquad \dfrac{\dfrac{\dfrac{\dfrac{}{Ma}{\scriptstyle(1)} \quad \dfrac{\forall x(Mx \to Sx)}{Ma \to Sa}}{Sa} \qquad \dfrac{\dfrac{\dfrac{\dfrac{\dfrac{}{Ma}{\scriptstyle(1)} \quad \dfrac{}{Pa}{\scriptstyle(2)}}{Ma \wedge Pa}}{\exists x(Mx \wedge Px)} \quad \neg\exists x(Mx \wedge Px)}{\bot}}{\neg Pa}{\scriptstyle(2)}}{Sa \wedge \neg Pa}}{\exists x(Sx \wedge \neg Px)}}{\exists x(Sx \wedge \neg Px)}{\scriptstyle(1)}$$

$\langle i, a, i\rangle$-3 (Disamis)

Some M is P	$\exists x(Mx \wedge Px)$
Every M is S	$\forall x(Mx \to Sx)$
Some S is P	$\exists x(Sx \wedge Px)$

$$\dfrac{\exists x(Mx \wedge Px) \qquad \dfrac{\dfrac{\dfrac{\dfrac{\dfrac{}{Ma \wedge Pa}{\scriptstyle(1)}}{Ma} \quad \dfrac{\forall x(Mx \to Sx)}{Ma \to Sa}}{Sa} \qquad \dfrac{\dfrac{}{Ma \wedge Pa}{\scriptstyle(1)}}{Pa}}{Sa \wedge Pa}}{\exists x(Sx \wedge Px)}}{\exists x(Sx \wedge Px)}{\scriptstyle(1)}$$

$\langle a, i, i\rangle$-3 (Datisi)

Every M is P	$\forall x(Mx \to Px)$
Some M is S	$\exists x(Mx \wedge Sx)$
Some S is P	$\exists x(Sx \wedge Px)$

$$\dfrac{\exists x(Mx \wedge Sx) \qquad \dfrac{\dfrac{\dfrac{\dfrac{}{Ma \wedge Sa}{\scriptstyle(1)}}{Sa} \qquad \dfrac{\dfrac{\dfrac{}{Ma \wedge Sa}{\scriptstyle(1)}}{Ma} \quad \dfrac{\forall x(Mx \to Px)}{Ma \to Pa}}{Pa}}{Sa \wedge Pa}}{\exists x(Sx \wedge Px)}}{\exists x(Sx \wedge Px)}{\scriptstyle(1)}$$

$\langle o, a, o\rangle$-3 (Bocardo)

Some M is not P	$\exists x(Mx \wedge \neg Px)$
Every M is S	$\forall x(Mx \to Sx)$
Some S is not P	$\exists x(Sx \wedge \neg Px)$

$$\dfrac{\exists x(Mx \wedge \neg Px) \qquad \dfrac{\dfrac{\dfrac{\dfrac{\dfrac{}{Ma \wedge \neg Pa}{}^{(1)}}{Ma} \quad \dfrac{\forall x(Mx \to Sx)}{Ma \to Sa}}{Sa} \qquad \dfrac{\dfrac{}{Ma \wedge \neg Pa}{}^{(1)}}{\neg Pa}}{Sa \wedge \neg Pa}}{\exists x(Sx \wedge \neg Px)}}{\exists x(Sx \wedge \neg Px)}{}^{(1)}$$

$\langle e, i, o\rangle$-3 (Ferison)

No M is P	$\neg\exists x(Mx \wedge Px)$
Some M is S	$\exists x(Mx \wedge Sx)$
Some S is not P	$\exists x(Sx \wedge \neg Px)$

$$\dfrac{\exists x(Mx \wedge Sx) \qquad \dfrac{\dfrac{\dfrac{\dfrac{}{Ma \wedge Sa}{}^{(1)}}{Sa} \qquad \dfrac{\dfrac{\dfrac{\dfrac{\dfrac{\dfrac{}{Ma \wedge Sa}{}^{(1)}}{Ma} \quad \dfrac{}{Pa}{}^{(2)}}{Ma \wedge Pa}}{\exists x(Mx \wedge Px)} \quad \neg\exists x(Mx \wedge Px)}{\bot}}{\neg Pa}{}^{(2)}}{Sa \wedge \neg Pa}}{\exists x(Sx \wedge \neg Px)}}{\exists x(Sx \wedge \neg Px)}{}^{(1)}$$

20.6 Syllogisms of the *Fourth Figure*

$\langle a, a, i\rangle$-4 (Bramantip)

*There is some P	$\exists x Px$
Every P is M	$\forall x(Px \to Mx)$
Every M is S	$\forall x(Mx \to Sx)$
Some S is P	$\exists x(Sx \wedge Px)$

$$\dfrac{\exists x Px \qquad \dfrac{\dfrac{\dfrac{\dfrac{\dfrac{}{Pa}{}^{(1)} \quad \dfrac{\forall x(Px \to Mx)}{Pa \to Ma}}{Ma} \quad \dfrac{\forall x(Mx \to Sx)}{Ma \to Sa}}{Sa} \qquad \dfrac{}{Pa}{}^{(1)}}{Sa \wedge Pa}}{\exists x(Sx \wedge Px)}}{\exists x(Sx \wedge Px)}{}^{(1)}$$

$\langle a, e, e\rangle$-4 (Camenes)

Every P is M	$\forall x(Px\to Mx)$
No M is S	$\neg\exists x(Mx\land Sx)$
No S is P	$\neg\exists x(Sx\land Px)$

$$\dfrac{\dfrac{[\exists x(Sx\land Px)]^{(1)} \quad \dfrac{\dfrac{\dfrac{\dfrac{\dfrac{[Sa\land Pa]^{(2)}}{Pa} \quad \dfrac{\forall x(Px\to Mx)}{Pa\to Ma}}{Ma} \quad \dfrac{[Sa\land Pa]^{(2)}}{Sa}}{Ma\land Sa}}{\exists x(Mx\land Sx)} \quad \neg\exists x(Mx\land Sx)}{\bot}}{\bot}\,(2)}{\neg\exists x(Sx\land Px)}\,(1)$$

$\langle i, a, i\rangle$-4 (Dimaris)

Some P is M	$\exists x(Px\land Mx)$
Every M is S	$\forall x(Mx\to Sx)$
Some S is P	$\exists x(Sx\land Px)$

$$\dfrac{\exists x(Px\land Mx) \quad \dfrac{\dfrac{\dfrac{\dfrac{[Pa\land Ma]^{(1)}}{Ma} \quad \dfrac{\forall x(Mx\to Sx)}{Ma\to Sa}}{Sa} \quad \dfrac{[Pa\land Ma]^{(1)}}{Pa}}{Sa\land Pa}}{\exists x(Sx\land Px)}}{\exists x(Sx\land Px)}\,(1)$$

$\langle a, e, o\rangle$-4 (Calemos)

*There is some S	$\exists xSx$
Every P is M	$\forall x(Px\to Mx)$
No M is S	$\neg\exists x(Mx\land Sx)$
Some S is not P	$\exists x(Sx\land\neg Px)$

$$\dfrac{\exists xSx \quad \dfrac{\dfrac{[Sa]^{(1)} \quad \dfrac{\dfrac{\dfrac{\dfrac{\dfrac{[Pa]^{(2)} \quad \dfrac{\forall x(Px\to Mx)}{Pa\to Ma}}{Ma} \quad [Sa]^{(1)}}{Ma\land Sa}}{\exists x(Mx\land Sx)} \quad \neg\exists x(Mx\land Sx)}{\bot}}{\neg Pa}\,(2)}{Sa\land\neg Pa}}{\exists x(Sx\land\neg Px)}}{\exists x(Sx\land\neg Px)}\,(1)$$

$\langle e, a, o\rangle$-4 (Fesapo)

*There is some M	$\exists x Mx$
No P is M	$\neg\exists x(Px \wedge Mx)$
Every M is S	$\forall x(Mx \to Sx)$
Some S is not P	$\exists x(Sx \wedge \neg Px)$

$$\dfrac{\exists x Mx \quad \dfrac{\dfrac{\dfrac{\overline{Ma}^{\,(1)} \quad \dfrac{\forall x(Mx\to Sx)}{Ma\to Sa}}{Sa} \quad \dfrac{\dfrac{\dfrac{\dfrac{\overline{Pa}^{\,(2)} \quad \overline{Ma}^{\,(1)}}{Pa\wedge Ma}}{\exists x(Px\wedge Mx)} \quad \neg\exists x(Px\wedge Mx)}{\bot}}{\neg Pa}{\scriptstyle(2)}}{Sa\wedge\neg Pa}}{\exists x(Sx\wedge\neg Px)}}{\exists x(Sx\wedge\neg Px)}{\scriptstyle(1)}$$

$\langle e, i, o\rangle$-4 (Fresison)

No P is M	$\neg\exists x(Px\wedge Mx)$
Some M is S	$\exists x(Mx\wedge Sx)$
Some S is not P	$\exists x(Sx\wedge\neg Px)$

$$\dfrac{\exists x(Mx\wedge Sx) \quad \dfrac{\dfrac{\dfrac{\overline{Ma\wedge Sa}^{\,(1)}}{Sa} \quad \dfrac{\dfrac{\dfrac{\dfrac{\overline{Pa}^{\,(2)} \quad \dfrac{\overline{Ma\wedge Sa}^{\,(1)}}{Ma}}{Pa\wedge Ma}}{\exists x(Px\wedge Mx)} \quad \neg\exists x(Px\wedge Mx)}{\bot}}{\neg Pa}{\scriptstyle(2)}}{Sa\wedge\neg Pa}}{\exists x(Sx\wedge\neg Px)}}{\exists x(Sx\wedge\neg Px)}{\scriptstyle(1)}$$

Note that we have made liberal use, in the foregoing natural deductions, of the serial forms of elimination for $\wedge$, $\to$, and $\forall$. These are *Gentzen–Prawitz* natural deductions.

Exercise 60. Turn each of the natural deductions provided above for the syllogisms into a natural deduction that uses only, for its eliminations, the parallelized forms of the elimination rules.

Exercise 61. Turn each of the natural deductions found in doing Exercise 60 (using only the parallelized forms of eliminations rules) into a proof in the sequent calculus of the same result.

Chapter 21

Proofs Galore

21.1 Preliminaries

A 'deductive problem', as the reader by now well knows, is a valid argument for which one is required to find a proof. Each *step* in a proof must be an application of a *primitive* rule of inference. The *length* of a proof (sometimes called its *complexity*) is measured by the number of steps it contains. The *level of difficulty* of a deductive problem is measured by the length of its shortest proof.

We furnish here some deductive problems of increasing levels of difficulty. The beginner is encouraged to work through these problems in their strictly increasing order of levels of difficulty. A *caveat* is needed: So-called '*levels* of difficulty' do not always line up directly with what one might call the '*degree* of difficulty' that a beginner (or even an expert) might regard a deductive problem as possessing. A problem of lower *level* of difficulty than another might be of higher *degree* of difficulty (in finding a proof for it). For example, the vast majority of students find that the blue problem of Level 7 below is easier to solve than any of the blue ones of Level 6. These four blue problems are known as the De Morgan dualities for the connectives; the similar quantifier dualities for $\forall$ and $\exists$ (see Levels 4 and 6) are in red.

We do not have any precisely definable measure of *degree* of difficulty of deductive problems.[1] The best we can do is classify deductive problems by their precisely definable *levels* of difficulty.

[1]The reader might wish to consult Kossak and Ording [2017].

Certain problems are *strictly classical*, in that their proofs require at least one application of one of the four strictly classical rules (Law of Excluded Middle; Dilemma; Classical Reductio; Double Negation Elimination). Exactly *which* of these rules yield the shortest possible classical proof is highly context-dependent. Some problems admit of *constructive* proofs (i.e., proofs that are not strictly classical) but also have strictly classical proofs that are shorter than their constructive ones. Thus, for example, there is a 2-step classical proof of

$$\frac{}{\neg\neg(A \vee \neg A)}\ ,$$

whose constructive proof takes 6 steps; and there is a 1-step classical proof of

$$\frac{\neg\neg\neg A}{\neg A}$$

whose constructive proof takes 4 steps.

Exercise 62. *Find the four proofs in question.*

The proofs called for in Exercise 62 will not appear in the 'levels of difficulty' list that is to follow.

One more *caveat* will be in order. Our measure of level of difficulty of a deductive problem is the *minimum number of steps in a proof that solves it.* But this number can be different for two 'essentially similar' solutions, one of which uses serial forms of elimination, and the other of which uses parallelized forms. Proofs using parallelized eliminations, however, tend to spread more widely across the page—even though they employ fewer steps than their 'serial' versions.

21.2 Problems in increasing levels of difficulty

21.2.1 Level of difficulty 1

All the following problems *are their own solutions.* Viewed as steps of inference, they are all instances of primitive rules of inference . They are their own 'one-step' proofs.

$$\frac{\neg A \quad A}{\bot} \qquad \frac{A}{A \vee B} \qquad \frac{B}{A \vee B} \qquad \frac{A \quad B}{A \wedge B} \qquad \frac{A \wedge B}{A}$$

$$\frac{A \wedge B}{B} \qquad \frac{B}{A \to B} \qquad \frac{F(t)}{\exists x F(x)} \qquad \frac{\forall x F(x)}{F(t)} \qquad \frac{\exists ! t}{t = t}$$

$$\frac{F(t)}{t = t}\ \ F \text{ atomic} \qquad \frac{}{A \vee \neg A} \qquad \frac{\neg\neg A}{A} \qquad \frac{A \to B \quad A}{B} \qquad \frac{t = u \quad u = v}{t = v}$$

Henceforth: problem on the left; solution on the right.

21.2.2 Level of difficulty 2

$$\frac{A}{\neg\neg A} \qquad \cfrac{\cfrac{{}^{(1)}\overline{\neg A} \quad A}{\bot}}{\neg\neg A}{\scriptstyle (1)}$$

$$\frac{A \vee B \quad \neg A}{B} \qquad \cfrac{A \vee B \quad \cfrac{{}^{(1)}\overline{A} \quad \neg A}{\bot} \quad \overline{B}^{(1)}}{B}{\scriptstyle (1)}$$

$$\frac{A \wedge B}{B \wedge A} \qquad \cfrac{A \wedge B \quad \cfrac{{}^{(1)}\overline{B} \quad \overline{A}^{(1)}}{B \wedge A}}{B \wedge A}{\scriptstyle (1)}$$

$$\frac{A \wedge B}{B \vee C} \qquad \cfrac{\cfrac{A \wedge B}{B}}{B \vee C}$$

$$\frac{\begin{matrix} A \\ \neg B \\ A \to B \end{matrix}}{\bot} \qquad \cfrac{\neg B \quad \cfrac{A \to B \quad A}{B}}{\bot}$$

$$\frac{A \wedge (B \wedge C)}{B} \qquad \cfrac{\cfrac{A \wedge (B \wedge C)}{B \wedge C}}{B}$$

$$\begin{array}{l} A \\ A\to B \\ \underline{B\to C} \\ C \end{array} \qquad \dfrac{B\to C \qquad \dfrac{A\to B \quad A}{B}}{C}$$

$$\dfrac{\forall x(Fx \wedge Gx)}{Ft} \qquad \dfrac{\dfrac{\forall x(Fx \wedge Gx)}{Ft \wedge Gt}}{Ft}$$

$$\dfrac{Ft}{\exists x(Fx \vee Gx)} \qquad \dfrac{\dfrac{Ft}{Ft \vee Gt}}{\exists x(Fx \vee Gx)}$$

$$\dfrac{A}{C\to(A\vee B)} \qquad \dfrac{\dfrac{A}{A\vee B}}{C\to(A\vee B)}$$

$$\dfrac{\begin{array}{l}A\\ A\to(B\wedge C)\end{array}}{C} \qquad\qquad \dfrac{\dfrac{A\to(B\wedge C)\quad A}{B\wedge C}}{C}$$

$$\dfrac{\neg A}{A\to B} \qquad\qquad \dfrac{\dfrac{\neg A \qquad \dfrac{}{A}{\scriptstyle(1)}}{\bot}}{A\to B}{\scriptstyle(1)}$$

$$\dfrac{}{\neg\neg A\to A} \qquad\qquad \dfrac{\dfrac{\dfrac{}{\neg\neg A}{\scriptstyle(1)}}{A}\,\text{DNE}}{\neg\neg A\to A}{\scriptstyle(1)}$$

$$\dfrac{\begin{array}{l}A\\ \neg(A\vee B)\end{array}}{\bot} \qquad\qquad \dfrac{\neg(A\vee B)\quad \dfrac{A}{A\vee B}}{\bot}$$

$$\dfrac{\begin{array}{l}B\\ \neg(A\to B)\end{array}}{\bot} \qquad\qquad \dfrac{\neg(A\to B)\quad \dfrac{B}{A\to B}}{\bot}$$

$$\dfrac{}{B\to(A\to B)} \qquad\qquad \dfrac{\dfrac{\dfrac{}{B}{\scriptstyle(1)}}{A\to B}}{B\to(A\to B)}{\scriptstyle(1)}$$

$$\dfrac{A}{(A\vee B)\vee C} \qquad\qquad \dfrac{\dfrac{A}{A\vee B}}{(A\vee B)\vee C}$$

$$\dfrac{A}{(C\to A)\vee B} \qquad\qquad \dfrac{\dfrac{A}{C\to A}}{(C\to A)\vee B}$$

$$\dfrac{B}{A\vee(C\to B)} \qquad\qquad \dfrac{\dfrac{B}{C\to B}}{A\vee(C\to B)}$$

$$\dfrac{A\wedge B}{C\to A} \qquad\qquad \dfrac{\dfrac{A\wedge B}{A}}{C\to A}$$

$$\dfrac{\forall x\exists yLxy}{\exists x\exists yLxy} \qquad \dfrac{\dfrac{\forall x\exists yLxy}{\exists yLay}}{\exists x\exists yLxy}$$

$$\dfrac{\forall x\forall yLxy}{\exists x\forall yLxy} \qquad \dfrac{\dfrac{\forall x\forall yLxy}{\forall yLay}}{\exists x\forall yLxy}$$

21.2.3 Level of difficulty 3

$$\dfrac{A\vee B}{B\vee A} \qquad \dfrac{A\vee B \quad \dfrac{\dfrac{}{A}{\scriptstyle(1)}}{B\vee A} \quad \dfrac{\dfrac{}{B}{\scriptstyle(1)}}{B\vee A}}{B\vee A}{\scriptstyle(1)}$$

$$\begin{array}{c} A\wedge B \\ \underline{B\wedge C} \\ A\wedge C \end{array} \qquad \dfrac{\dfrac{A\wedge B}{A} \quad \dfrac{B\wedge C}{C}}{A\wedge C}$$

$$\begin{array}{c} A\to B \\ \underline{B\to C} \\ A\to C \end{array} \qquad \dfrac{\dfrac{B\to C \quad \dfrac{A\to B \quad \dfrac{}{A}{\scriptstyle(1)}}{B}}{C}}{A\to C}{\scriptstyle(1)}$$

$$\dfrac{A\to(B\wedge C)}{A\to C} \qquad \dfrac{\dfrac{\dfrac{A\to(B\wedge C) \quad \dfrac{}{A}{\scriptstyle(1)}}{B\wedge C}}{C}}{A\to C}{\scriptstyle(1)}$$

$$\dfrac{}{(A\wedge B)\to(C\vee B)} \qquad \dfrac{\dfrac{\dfrac{\dfrac{}{A\wedge B}{\scriptstyle(1)}}{B}}{C\vee B}}{(A\wedge B)\to(C\vee B)}{\scriptstyle(1)}$$

$$\dfrac{}{A\to\neg\neg A} \qquad \dfrac{\dfrac{\dfrac{{\scriptstyle(2)}\dfrac{}{\neg A} \quad \dfrac{}{A}{\scriptstyle(1)}}{\bot}}{\neg\neg A}{\scriptstyle(2)}}{A\to\neg\neg A}{\scriptstyle(1)}$$

$\dfrac{A\to\neg A}{\neg A}$	$\dfrac{\dfrac{\dfrac{A\to\neg A \qquad \overline{A}^{(1)}}{\neg A} \qquad \overline{A}^{(1)}}{\bot}}{\neg A}{(1)}$
$\dfrac{\neg A}{\neg(A\wedge B)}$	$\dfrac{\dfrac{\neg A \qquad \dfrac{\overline{A\wedge B}^{(1)}}{A}}{\bot}}{\neg(A\wedge B)}{(1)}$
$\dfrac{\neg(A\to B)}{\neg B}$	$\dfrac{\dfrac{\neg(A\to B) \qquad \dfrac{\overline{B}^{(1)}}{A\to B}}{\bot}}{\neg B}{(1)}$
$\dfrac{\neg(A\vee B)}{\neg A}$	$\dfrac{\dfrac{\neg(A\vee B) \qquad \dfrac{\overline{A}^{(1)}}{A\vee B}}{\bot}}{\neg A}{(1)}$
$\dfrac{\forall x\forall y Lxy}{\forall z Lzz}$	$\dfrac{\dfrac{\dfrac{\forall x\forall y Lxy}{\forall y Lay}}{Laa}}{\forall z Lzz}$
$\dfrac{\exists z Lzz}{\exists x\exists y Lxy}$	$\dfrac{\exists z Lzz \qquad \dfrac{\dfrac{\overline{Laa}^{(1)}}{\exists y Lay}}{\exists x\exists y Lxy}}{\exists x\exists y Lxy}{(1)}$
$\neg B$ $\dfrac{A\to B}{\neg A}$	$\dfrac{\dfrac{\neg B \qquad \dfrac{A\to B \qquad \overline{A}^{(1)}}{B}}{\bot}}{\neg A}{(1)}$
B $\dfrac{\neg A\to\neg B}{A}$	$\dfrac{\dfrac{\dfrac{\neg A\to\neg B \qquad \overline{\neg A}^{(1)}}{\neg B} \qquad B}{\bot}}{A}{(1)\ \text{CR}}$

$$\frac{\neg A \lor B}{A \to B}$$

$$\dfrac{\dfrac{\neg A \lor B \qquad \dfrac{\overline{\neg A}^{\,(2)} \quad \overline{A}^{\,(1)}}{\bot} \qquad \overline{B}^{\,(2)}}{B}_{(2)}}{A \to B}_{(1)}$$

$$\frac{\exists x \forall y Lyx}{\exists x Lxx}$$

$$\dfrac{\exists x \forall y Lyx \qquad \dfrac{\dfrac{\overline{\forall y Lya}^{\,(1)}}{Laa}}{\exists x Lxx}}{\exists x Lxx}_{(1)}$$

$$\frac{\forall x Lxx}{\exists x \exists y Lxy}$$

$$\dfrac{\dfrac{\dfrac{\forall x Lxx}{Laa}}{\exists y Lay}}{\exists x \exists y Lxy}$$

$$\frac{\forall x Lxx}{\forall x \exists y Lxy}$$

$$\dfrac{\dfrac{\dfrac{\forall x Lxx}{Laa}}{\exists y Lay}}{\forall x \exists y Lxy}$$

$$\frac{A \lor B}{\neg A \to B}$$

$$\dfrac{\dfrac{A \lor B \qquad \dfrac{\overline{\neg A}^{\,(1)} \quad \overline{A}^{\,(2)}}{\bot} \qquad \overline{B}^{\,(2)}}{B}_{(2)}}{\neg A \to B}_{(1)}$$

$$\frac{\forall x (Fx \land Gx)}{\forall z Gz}$$

$$\dfrac{\dfrac{\dfrac{\forall x (Fx \land Gx)}{Fa \land Ga}}{Ga}}{\forall z Gz}$$

$$\frac{\exists z Gz}{\exists x (Fx \lor Gx)}$$

$$\dfrac{\exists z Gz \qquad \dfrac{\dfrac{\overline{Ga}^{\,(1)}}{Fa \lor Ga}}{\exists x (Fx \lor Gx)}}{\exists x (Fx \lor Gx)}_{(1)}$$

21.2.4 Level of difficulty 4

$$\frac{A\vee\neg B}{B\to A}$$

$$\dfrac{A\vee\neg B \qquad \dfrac{\overline{A}^{(1)}}{B\to A} \qquad \dfrac{\dfrac{{}^{(1)}\overline{\neg B} \quad \overline{B}^{(2)}}{\bot}}{B\to A}{\scriptstyle(2)}}{B\to A}{\scriptstyle(1)}$$

$$\frac{\neg(A\to B)}{A}$$

$$\dfrac{\dfrac{\neg(A\to B) \qquad \dfrac{\dfrac{{}^{(1)}\overline{\neg A} \quad \overline{A}^{(2)}}{\bot}}{A\to B}{\scriptstyle(2)}}{\bot}}{A}{\scriptstyle(1)\ \mathrm{CR}}$$

$$\frac{\exists x\forall yLxy}{\forall z\exists wLwz}$$

$$\dfrac{\dfrac{\exists x\forall yLxy \qquad \dfrac{\dfrac{\overline{\forall yLay}^{(1)}}{Lab}}{\exists wLwb}}{\exists wLwb}{\scriptstyle(1)}}{\forall z\exists wLwz}$$

$$\frac{A\to B}{\neg B\to\neg A}$$

$$\dfrac{\dfrac{\dfrac{{}^{(1)}\overline{\neg B} \qquad \dfrac{A\to B \quad \overline{A}^{(2)}}{B}}{\bot}}{\neg A}{\scriptstyle(2)}}{\neg B\to\neg A}{\scriptstyle(1)}$$

$$\frac{\neg B\to\neg A}{A\to B}$$

$$\dfrac{\dfrac{\dfrac{\dfrac{\neg B\to\neg A \quad \overline{\neg B}^{(2)}}{\neg A} \qquad \overline{A}^{(1)}}{\bot}}{B}{\scriptstyle(2)}}{A\to B}{\scriptstyle(1)}$$

$$\frac{A\vee B}{\neg B\to A}$$

$$\dfrac{A\vee B \qquad \dfrac{\overline{A}^{(1)}}{\neg B\to A} \qquad \dfrac{\dfrac{{}^{(2)}\overline{\neg B} \quad \overline{B}^{(1)}}{\bot}}{\neg B\to A}{\scriptstyle(2)}}{\neg B\to A}{\scriptstyle(1)}$$

$$\begin{array}{c} A\to B \\ A\to C \\ \hline A\to(B\wedge C)\end{array} \qquad \dfrac{\dfrac{\dfrac{A\to B \quad \overline{A}^{(1)}}{B} \quad \dfrac{A\to C \quad \overline{A}^{(1)}}{C}}{B\wedge C}}{A\to(B\wedge C)}{\scriptstyle (1)}$$

$$\begin{array}{c} A\to C \\ B\to C \\ \hline (A\vee B)\to C\end{array} \qquad \dfrac{\dfrac{{}^{(1)}\overline{A\vee B} \quad \dfrac{A\to C \quad \overline{A}^{(2)}}{C} \quad \dfrac{B\to C \quad \overline{B}^{(2)}}{C}}{C}{\scriptstyle (2)}}{(A\vee B)\to C}{\scriptstyle (1)}$$

$$\dfrac{A\to B}{\neg A\vee B} \qquad \dfrac{\dfrac{\dfrac{A\to B \quad \overline{A}^{(1)}}{B}}{\neg A\vee B} \quad \dfrac{\overline{\neg A}^{(1)}}{\neg A\vee B}}{\neg A\vee B}{\scriptstyle (1)}$$

$$\dfrac{\exists x\forall y Lyx}{\exists x\exists y Lxy} \qquad \dfrac{\dfrac{\exists x\forall y Lyx \quad \dfrac{\dfrac{\overline{\forall y Lya}^{(1)}}{Lba}}{\exists y Lby}}{\exists y Lby}{\scriptstyle (1)}}{\exists x\exists y Lxy}$$

$$\dfrac{\forall x\exists y Lyx}{\exists x\exists y Lxy} \qquad \dfrac{\dfrac{\forall x\exists y Lyx}{\exists y Lya} \quad \dfrac{\dfrac{\overline{Lba}^{(1)}}{\exists y Lby}}{\exists x\exists y Lxy}}{\exists x\exists y Lxy}{\scriptstyle (1)}$$

$$\dfrac{\forall x\forall y Lyx}{\exists x\exists y Lxy} \qquad \dfrac{\dfrac{\dfrac{\dfrac{\forall x\forall y Lyx}{\forall y Lya}}{Lba}}{\exists y Lby}}{\exists x\exists y Lxy}$$

$$\dfrac{\forall x\forall y Lyx}{\exists x\forall y Lxy} \qquad \dfrac{\dfrac{\dfrac{\dfrac{\forall x\forall y Lyx}{\forall y Lya}}{Lba}}{\forall y Lby}}{\exists x\forall y Lxy}$$

$\dfrac{\forall x\neg Fx}{\neg\exists xFx}$

$$\dfrac{\dfrac{\overline{\exists xFx}^{\,(1)} \qquad \dfrac{\dfrac{\forall x\neg Fx}{\neg Fa} \quad \overline{Fa}^{\,(2)}}{\bot}}{\bot}\,{\scriptstyle(2)}}{\neg\exists xFx}\,{\scriptstyle(1)}$$

This problem and the next two are the *constructively* provable De Morgan dualities for the quantifiers.

$\dfrac{\neg\exists xFx}{\forall x\neg Fx}$

$$\dfrac{\dfrac{\dfrac{\neg\exists xFx \qquad \dfrac{\overline{Fa}^{\,(1)}}{\exists xFx}}{\bot}}{\neg Fa}\,{\scriptstyle(1)}}{\forall x\neg Fx}$$

$\dfrac{\exists x\neg Fx}{\neg\forall xFx}$

$$\dfrac{\dfrac{\exists x\neg Fx \qquad \dfrac{\overline{\neg Fa}^{\,(2)} \quad \dfrac{\overline{\forall xFx}^{\,(1)}}{Fa}}{\bot}}{\bot}\,{\scriptstyle(2)}}{\neg\forall xFx}\,{\scriptstyle(1)}$$

$\overline{(A\to\neg A)\to\neg A}$

$$\dfrac{\dfrac{\dfrac{\dfrac{\overline{A\to\neg A}^{\,(1)} \quad \overline{A}^{\,(2)}}{\neg A} \qquad \overline{A}^{\,(2)}}{\bot}}{\neg A}\,{\scriptstyle(2)}}{(A\to\neg A)\to\neg A}\,{\scriptstyle(1)}$$

$\dfrac{A\to B}{\neg(A\wedge\neg B)}$

$$\dfrac{\dfrac{\dfrac{\overline{A\wedge\neg B}^{\,(1)}}{\neg B} \qquad \dfrac{A\to B \quad \dfrac{\overline{A\wedge\neg B}^{\,(1)}}{A}}{B}}{\bot}}{\neg(A\wedge\neg B)}\,{\scriptstyle(1)}$$

$\dfrac{\neg(A\wedge\neg B)}{A\to B}$

$$\dfrac{\dfrac{\dfrac{\neg(A\wedge\neg B) \qquad \dfrac{\overline{A}^{\,(1)} \quad \overline{\neg B}^{\,(2)}}{A\wedge\neg B}}{\bot}}{B}\,{\scriptstyle(2)}}{A\to B}\,{\scriptstyle(1)}$$

$$\dfrac{A \wedge B}{\neg(A \to \neg B)}$$

$$\dfrac{\dfrac{A \to \neg B \qquad \dfrac{\overline{A \wedge B}^{\,(1)}}{A}}{\neg B} \qquad \dfrac{\overline{A \wedge B}^{\,(1)}}{B}}{\dfrac{\bot}{\neg(A \to \neg B)}\,(1)}$$

$$\dfrac{(A \to C) \wedge (B \to C)}{(A \wedge B) \to C}$$

$$\dfrac{\dfrac{\dfrac{(A \to C) \wedge (B \to C)}{A \to C} \qquad \dfrac{\overline{A \wedge B}^{\,(1)}}{A}}{C}}{(A \wedge B) \to C}\,(1)$$

$$\dfrac{(A \to B) \wedge (A \to C)}{A \to (B \vee C)}$$

$$\dfrac{\dfrac{\dfrac{\dfrac{(A \to B) \wedge (A \to C)}{A \to B} \qquad \overline{A}^{\,(1)}}{B}}{B \vee C}}{A \to (B \vee C)}\,(1)$$

$$\dfrac{A \to (B \to C)}{B \to (A \to C)}$$

$$\dfrac{\dfrac{\dfrac{\dfrac{A \to (B \to C) \qquad \overline{A}^{\,(2)}}{B \to C} \qquad \overline{B}^{\,(1)}}{C}\,(2)}{A \to C}}{B \to (A \to C)}\,(1)$$

$$\dfrac{(A \wedge B) \to C}{A \to (B \to C)}$$

$$\dfrac{\dfrac{\dfrac{(A \wedge B) \to C \qquad \dfrac{{}^{(1)}\overline{A} \qquad \overline{B}^{\,(2)}}{A \wedge B}}{C}\,(2)}{B \to C}}{A \to (B \to C)}\,(1)$$

$$\dfrac{\begin{array}{c}\forall x F x \\ \forall x G x\end{array}}{\forall x (Fx \wedge Gx)}$$

$$\dfrac{\dfrac{\dfrac{\forall x F x}{Fa} \qquad \dfrac{\forall x G x}{Ga}}{Fa \wedge Ga}}{\forall x (Fx \wedge Gx)}$$

$$\dfrac{A \to (B \to C)}{(A \wedge B) \to C}$$

$$\dfrac{\dfrac{\dfrac{A \to (B \to C) \qquad \dfrac{\overline{A \wedge B}^{\,(1)}}{A}}{B \to C} \qquad \dfrac{\overline{A \wedge B}^{\,(1)}}{B}}{C}}{(A \wedge B) \to C}\,(1)$$

$$\dfrac{(A\wedge B)\to C}{A\to(B\to C)}$$

$$\dfrac{\dfrac{\dfrac{(A\wedge B)\to C \qquad \dfrac{{\scriptstyle(1)}\dfrac{}{A}\quad \dfrac{}{B}{\scriptstyle(2)}}{A\wedge B}}{C}{\scriptstyle(2)}}{B\to C}{\scriptstyle(1)}}{A\to(B\to C)}$$

$$\dfrac{}{(\neg A\to A)\to A}$$

$$\dfrac{\dfrac{\dfrac{{\scriptstyle(2)}\dfrac{}{\neg A}\qquad \dfrac{\neg A\to A \quad \dfrac{}{\neg A}{\scriptstyle(2)}}{A}}{\bot}{\scriptstyle(2)}}{A}{\scriptstyle(1)}}{(\neg A\to A)\to A}$$

21.2.5 Level of difficulty 5

$$\dfrac{\begin{array}{l}\neg(A\to\neg A)\\ \neg(\neg A\to A)\end{array}}{\bot}$$

$$\dfrac{\neg(A\to\neg A)\qquad \dfrac{\dfrac{\dfrac{\neg(\neg A\to A)\quad \dfrac{\dfrac{}{A}{\scriptstyle(1)}}{\neg A\to A}}{\bot}{\scriptstyle(1)}}{\neg A}}{A\to\neg A}}{\bot}$$

$$\dfrac{\begin{array}{l}\forall xFx\\ \exists x(\neg Fx\vee Gx)\end{array}}{\exists xGx}$$

$$\dfrac{\exists x(\neg Fx\vee Gx)\qquad \dfrac{\dfrac{{\scriptstyle(1)}\dfrac{}{\neg Fa\vee Ga}\quad \dfrac{{\scriptstyle(2)}\dfrac{}{\neg Fa}\quad \dfrac{\forall xFx}{Fa}}{\bot}\quad \dfrac{}{Ga}{\scriptstyle(2)}}{Ga}{\scriptstyle(2)}}{\exists xGx}}{\exists xGx}{\scriptstyle(1)}$$

$$\dfrac{}{\neg\neg\neg A\to\neg A}$$

$$\dfrac{\dfrac{\dfrac{{\scriptstyle(1)}\dfrac{}{\neg\neg\neg A}\qquad \dfrac{\dfrac{{\scriptstyle(3)}\dfrac{}{\neg A}\quad \dfrac{}{A}{\scriptstyle(2)}}{\bot}{\scriptstyle(3)}}{\neg\neg A}}{\bot}{\scriptstyle(2)}}{\neg A}{\scriptstyle(1)}}{\neg\neg\neg A\to\neg A}$$

$$\dfrac{A\to\neg(B\vee C)}{B\to\neg A}$$

$$\dfrac{\dfrac{\dfrac{\dfrac{A\to\neg(B\vee C)\quad \overline{A}^{(2)}}{\neg(B\vee C)}\qquad \dfrac{\overline{B}^{(1)}}{B\vee C}}{\bot}}{\neg A}{\scriptstyle(2)}}{B\to\neg A}{\scriptstyle(1)}$$

$$\dfrac{A\to(B\wedge C)}{\neg B\to\neg A}$$

$$\dfrac{\dfrac{\dfrac{{\scriptstyle(1)}\overline{\neg B}\qquad \dfrac{\dfrac{A\to(B\wedge C)\quad \overline{A}^{(2)}}{B\wedge C}}{B}}{\bot}}{\neg A}{\scriptstyle(2)}}{\neg B\to\neg A}{\scriptstyle(1)}$$

$$\dfrac{(A\to C)\vee(B\to C)}{(A\wedge B)\to C}$$

$$\dfrac{\dfrac{(A\to C)\vee(B\to C)\qquad \dfrac{{\scriptstyle(2)}\overline{A\to C}\quad \dfrac{\overline{A\wedge B}^{(1)}}{A}}{C}\qquad \dfrac{{\scriptstyle(2)}\overline{B\to C}\quad \dfrac{\overline{A\wedge B}^{(1)}}{B}}{C}}{C}{\scriptstyle(2)}}{(A\wedge B)\to C}{\scriptstyle(1)}$$

$$\dfrac{(A\to B)\wedge(A\to C)}{A\to(B\wedge C)}$$

$$\dfrac{\dfrac{\dfrac{\dfrac{(A\to B)\wedge(A\to C)}{A\to B}\quad \overline{A}^{(1)}}{B}\qquad \dfrac{\dfrac{(A\to B)\wedge(A\to C)}{A\to C}\quad \overline{A}^{(1)}}{C}}{B\wedge C}}{A\to(B\wedge C)}{\scriptstyle(1)}$$

$$\dfrac{\forall x(Fx\to Gx)}{\forall xFx\to\forall xGx}$$

$$\dfrac{\dfrac{\dfrac{\dfrac{\forall x(Fx\to Gx)}{Fa\to Ga}\quad \dfrac{\overline{\forall xFx}^{(1)}}{Fa}}{Ga}}{\forall xGx}}{\forall xFx\to\forall xGx}{\scriptstyle(1)}$$

$$\dfrac{\forall x(Fx\to Gx)}{\exists xFx\to\exists xGx}$$

$$\dfrac{\dfrac{{\scriptstyle(1)}\overline{\exists xFx}\qquad \dfrac{\dfrac{\dfrac{\forall x(Fx\to Gx)}{Fa\to Ga}\quad \overline{Fa}^{(2)}}{Ga}}{\exists xGx}}{\exists xGx}{\scriptstyle(2)}}{\exists xFx\to\exists xGx}{\scriptstyle(1)}$$

$$\dfrac{\forall xFx \quad \exists x(\neg Fx \vee Gx)}{\exists xGx}$$

$$\dfrac{\exists x(\neg Fx \vee Gx) \qquad \dfrac{\dfrac{\dfrac{}{\neg Fa \vee Ga}{\scriptstyle(1)} \qquad \dfrac{\dfrac{}{\neg Fa}{\scriptstyle(2)} \quad \dfrac{\forall xFx}{Fa}}{\bot} \qquad \dfrac{}{Ga}{\scriptstyle(2)}}{Ga}{\scriptstyle(2)}}{\exists xGx}}{\exists xGx}{\scriptstyle(1)}$$

(6 steps if you insert a step of EFQ)

21.2.6 Level of difficulty 6

$$\dfrac{\neg(A \wedge B)}{\neg A \vee \neg B}$$

$$\dfrac{\dfrac{\dfrac{\dfrac{\neg(A \wedge B) \qquad \dfrac{\dfrac{}{A}{\scriptstyle(1)} \quad \dfrac{}{B}{\scriptstyle(1)}}{A \wedge B}}{\bot}}{\neg B}{\scriptstyle(2)}}{\neg A \vee \neg B} \qquad \dfrac{\dfrac{}{\neg A}{\scriptstyle(1)}}{\neg A \vee \neg B}}{\neg A \vee \neg B}{\scriptstyle(1)}$$

This is the only De Morgan duality for the connectives that is *strictly classical.* The next two are *constructive.*

$$\dfrac{\neg A \vee \neg B}{\neg(A \wedge B)}$$

$$\dfrac{\dfrac{\neg A \vee \neg B \qquad \dfrac{\dfrac{}{\neg A}{\scriptstyle(2)} \quad \dfrac{\dfrac{}{A \wedge B}{\scriptstyle(1)}}{A}}{\bot} \qquad \dfrac{\dfrac{}{\neg B}{\scriptstyle(2)} \quad \dfrac{\dfrac{}{A \wedge B}{\scriptstyle(1)}}{B}}{\bot}}{\bot}{\scriptstyle(2)}}{\neg(A \wedge B)}{\scriptstyle(1)}$$

$$\dfrac{\neg A \wedge \neg B}{\neg(A \vee B)}$$

$$\dfrac{\dfrac{\dfrac{}{A \vee B}{\scriptstyle(1)} \qquad \dfrac{\dfrac{\neg A \wedge \neg B}{\neg A} \quad \dfrac{}{A}{\scriptstyle(2)}}{\bot} \qquad \dfrac{\dfrac{\neg A \wedge \neg B}{\neg B} \quad \dfrac{}{B}{\scriptstyle(2)}}{\bot}}{\bot}{\scriptstyle(2)}}{\neg(A \vee B)}{\scriptstyle(1)}$$

$$\dfrac{\forall x(Fx \to Gx) \quad \exists x(Fx \wedge \neg Gx)}{\bot}$$

$$\dfrac{\exists x(Fx \wedge \neg Gx) \qquad \dfrac{\dfrac{\dfrac{}{Fa \wedge \neg Ga}{\scriptstyle(1)}}{\neg Ga} \qquad \dfrac{\dfrac{\forall x(Fx \to Gx)}{Fa \to Ga} \quad \dfrac{\dfrac{}{Fa \wedge \neg Ga}{\scriptstyle(1)}}{Fa}}{Ga}}{\bot}}{\bot}{\scriptstyle(1)}$$

$$\dfrac{(A\to C)\wedge(B\to C)}{(A\vee B)\to C}$$

$$\dfrac{\dfrac{{}^{(1)}\dfrac{}{A\vee B}\quad \dfrac{\dfrac{(A\to C)\wedge(B\to C)}{A\to C}\quad \dfrac{}{A}{}^{(2)}}{C}\quad \dfrac{\dfrac{(A\to C)\wedge(B\to C)}{B\to C}\quad \dfrac{}{B}{}^{(2)}}{C}}{C}{}^{(2)}}{(A\vee B)\to C}{}^{(1)}$$

This problem would be Level of Difficulty 5 if one were to use *parallelized* $\wedge$-Elimination:

$$\dfrac{(A\to C)\wedge(B\to C)\quad \dfrac{\dfrac{{}^{(2)}\dfrac{}{A\vee B}\quad \dfrac{{}^{(2)}\dfrac{}{A\to C}\quad \dfrac{}{A}{}^{(3)}}{C}\quad \dfrac{{}^{(2)}\dfrac{}{B\to C}\quad \dfrac{}{B}{}^{(3)}}{C}}{C}{}^{(3)}}{(A\vee B)\to C}{}^{(2)}}{(A\vee B)\to C}{}^{(1)}$$

$$\dfrac{\exists x(Fx\wedge\neg Gx)}{\neg\forall x(Fx\to Gx)}$$

$$\dfrac{\dfrac{\exists x(Fx\wedge\neg Gx)\quad \dfrac{\dfrac{{}^{(2)}\dfrac{}{Fa\wedge\neg Ga}}{\neg Ga}\quad \dfrac{\dfrac{{}^{(2)}\dfrac{}{Fa\wedge\neg Ga}}{Fa}\quad \dfrac{\dfrac{}{\forall x(Fx\to Gx)}{}^{(1)}}{Fa\to Ga}}{Ga}}{\bot}}{\bot}{}^{(2)}}{\neg\forall x(Fx\to Gx)}{}^{(1)}$$

This proof could be shortened by one step by using parallelized $\wedge$-Elimination.

$$\dfrac{\begin{array}{c}A\to\neg A\\ \neg A\to A\end{array}}{\bot}\ **$$

$$\dfrac{\neg A\to A\quad \dfrac{\dfrac{A\to\neg A\quad {}^{(2)}\dfrac{}{A}\quad \dfrac{{}^{(3)}\dfrac{}{\neg A}\quad \dfrac{}{A}{}^{(2)}}{\bot}}{\bot}{}^{(3)}}{\neg A}{}^{(2)}\quad \dfrac{A\to\neg A\quad \dfrac{}{A}{}^{(1)}\quad \dfrac{{}^{(4)}\dfrac{}{\neg A}\quad \dfrac{}{A}{}^{(1)}}{\bot}}{\bot}{}^{(4)}}{\bot}{}^{(1)}$$

** Level of difficulty 6 if one uses *parallelized* $\to$E, as in this solution. Level of difficulty 10 if one uses *serial* $\to$E.

$$\dfrac{\begin{array}{c}\forall x(Mx\to Px)\\ \forall x(Sx\to Mx)\end{array}}{\forall x(Sx\to Px)}$$

$$\dfrac{\dfrac{\dfrac{\dfrac{\forall x(Mx\to Px)}{Ma\to Pa}\quad \dfrac{\dfrac{\forall x(Sx\to Mx)}{Sa\to Ma}\quad \dfrac{}{Sa}{}^{(1)}}{Ma}}{Pa}}{Sa\to Pa}{}^{(1)}}{\forall x(Sx\to Px)}$$

$$\dfrac{}{\neg\neg(A\lor\neg A)}$$

$$\dfrac{\dfrac{}{\neg(A\lor\neg A)}{\scriptstyle(1)} \qquad \dfrac{\dfrac{\dfrac{\dfrac{}{\neg(A\lor\neg A)}{\scriptstyle(1)}\quad\dfrac{\dfrac{}{A}{\scriptstyle(2)}}{A\lor\neg A}}{\bot}}{\neg A}{\scriptstyle(2)}}{A\lor\neg A}}{\dfrac{\bot}{\neg\neg(A\lor\neg A)}{\scriptstyle(1)}}$$

$$\dfrac{\forall xFx\land\forall xGx}{\forall x(Fx\land Gx)}$$

$$\dfrac{\dfrac{\dfrac{\dfrac{\forall xFx\land\forall xGx}{\forall xFx}}{Fa}\qquad\dfrac{\dfrac{\forall xFx\land\forall xGx}{\forall xGx}}{Ga}}{Fa\land Ga}}{\forall x(Fx\land Gx)}$$

$$\dfrac{\exists x(Fx\land Gx)}{\exists xFx\land\exists xGx}$$

$$\dfrac{\exists x(Fx\land Gx)\qquad \dfrac{\dfrac{\dfrac{\dfrac{}{Fa\land Ga}{\scriptstyle(1)}}{Fa}}{\exists xFx}\qquad \dfrac{\dfrac{\dfrac{}{Fa\land Ga}{\scriptstyle(1)}}{Ga}}{\exists xGx}}{\exists xFx\land\exists xGx}}{\exists xFx\land\exists xGx}{\scriptstyle(1)}$$

$$\dfrac{\exists x(Fx\lor Gx)}{\exists xFx\lor\exists xGx}$$

$$\dfrac{\exists x(Fx\lor Gx)\qquad \dfrac{\dfrac{}{Fa\lor Ga}{\scriptstyle(1)}\quad \dfrac{\dfrac{\dfrac{}{Fa}{\scriptstyle(2)}}{\exists xFx}}{\exists xFx\lor\exists xGx}\quad \dfrac{\dfrac{\dfrac{}{Ga}{\scriptstyle(2)}}{\exists xGx}}{\exists xFx\lor\exists xGx}}{\exists xFx\lor\exists xGx}{\scriptstyle(2)}}{\exists xFx\lor\exists xGx}{\scriptstyle(1)}$$

$$\dfrac{\forall x(Fx\to Gx)}{\forall x(\neg Gx\to\neg Fx)}$$

$$\dfrac{\dfrac{\dfrac{\dfrac{\dfrac{}{\neg Ga}{\scriptstyle(1)}\quad \dfrac{\dfrac{\forall x(Fx\to Gx)}{Fa\to Ga}\quad \dfrac{}{Fa}{\scriptstyle(2)}}{Ga}}{\bot}}{\neg Fa}{\scriptstyle(2)}}{\neg Ga\to\neg Fa}{\scriptstyle(1)}}{\forall x(\neg Gx\to\neg Fx)}$$

$$\dfrac{\begin{array}{l}\exists x\neg Gx\\ \forall x(Fx\to Gx)\end{array}}{\exists x\neg Fx}$$

$$\dfrac{\exists x\neg Gx\qquad \dfrac{\dfrac{\dfrac{\dfrac{}{\neg Ga}{\scriptstyle(1)}\quad\dfrac{\dfrac{\forall x(Fx\to Gx)}{Fa\to Ga}\quad\dfrac{}{Fa}{\scriptstyle(2)}}{Ga}}{\bot}}{\neg Fa}{\scriptstyle(2)}}{\exists x\neg Fx}}{\exists x\neg Fx}{\scriptstyle(1)}$$

$$\frac{\forall x\neg Gx \qquad \forall x(Fx\to Gx)}{\forall x\neg Fx}$$

$$\dfrac{\dfrac{\dfrac{\dfrac{\forall x\neg Gx}{\neg Ga}\qquad \dfrac{\dfrac{\forall x(Fx\to Gx)}{Fa\to Ga}\qquad \dfrac{}{Fa}{\scriptstyle(1)}}{Ga}}{\bot}{\scriptstyle(1)}}{\neg Fa}}{\forall x\neg Fx}$$

$$\frac{\forall x\forall y(Lxy\to\neg Lyx)}{\forall x\neg Lxx}$$

$$\dfrac{\dfrac{\dfrac{\dfrac{\dfrac{\dfrac{\forall x\forall y(Lxy\to\neg Lyx)}{\forall y(Lay\to\neg Lya)}}{Laa\to\neg Laa}\qquad \dfrac{}{Laa}{\scriptstyle(1)}}{\neg Laa}\qquad \dfrac{}{Laa}{\scriptstyle(1)}}{\bot}{\scriptstyle(1)}}{\neg Laa}}{\forall x\neg Lxx}$$

$$\frac{\neg\forall xFx}{\exists x\neg Fx}$$

$$\dfrac{\dfrac{\neg\forall xFx \qquad \dfrac{\dfrac{\dfrac{{\scriptstyle(1)}\dfrac{}{\neg\exists x\neg Fx}\qquad \dfrac{\dfrac{}{\neg Fa}{\scriptstyle(2)}}{\exists x\neg Fx}}{\bot}{\scriptstyle(2)\ \mathrm{CR}}}{Fa}}{\forall xFx}}{\bot}{\scriptstyle(1)}}{\exists x\neg Fx}$$

This is the *strictly classical* De Morgan duality for the quantifiers

---

$$\frac{A\wedge(B\vee C)}{(A\wedge B)\vee C}$$

$$\dfrac{\dfrac{A\wedge(B\vee C)}{B\vee C}\qquad \dfrac{\dfrac{\dfrac{A\wedge(B\vee C)}{A}\qquad \dfrac{}{B}{\scriptstyle(1)}}{A\wedge B}}{(A\wedge B)\vee C}\qquad \dfrac{\dfrac{}{C}{\scriptstyle(1)}}{(A\wedge B)\vee C}}{(A\wedge B)\vee C}{\scriptstyle(1)}$$

$$\frac{}{(A\to B)\vee(B\to A)}$$

$$\dfrac{\dfrac{\dfrac{{\scriptstyle(1)}\dfrac{}{A}}{B\to A}}{(A\to B)\vee(B\to A)}\qquad \dfrac{\dfrac{\dfrac{{\scriptstyle(1)}\dfrac{}{\neg A}\qquad \dfrac{}{A}{\scriptstyle(2)}}{\bot}{\scriptstyle(2)}}{A\to B}}{(A\to B)\vee(B\to A)}}{(A\to B)\vee(B\to A)}{\scriptstyle(1)}$$

(7 steps if you insert a step of EFQ)

$$\frac{\forall x\neg\exists yLxy}{\forall x\neg\exists yLyx}$$

$$\dfrac{\dfrac{\dfrac{{\scriptstyle(1)}\dfrac{}{\exists yLya}\qquad \dfrac{\dfrac{\forall x\neg\exists yLxy}{\neg\exists yLby}\qquad \dfrac{\dfrac{}{Lba}{\scriptstyle(2)}}{\exists yLby}}{\bot}}{\bot}{\scriptstyle(2)}}{\neg\exists yLya}{\scriptstyle(1)}}{\forall x\neg\exists yLyx}$$

$$\frac{(A\to B)\vee(A\to C)}{A\to(B\vee C)}$$

$$\dfrac{\dfrac{(A\to B)\vee(A\to C) \qquad \dfrac{\dfrac{\overline{A\to B}^{\,(2)} \quad \overline{A}^{\,(1)}}{B}}{B\vee C} \qquad \dfrac{\dfrac{\overline{A\to C}^{\,(2)} \quad \overline{A}^{\,(1)}}{C}}{B\vee C}}{B\vee C}\,(2)}{A\to(B\vee C)}\,(1)$$

21.2.7 Level of difficulty 7

$$\frac{\neg(A\vee B)}{\neg A\wedge\neg B}$$

$$\dfrac{\dfrac{\dfrac{\neg(A\vee B) \quad \dfrac{\overline{A}^{\,(1)}}{A\vee B}}{\bot}\,(1)}{\neg A} \qquad \dfrac{\dfrac{\neg(A\vee B) \quad \dfrac{\overline{B}^{\,(2)}}{A\vee B}}{\bot}\,(2)}{\neg B}}{\neg A\wedge\neg B}$$

This is the propositional De Morgan duality that most students find easier than the other ones

$$\frac{\begin{array}{c}\forall x(Mx\to Px)\\ \exists x(Sx\wedge Mx)\end{array}}{\exists x(Sx\wedge Px)}$$

$$\dfrac{\exists x(Sx\wedge Mx) \qquad \dfrac{\dfrac{\dfrac{{}^{(1)}\overline{Sa\wedge Ma}}{Sa} \qquad \dfrac{\dfrac{\forall x(Mx\to Px)}{Ma\to Pa} \quad \dfrac{{}^{(1)}\overline{Sa\wedge Ma}}{Ma}}{Pa}}{Sa\wedge Pa}}{\exists x(Sx\wedge Px)}}{\exists x(Sx\wedge Px)}\,(1)$$

$$\frac{\begin{array}{c}\exists xSx\\ \forall x(Mx\to Px)\\ \forall x(Sx\to Mx)\end{array}}{\exists x(Sx\wedge Px)}$$

$$\dfrac{\exists xSx \qquad \dfrac{\dfrac{{}^{(1)}\overline{Sa} \qquad \dfrac{\dfrac{\forall x(Mx\to Px)}{Ma\to Pa} \quad \dfrac{\dfrac{\forall x(Sx\to Mx)}{Sa\to Ma} \quad \overline{Sa}^{\,(1)}}{Ma}}{Pa}}{Sa\wedge Pa}}{\exists x(Sx\wedge Px)}}{\exists x(Sx\wedge Px)}\,(1)$$

$$\frac{\begin{array}{c}\exists xMx\\ \forall x(Mx\to Px)\\ \forall x(Mx\to Sx)\end{array}}{\exists x(Sx\wedge Px)}$$

$$\dfrac{\exists xMx \qquad \dfrac{\dfrac{\dfrac{\dfrac{\forall x(Mx\to Sx)}{Ma\to Sa} \quad \overline{Ma}^{\,(1)}}{Sa} \qquad \dfrac{\dfrac{\forall x(Mx\to Px)}{Ma\to Pa} \quad \overline{Ma}^{\,(1)}}{Pa}}{Sa\wedge Pa}}{\exists x(Sx\wedge Px)}}{\exists x(Sx\wedge Px)}\,(1)$$

$$\frac{(A\vee B)\to C}{(A\to C)\wedge(B\to C)}$$

$$\dfrac{\dfrac{\dfrac{(A\vee B)\to C \quad \dfrac{\overline{A}^{\,(1)}}{A\vee B}}{C}\,(1)}{A\to C} \qquad \dfrac{\dfrac{(A\vee B)\to C \quad \dfrac{\overline{B}^{\,(2)}}{A\vee B}}{C}\,(2)}{B\to C}}{(A\to C)\wedge(B\to C)}$$

$$\begin{array}{l}\exists x(Mx\wedge Px)\\ \underline{\forall x(Mx\to Sx)}\\ \exists x(Sx\wedge Px)\end{array}$$

$$\dfrac{\exists x(Mx\wedge Px)\qquad \dfrac{\dfrac{\dfrac{\dfrac{\forall x(Mx\to Sx)}{Ma\to Sa}\quad \dfrac{\overline{Ma\wedge Pa}^{(1)}}{Ma}}{Sa}\qquad \dfrac{\overline{Ma\wedge Pa}^{(1)}}{Pa}}{Sa\wedge Pa}}{\exists x(Sx\wedge Px)}}{\exists x(Sx\wedge Px)}(1)$$

$$\begin{array}{l}\forall x(Mx\to Px)\\ \underline{\exists x(Mx\wedge Sx)}\\ \exists x(Sx\wedge Px)\end{array}$$

$$\dfrac{\exists x(Mx\wedge Sx)\qquad \dfrac{\dfrac{\dfrac{{}^{(1)}\overline{Ma\wedge Sa}}{Sa}\qquad \dfrac{\dfrac{\forall x(Mx\to Px)}{Ma\to Pa}\quad \dfrac{\overline{Ma\wedge Sa}^{(1)}}{Ma}}{Pa}}{Sa\wedge Pa}}{\exists x(Sx\wedge Px)}}{\exists x(Sx\wedge Px)}(1)$$

$$\begin{array}{l}\exists x(Mx\wedge \neg Px)\\ \underline{\forall x(Mx\to Sx)}\\ \exists x(Sx\wedge \neg Px)\end{array}$$

For $\neg P$ read Q!
The negation operator plays no role.

$$\dfrac{\exists x(Mx\wedge \neg Px)\qquad \dfrac{\dfrac{\dfrac{\dfrac{\forall x(Mx\to Sx)}{Ma\to Sa}\quad \dfrac{\overline{Ma\wedge \neg Pa}^{(1)}}{Ma}}{Sa}\qquad \dfrac{\overline{Ma\wedge \neg Pa}^{(1)}}{\neg Pa}}{Sa\wedge \neg Pa}}{\exists x(Sx\wedge \neg Px)}}{\exists x(Sx\wedge \neg Px)}(1)$$

$$\begin{array}{l}\exists xPx\\ \forall x(Px\to Mx)\\ \underline{\forall x(Mx\to Sx)}\\ \exists x(Sx\wedge Px)\end{array}$$

$$\dfrac{\exists xPx\qquad \dfrac{\dfrac{\dfrac{\dfrac{\forall x(Mx\to Sx)}{Ma\to Sa}\quad \dfrac{\dfrac{\forall x(Px\to Mx)}{Pa\to Ma}\quad \overline{Pa}^{(1)}}{Ma}}{Sa}\qquad \overline{Pa}^{(1)}}{Sa\wedge Pa}}{\exists x(Sx\wedge Px)}}{\exists x(Sx\wedge Px)}(1)$$

$$\begin{array}{l}\exists x(Px\wedge Mx)\\ \underline{\forall x(Mx\to Sx)}\\ \exists x(Sx\wedge Px)\end{array}$$

$$\dfrac{\exists x(Px\wedge Mx)\qquad \dfrac{\dfrac{\dfrac{\dfrac{\forall x(Mx\to Sx)}{Ma\to Sa}\quad \dfrac{\overline{Pa\wedge Ma}^{(1)}}{Ma}}{Sa}\qquad \dfrac{\overline{Pa\wedge Ma}^{(1)}}{Pa}}{Sa\wedge Pa}}{\exists x(Sx\wedge Px)}}{\exists x(Sx\wedge Px)}(1)$$

$$\frac{A \rightarrow (B \rightarrow C)}{(A \rightarrow \neg C) \rightarrow (A \rightarrow \neg B)}$$

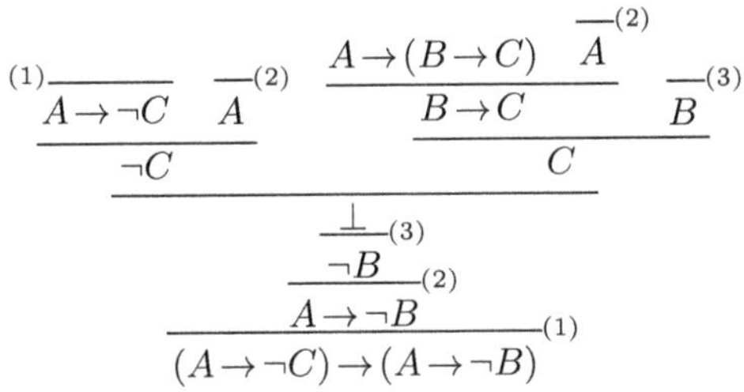

$$\frac{A \rightarrow (B \vee C)}{\neg C \rightarrow (\neg B \rightarrow \neg A)}$$

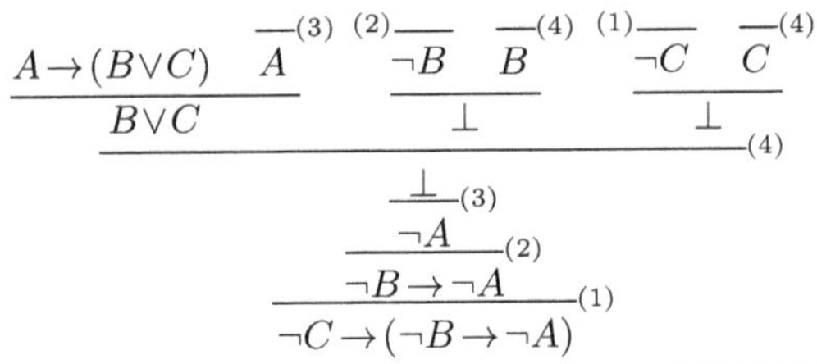

$$\frac{A \rightarrow (B \wedge C)}{(A \rightarrow B) \wedge (A \rightarrow C)}$$

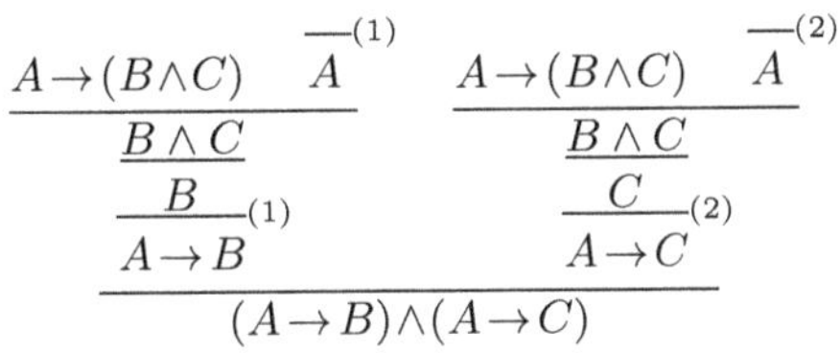

$$\frac{(A \vee B) \wedge (\neg A \vee C)}{B \vee C}$$

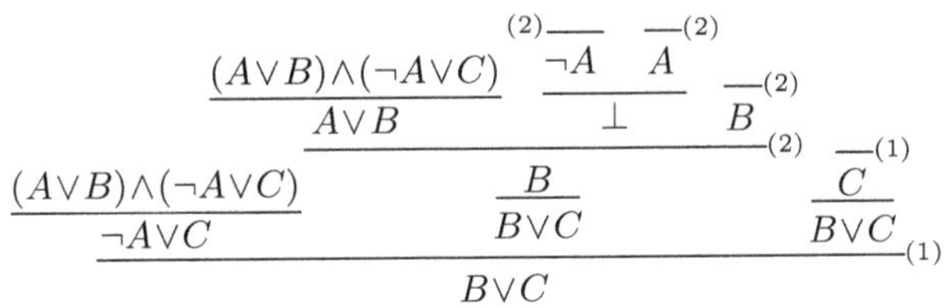

21.2.8 Level of difficulty 8

$$\frac{\neg \forall x(Fx \rightarrow Gx)}{\exists x(Fx \wedge \neg Gx)}$$

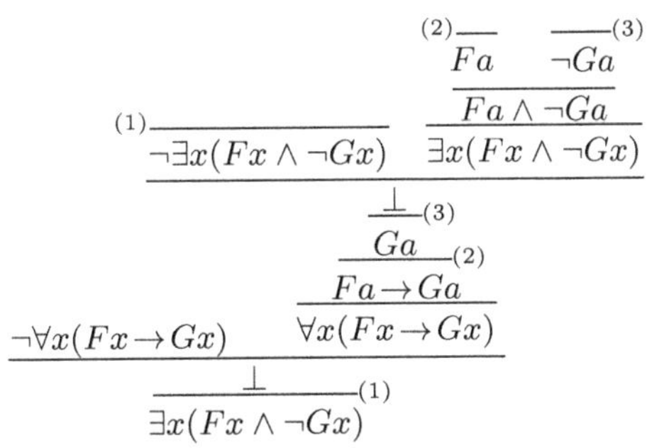

$$\frac{\forall x(Fx \lor Gx)}{\exists xFx \lor \forall xGx}$$

$$\dfrac{\dfrac{\dfrac{\forall x(Fx \lor Gx)}{Fa \lor Ga} \quad \dfrac{\overline{\neg\exists xFx}^{\,(1)} \quad \dfrac{\overline{Fa}^{\,(2)}}{\exists xFx}}{\bot} \quad \overline{Ga}^{\,(2)}}{\dfrac{\dfrac{Ga}{\forall xGx}}{\exists xFx \lor \forall xGx}}{\scriptstyle (2)} \qquad \dfrac{\overline{\exists xFx}^{\,(1)}}{\exists xFx \lor \forall xGx}}{\exists xFx \lor \forall xGx}{\scriptstyle (1)\ \text{Dil}}$$

$$\frac{\forall x\forall y(\neg x=y \to Rxy)}{\forall x\forall y(x=y \lor Rxy)}$$

$$\dfrac{\dfrac{\dfrac{\overline{a=b}^{\,(1)}}{a=b \lor Rab} \qquad \dfrac{\dfrac{\dfrac{\dfrac{\forall x\forall y(\neg x=y \to Rxy)}{\forall y(\neg a=y \to Ray)}}{\neg a=b \to Rab} \quad \overline{\neg a=b}^{\,(1)}}{Rab}}{a=b \lor Rab}}{a=b \lor Rab}{\scriptstyle (1)\ \text{Dil}}}{\dfrac{\forall y(a=y \lor Ray)}{\forall x\forall y(x=y \lor Rxy)}}$$

$$\frac{\begin{array}{l}\neg\neg A\\ \neg\neg B\\ A \to (B \to C)\end{array}}{\neg\neg C}$$

$$\dfrac{\dfrac{\neg\neg A \qquad \dfrac{\dfrac{\neg\neg B \qquad \dfrac{\dfrac{\overline{\neg C}^{\,(1)} \qquad \dfrac{\dfrac{A \to (B \to C) \quad \overline{A}^{\,(2)}}{B \to C} \quad \overline{B}^{\,(3)}}{C}}{\bot}}{\neg B}{\scriptstyle (3)}}{\bot}}{\neg A}{\scriptstyle (2)}}{\bot}}{\neg\neg C}{\scriptstyle (1)}$$

$$\frac{\begin{array}{l}\forall x(Fx \to Gx)\\ \neg\exists x(Gx \land Hx)\end{array}}{\forall x(Hx \to \neg Fx)}$$

$$\dfrac{\dfrac{\dfrac{\dfrac{\neg\exists x(Gx \land Hx) \qquad \dfrac{\dfrac{\dfrac{\dfrac{\forall x(Fx \to Gx)}{Fa \to Ga} \quad \overline{Fa}^{\,(2)}}{Ga} \quad \overline{Ha}^{\,(1)}}{Ga \land Ha}}{\exists x(Gx \land Hx)}}{\bot}}{\neg Fa}{\scriptstyle (2)}}{Ha \to \neg Fa}{\scriptstyle (1)}}{\forall x(Hx \to \neg Fx)}$$

$$\frac{\neg(A \to B)}{A \wedge \neg B}$$

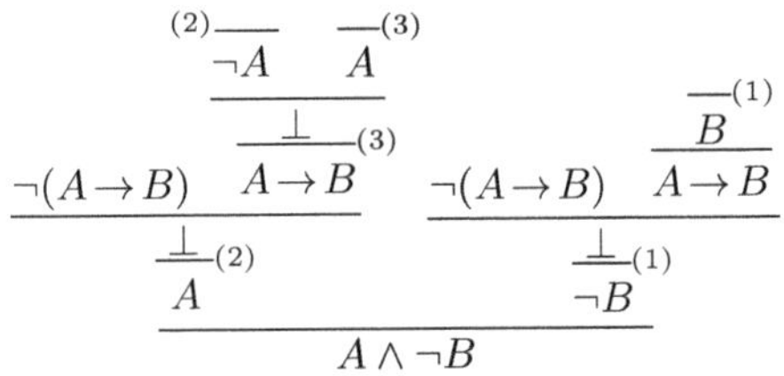

$$\frac{(A \wedge B) \to C}{(A \to C) \vee (B \to C)}$$

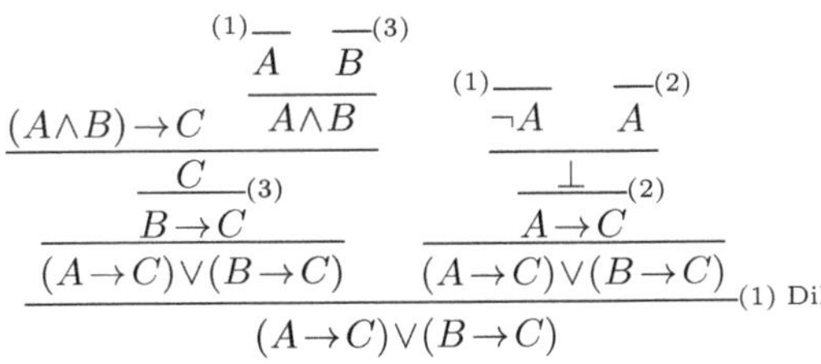

21.2.9 Level of difficulty 9

$$\frac{\neg\exists x(Mx \wedge Px) \quad \forall x(Sx \to Mx)}{\neg\exists x(Sx \wedge Px)}$$

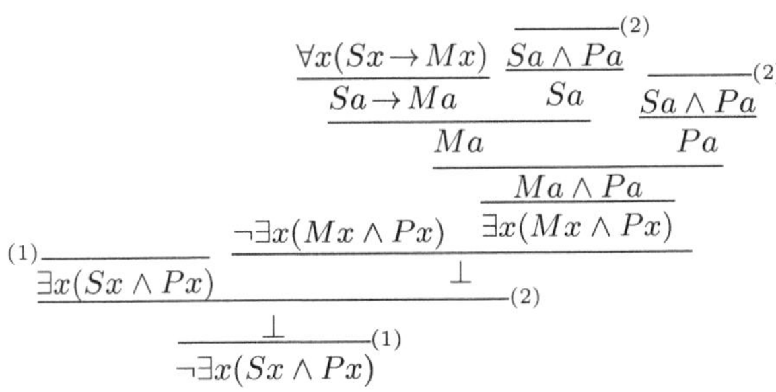

$$\frac{\neg\exists x(Mx \wedge Px) \quad \exists x(Sx \wedge Mx)}{\exists x(Sx \wedge \neg Px)}$$

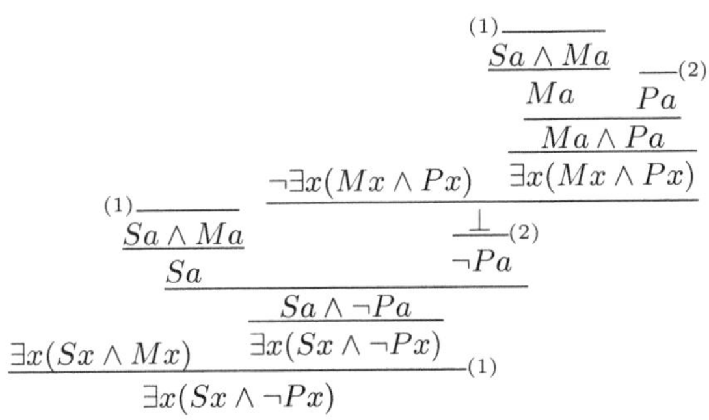

$$\begin{array}{l} \exists xSx \\ \neg\exists x(Mx\wedge Px) \\ \forall x(Sx\to Mx) \\ \hline \exists x(Sx\wedge\neg Px) \end{array}$$

$$\dfrac{\exists xSx \qquad \dfrac{\dfrac{\overline{Sa}^{\,(1)} \qquad \dfrac{\dfrac{\neg\exists x(Mx\wedge Px) \qquad \dfrac{\dfrac{\dfrac{\dfrac{\forall x(Sx\to Mx)}{Sa\to Ma} \quad \overline{Sa}^{\,(1)}}{Ma} \quad \overline{Pa}^{\,(2)}}{Ma\wedge Pa}}{\exists x(Mx\wedge Px)}}{\bot}}{\neg Pa}{\scriptstyle(2)}}{Sa\wedge\neg Pa}}{\exists x(Sx\wedge\neg Px)}}{\exists x(Sx\wedge\neg Px)}{\scriptstyle(1)}$$

$$\begin{array}{l} (A\to B)\wedge(B\to C) \\ (B\to A)\wedge(C\to B) \\ \hline (A\to C)\wedge(C\to A) \end{array}$$

$$\dfrac{\dfrac{\dfrac{\dfrac{(A\to B)\wedge(B\to C)}{B\to C} \qquad \dfrac{\dfrac{(A\to B)\wedge(B\to C)}{A\to B} \quad \overline{A}^{\,(1)}}{B}}{C}}{A\to C}{\scriptstyle(1)} \qquad \dfrac{\dfrac{\dfrac{(B\to A)\wedge(C\to B)}{B\to A} \qquad \dfrac{\dfrac{(B\to A)\wedge(C\to B)}{C\to B} \quad \overline{C}^{\,(2)}}{B}}{A}}{C\to A}{\scriptstyle(2)}}{(A\to C)\wedge(C\to A)}$$

$$\begin{array}{l} \exists xFx\to\exists xGx \\ \hline \exists x(Fx\to Gx) \end{array}$$

$$\dfrac{\dfrac{\dfrac{\dfrac{\overline{Fa}^{\,(2)}}{\exists xFx} \quad \exists xFx\to\exists xGx}{\exists xGx} \qquad \dfrac{\dfrac{\overline{Gb}^{\,(3)}}{Fb\to Gb}}{\exists x(Fx\to Gx)}}{\exists x(Fx\to Gx)}{\scriptstyle(3)} \qquad \dfrac{\dfrac{\dfrac{{}^{(2)}\overline{\neg Fa} \quad \overline{Fa}^{\,(1)}}{\bot}}{Fa\to Ga}{\scriptstyle(1)}}{\exists x(Fx\to Gx)}}{\exists x(Fx\to Gx)}{\scriptstyle(2)\ \text{Dil}}$$

$$\begin{array}{l} \neg\neg(A\to B) \\ \neg\neg(B\to C) \\ \hline \neg\neg(A\to C) \end{array}$$

$$\dfrac{\dfrac{\neg\neg(A\to B) \qquad \dfrac{\dfrac{\neg\neg(B\to C) \qquad \dfrac{\dfrac{{}^{(1)}\overline{\neg(A\to C)} \qquad \dfrac{\dfrac{{}^{(3)}\overline{B\to C} \qquad \dfrac{{}^{(2)}\overline{A\to B} \quad \overline{A}^{\,(4)}}{B}}{C}}{A\to C}{\scriptstyle(4)}}{\bot}}{\neg(B\to C)}{\scriptstyle(3)}}{\bot}}{\neg(A\to B)}{\scriptstyle(2)}}{\bot}}{\neg\neg(A\to C)}{\scriptstyle(1)}$$

21.2.10 Level of difficulty 10

$$\frac{(A\to B)\wedge(B\to A)}{((A\to C)\to(B\to C))\wedge((B\to C)\to(A\to C))}$$

$$\dfrac{\dfrac{\dfrac{\dfrac{\dfrac{}{A\to C}{\scriptstyle(1)}\quad \dfrac{\dfrac{(A\to B)\wedge(B\to A)}{B\to A}\quad \dfrac{}{B}{\scriptstyle(2)}}{A}}{C}}{B\to C}{\scriptstyle(2)}}{(A\to C)\to(B\to C)}{\scriptstyle(1)}\qquad \dfrac{\dfrac{\dfrac{\dfrac{}{B\to C}{\scriptstyle(3)}\quad \dfrac{\dfrac{(A\to B)\wedge(B\to A)}{A\to B}\quad \dfrac{}{A}{\scriptstyle(2)}}{B}}{C}}{A\to C}{\scriptstyle(2)}}{(B\to C)\to(A\to C)}{\scriptstyle(3)}}{((A\to C)\to(B\to C))\wedge((B\to C)\to(A\to C))}$$

This proofs has 11 steps because it uses the serial form of $\wedge$-E.
If one were to use the parallelized form instead, it would have 10 steps.

$$\frac{A\to(B\vee C)}{(A\to B)\vee(A\to C)}$$

$$\dfrac{\dfrac{\dfrac{A\to(B\vee C)\quad \dfrac{}{A}{\scriptstyle(1)}}{B\vee C}\quad \dfrac{\dfrac{\dfrac{}{B}{\scriptstyle(2)}}{A\to B}}{(A\to B)\vee(A\to C)}\quad \dfrac{\dfrac{\dfrac{}{C}{\scriptstyle(2)}}{A\to C}}{(A\to B)\vee(A\to C)}}{(A\to B)\vee(A\to C)}{\scriptstyle(2)}\qquad \dfrac{\dfrac{\dfrac{\dfrac{}{\neg A}{\scriptstyle(1)}\quad \dfrac{}{A}{\scriptstyle(3)}}{\bot}}{A\to B}{\scriptstyle(3)}}{(A\to B)\vee(A\to C)}}{(A\to B)\vee(A\to C)}{\scriptstyle(1)\ \text{Dil}}$$

$$\frac{(A\to B)\wedge(B\to A)}{((C\to A)\to(C\to B))\wedge((C\to B)\to(C\to A))}$$

$$\dfrac{\dfrac{\dfrac{\dfrac{\dfrac{(A\to B)\wedge(B\to A)}{A\to B}\quad \dfrac{\dfrac{}{C\to A}{\scriptstyle(1)}\quad \dfrac{}{C}{\scriptstyle(2)}}{A}}{B}}{C\to B}{\scriptstyle(2)}}{(C\to A)\to(C\to B)}{\scriptstyle(1)}\qquad \dfrac{\dfrac{\dfrac{\dfrac{(A\to B)\wedge(B\to A)}{B\to A}\quad \dfrac{\dfrac{}{C\to B}{\scriptstyle(1)}\quad \dfrac{}{C}{\scriptstyle(2)}}{B}}{A}}{C\to A}{\scriptstyle(2)}}{(C\to B)\to(C\to A)}{\scriptstyle(1)}}{((C\to A)\to(C\to B))\wedge((C\to B)\to(C\to A))}$$

This proofs has 11 steps because it uses the serial form of $\wedge$-E.
If one were to use the parallelized form instead, it would have 10 steps.

$$\frac{\forall xFx\to\forall xGx}{\exists x\neg Fx\vee\exists xGx}$$

$$\dfrac{\dfrac{{\scriptstyle(1)}\dfrac{}{\exists x\neg Fx}}{\exists x\neg Fx\vee\exists xGx} \qquad \dfrac{\dfrac{\dfrac{\dfrac{\forall xFx\to\forall xGx \qquad \dfrac{\dfrac{\dfrac{{\scriptstyle(1)}\dfrac{}{\neg\exists x\neg Fx} \quad \dfrac{\dfrac{}{\neg Fb}{\scriptstyle(2)}}{\exists x\neg Fx}}{\bot}{\scriptstyle(2)}}{Fb}}{\forall xFx}}{\forall xGx}}{Ga}}{\exists xGx}}{\exists x\neg Fx\vee\exists xGx}}{\exists x\neg Fx\vee\exists xGx}{\scriptstyle(1)\ \text{Dil}}$$

21.2.11 Level of difficulty 11

$$\dfrac{}{((A\to\neg B)\to\neg(A\wedge B))\wedge(\neg(A\wedge B)\to(A\to\neg B))}$$

$$\dfrac{\dfrac{\dfrac{\dfrac{\dfrac{{\scriptstyle(1)}\dfrac{}{A\to\neg B} \quad \dfrac{\dfrac{}{A\wedge B}{\scriptstyle(2)}}{A}}{\neg B} \quad \dfrac{\dfrac{}{A\wedge B}{\scriptstyle(2)}}{B}}{\bot}{\scriptstyle(2)}}{\neg(A\wedge B)}{\scriptstyle(1)}}{(A\to\neg B)\to\neg(A\wedge B)} \qquad \dfrac{\dfrac{\dfrac{\dfrac{{\scriptstyle(3)}\dfrac{}{\neg(A\wedge B)} \quad \dfrac{{\scriptstyle(4)}\dfrac{}{A} \quad \dfrac{}{B}{\scriptstyle(5)}}{A\wedge B}}{\bot}{\scriptstyle(5)}}{\neg B}{\scriptstyle(4)}}{A\to\neg B}{\scriptstyle(3)}}{\neg(A\wedge B)\to(A\to\neg B)}}{((A\to\neg B)\to\neg(A\wedge B))\wedge(\neg(A\wedge B)\to(A\to\neg B))}$$

This proof has 12 steps because it uses the serial form of $\wedge$-E.
If one were to use the parallelized form instead, it would have 11 steps.

$\dfrac{\forall x(\neg Px\to\exists y(\neg Py\wedge\forall z(z<y\to Pz)))}{\forall x(\forall y(y<x\to Py)\to Px)\to\forall zPz}$ has the overall form $\dfrac{\forall x(\neg Px\to\exists y(\neg Py\wedge Qy))}{\forall x(Qx\to Px)\to\forall zPz}$

$$\dfrac{\dfrac{CR\ \dfrac{\dfrac{\dfrac{\forall x(\neg Px\to\exists y(\neg Py\wedge Qy))}{\neg Pa\to\exists y(\neg Py\wedge Qy)} \quad \dfrac{}{\neg Pa}{\scriptstyle(1)}}{\exists y(\neg Py\wedge Qy))} \qquad \dfrac{\dfrac{{\scriptstyle(2)}\dfrac{}{\neg Pb\wedge Qb}}{\neg Pb} \quad \dfrac{\dfrac{\dfrac{}{\forall x(Qx\to Px)}{\scriptstyle(1)}}{Qb\to Pb} \quad \dfrac{\dfrac{}{\neg Pb\wedge Qb}{\scriptstyle(2)}}{Qb}}{Pb}}{\bot}}{\bot}{\scriptstyle(2)}}{\dfrac{\bot}{Pa}{\scriptstyle(1)}}}{\forall zPz}$$

$$\dfrac{\forall zPz}{\forall x(Qx\to Px)\to\forall zPz}{\scriptstyle(1)}$$

We have followed here Quine's famous *Maxim of Shallow Analysis*:

Do not expose more logical structure than is needed to carry out the proof.

With the coarser structure involving Q, one loses sight of the *dyadic* predicate $<$, and the problem becomes one in *monadic* predicate logic, which is decidable.

21.2.12 Level of difficulty 12

$$\dfrac{\forall xFx\to\forall xGx}{\exists x(Fx\to Gx)}$$

$$\dfrac{\dfrac{[\neg\exists x(Fx\to Gx)]^{(1)} \qquad \dfrac{\dfrac{\dfrac{\dfrac{\forall xFx\to\forall xGx \qquad \dfrac{\dfrac{\dfrac{[\neg\exists x(Fx\to Gx)]^{(1)} \qquad \dfrac{\dfrac{\dfrac{[\neg Fa]^{(2)} \quad [Fa]^{(3)}}{\bot}}{Fa\to Ga}\,(3)}{\exists x(Fx\to Gx)}}{\bot}}{Fa}\,(2)}{\forall xFx}}{\forall xGx}}{Ga}}{Fa\to Ga}}{\exists x(Fx\to Gx)}}{\bot}}{\exists x(Fx\to Gx)}\,(1)\ \text{CR}$$

$$\dfrac{\neg\exists x(Fx\land\forall z(Gzx\to Hzx))}{\forall x(Fx\to\exists z(Gzx\land\neg Hzx))}$$

$$\dfrac{\dfrac{\dfrac{\dfrac{\neg\exists x(Fx\land\forall z(Gzx\to Hzx)) \qquad \dfrac{\dfrac{[Fa]^{(1)} \qquad \dfrac{\dfrac{\dfrac{\dfrac{[\neg\exists z(Gza\land\neg Hza)]^{(2)} \qquad \dfrac{\dfrac{[Gba]^{(3)} \quad [\neg Hba]^{(4)}}{Gba\land\neg Hba}}{\exists z(Gza\land\neg Hza)}}{\bot}}{Hba}\,(4)}{Gba\to Hba}\,(3)}{\forall z(Gza\to Hza)}}{Fa\land\forall z(Gza\to Hza)}}{\exists x(Fx\land\forall z(Gzx\to Hzx))}}{\bot}}{\exists z(Gza\land\neg Hza)}\,(2)\ \text{CR}}{Fa\to\exists z(Gza\land\neg Hza)}\,(1)}{\forall x(Fx\to\exists z(Gzx\land\neg Hzx))}$$

21.2.13 Level of difficulty 13

We define $A \leftrightarrow B$ to be short for $(A \to B) \wedge (B \to A)$.

$$\dfrac{\exists x \forall y(Rxy \leftrightarrow \neg Ryy)}{\bot} \qquad \dfrac{\exists x \forall y(Rxy \leftrightarrow \neg Ryy) \qquad \begin{array}{c} \underbrace{\dfrac{\dfrac{\dfrac{}{\forall y(Ray \leftrightarrow \neg Ray)}{\scriptstyle (1)}}{Raa \leftrightarrow \neg Raa}}{Raa \to \neg Raa} \; , \; \dfrac{\dfrac{\dfrac{}{\forall y(Ray \leftrightarrow \neg Ray)}{\scriptstyle (1)}}{Raa \leftrightarrow \neg Raa}}{\neg Raa \to Raa}} \\ \Pi \\ \bot \end{array}}{\bot}{\scriptstyle (1)}$$

where the disproof Π is obtained by substituting Raa for A in the following disproof:

$$\dfrac{\dfrac{A \to \neg A \qquad \dfrac{\neg A \to A \qquad \dfrac{\dfrac{\dfrac{A \to \neg A \quad \overline{A}^{(1)}}{\neg A} \quad \overline{A}^{(1)}}{\bot}}{\neg A}{\scriptstyle (1)}}{A}}{\neg A} \qquad \dfrac{\neg A \to A \qquad \dfrac{\dfrac{\dfrac{A \to \neg A \quad \overline{A}^{(1)}}{\neg A} \quad \overline{A}^{(1)}}{\bot}}{\neg A}{\scriptstyle (1)}}{A}}{\bot}$$

That makes for 15 steps overall, reducible to 13 steps if one parallelizes the applications of $\forall$E and parallelizes the applications of $\wedge$E that are implicit in the inferential moves from the biconditional to the conditionals in each direction.

21.2.14 Level of difficulty 14

$$\dfrac{\forall x(Qx \to Px) \to \forall z Pz}{\forall z(\neg Pz \to \exists y(\neg Py \wedge Qy))}$$

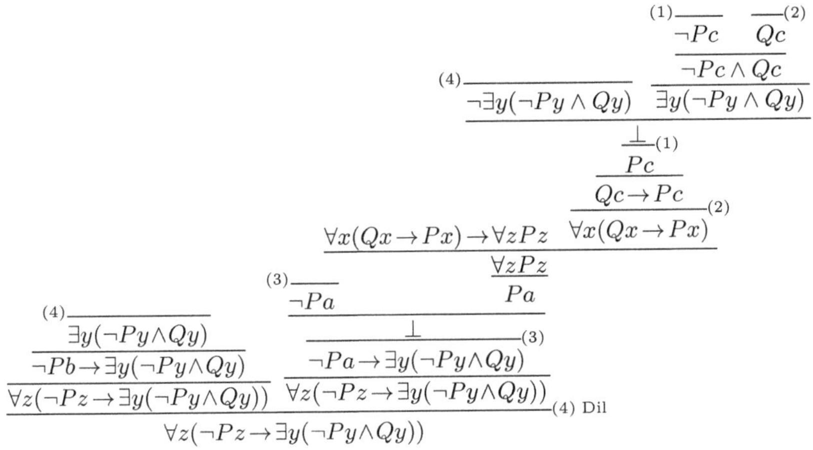

21.3 Note on the problem $A \to \neg A, \neg A \to A : \bot$

With serial $\to$E ('modus ponens'), in the Gentzen–Prawitz system of natural deduction, the normal-form constructive disproof for the titular problem has 10 steps:

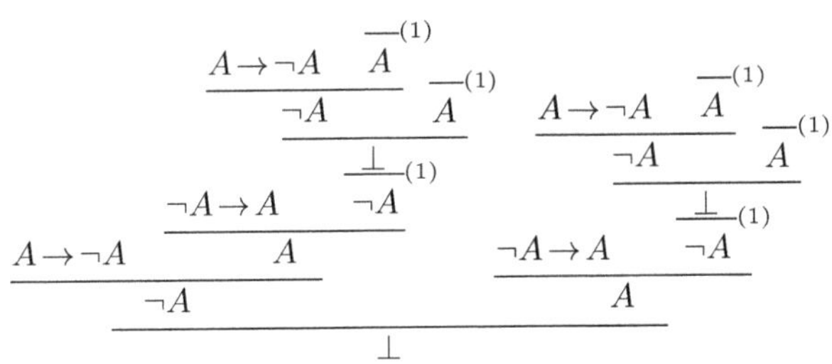

With parallelized $\to$E, the disproof has only 6 steps:

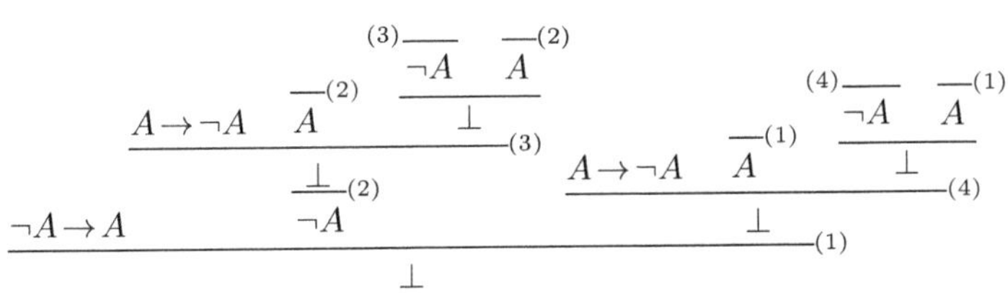

21.4 The natural-deduction disproof of $A \to \neg A, \neg A \to A : \bot$ along with its corresponding sequent disproof

Here again is the 6-step constructive *natural-deduction* disproof:

$$\dfrac{\neg A \to A \qquad \dfrac{\dfrac{A \to \neg A \quad \overline{A}^{\,(2)} \quad \dfrac{\overline{\neg A}^{\,(3)} \quad \overline{A}^{\,(2)}}{\bot}}{\bot}\,{\scriptstyle(3)}}{\neg A}\,{\scriptstyle(2)} \qquad \dfrac{A \to \neg A \quad \overline{A}^{\,(1)} \quad \dfrac{\overline{\neg A}^{\,(4)} \quad \overline{A}^{\,(1)}}{\bot}}{\bot}\,{\scriptstyle(4)}}{\bot}\,{\scriptstyle(1)}$$

And here is the corresponding 6-step constructive *sequent* disproof:

$$\dfrac{\dfrac{\dfrac{A : A \qquad \dfrac{A : A}{\neg A,\ A : \bot}}{A \to \neg A,\ A : \bot}}{A \to \neg A : \neg A} \qquad \dfrac{A : A \qquad \dfrac{A : A}{\neg A,\ A : \bot}}{A \to \neg A,\ A : \bot}}{\neg A \to A,\ A \to \neg A : \bot}$$

Intuition: One and the same abstract (dis)proof is being represented here by two distinct but 'equivalent' (or 'macro-structurally isomorphic') derivations. One is a constructive natural deduction employing only parallelized elimination rules; the other is a constructive sequent (dis)proof. Each of them regiments the same (constructive) 'line of argument'. This cannot be said of the Gentzen–Prawitz natural deduction and any Gentzenian version of its (too loosely!) corresponding sequent (dis)proof. The insight comes only with the appropriately relevantized systems, *and* by imposing the requirement that all eliminations in natural deduction be in parallelized form with their major premises standing proud with no non-trivial proof-work above them.
Thesis: The identity conditions for 'abstract (classical) proof' are those for the classical proofs in the logical system for relevant reasoning that has been set forth in this primer, whether in the form of a (classical) natural deduction using parallelized elimination rules or in the form of its corresponding and genuinely equivalent (classical) sequent proof.

Exercise 63. *Turn all the foregoing natural deductions in this chapter into ones that have all their eliminations parallelized. Then turn the latter into sequent proofs.*

Chapter 22

Summary of Logical Skills

We have now reached a point in our laying out of concepts and constructs in Logic where we can re-visit the list of skills provided in §1.8 and supply some more details about the kinds of problems whose solutions test one's mastery of those skills. The following is a more complete list of the skills involved, whose description the reader will now be in a better position to understand. All models M mentioned below are presumed to be finite. They are also presumed to provide interpretations for all the extra-logical vocabulary in the sentences involved.

1. *Translate* English sentences into logical notation—that is, provide them with their 'logical forms'.

2. Give an *inverted tree-analysis* of the structure of a formal sentence.

3. Given a finite sequence of symbols, determine whether it is a well-formed *term*; and, if it is well-formed, identify which variable occurrences within it, if any, are free.

4. Given a finite sequence of symbols, determine whether it is a well-formed *formula*; and, if it is well-formed, identify which variable occurrences within it, if any, are free.

5. Using the truth tables for the connectives, *work out the truth values* of a formal sentence of propositional logic under an assignment of truth values to its atoms.

6. Using the atomic diagram of an assignment τ of truth values to its atoms, *construct a* τ*-relative verification* of a formal sentence of propositional logic that *you are told* is true under τ.

7. Using the atomic diagram of an assignment τ of truth values to its atoms, *construct a* τ*-relative falsification* of a formal sentence of propositional logic that *you are told* is false under τ.

8. Using the atomic diagram of an assignment τ of truth values to its atoms, *construct the appropriate* τ*-relative evaluation* (a verification or a falsification) of a formal sentence of propositional logic, without being told what its truth value is.

9. Given what purports to be an evaluation of a sentence of *propositional* logic relative to some assignment of truth values to its atoms, tell whether it is indeed one; and whether it is a verification or a falsification.

10. Given what purports to be an evaluation of a sentence of *first-order* logic relative to some model M that interprets all its extra-logical vocabulary, tell whether it is indeed one; and whether it is a verification or a falsification.

11. Given a model M and a formal sentence φ of first-order logic *determine the truth values* in M of φ.

12. Using the atomic diagram of a model M, *construct an* M*-relative verification* of a formal sentence of first-order logic that *you are told* is true in M.

13. Using the atomic diagram of a model M, *construct an* M*-relative falsification* of a formal sentence of first-order logic that *you are told* is false in M.

14. Using the atomic diagram of a model M, *construct the appropriate* M*-relative evaluation* (a verification or a falsification) of a formal sentence of first-order logic that you are given, without being told what its truth values is.

15. Given a sentence of propositional logic, *construct a truth table* dealing with every possible assignment of truth values to its atoms,

in which every subsentence receives its proper truth values. Read off from the completed truth table whether the given sentence is logically true, logically false, or contingent.

16. Given a sentence φ of first-order logic that you are told is contingent,

    (a) *describe a model M* that interprets the extra-logical vocabulary in φ, and *construct* an M-relative verification for φ; and

    (b) *describe a model N* that interprets the extra-logical vocabulary in φ, and *construct* an N-relative falsification for φ.

17. Given a finite model M, *construct a sentence* using the extra-logical vocabulary interpreted in M *that categorically describes* M.

18. Given an argument $\varphi_1, \ldots, \varphi_n : \psi$ that you are told is invalid, *describe a countermodel M*—that is, a model M that makes all the premises true but the conclusion false. If your countermodel M is finite, construct M-relative verifications for $\varphi_1, \ldots, \varphi_n$, and construct an M-relative falsification for ψ.

19. Given an argument $\varphi_1, \ldots, \varphi_n : \psi$ that you are told is valid, *construct a natural deduction* whose conclusion is ψ, and whose undischarged assumptions are among $\varphi_1, \ldots, \varphi_n$.

20. Given an argument $\varphi_1, \ldots, \varphi_n : \psi$, *determine whether it is valid.* If it is *valid*, construct a *natural deduction* whose conclusion is ψ, and whose undischarged assumptions are among $\varphi_1, \ldots, \varphi_n$. If it is *invalid*, describe a *countermodel M* to it, and (if M is finite) construct M-relative verifications for $\varphi_1, \ldots, \varphi_n$ and an M-relative falsification for ψ.

21. Given an argument in English, translate it into logical notation as a formal argument $\varphi_1, \ldots, \varphi_n : \psi$. Determine whether the formal argument is valid. If it is valid, construct a natural deduction whose conclusion is ψ, and whose undischarged assumptions are among $\varphi_1, \ldots, \varphi_n$. If it is invalid, describe a countermodel M to it; and, if M is finite, construct M-relative verifications for $\varphi_1, \ldots, \varphi_n$ and an M-relative falsification for ψ.

Chapter 23

Some accessible metatheory

23.1 Strength of argument

Let Δ and Γ be two sets of sentences. Suppose that Δ logically implies every sentence in Γ. Then we say that Δ is *at least as strong as* Γ. If moreover Γ is not at least as strong as Δ—that is, if some sentence in Δ does not follow logically from Γ—then we say that Δ is *stronger than* Γ.

If Δ is at least as strong as Γ, and *vice versa*, then we say that Δ has *the same logical strength* as Γ.

If Δ is at least as strong as Γ, then the *argument* $\Gamma:\varphi$ is said to be *at least as strong as* the *argument* $\Delta:\varphi$. And if Δ is *stronger* than Γ, then the argument $\Gamma:\varphi$ is said to be *stronger than* the argument $\Delta:\varphi$. Note the reversal in order. The weaker the set of premises, the stronger the argument. An apt slogan might be that *more is less*—in the sense that the more (logical strength) one has in the *premises*, the less (logical strength) one has in any *argument from* those premises.

If Γ is a subset of Δ, then of course Δ logically implies every sentence in Γ; whence Δ is at least as strong as Γ, and the argument $\Gamma:\varphi$ is at least as strong as $\Delta:\varphi$.

Suppose that Γ is a *proper* subset of Δ—that is, every member of Γ is in Δ, but some member of Δ is not in Γ. If Γ logically implies all those members of Δ not in Γ, then Γ has the same logical strength as Δ; otherwise, Δ is stronger than Γ.

Any member of a set of sentences that follows logically from the other members is called *redundant* in that set. A set that contains no redundant members is called *irredundant*. Any irredundant set is stronger than each of its proper subsets. If Δ is irredundant and Γ is a proper subset of Δ, then the argument $\Gamma : \varphi$ is stronger than the argument $\Delta : \varphi$.

It is desirable, when stating an argument $\Delta : \varphi$, to have Δ irredundant. Mathematicians and logicians will not knowingly state an argument $\Delta : \varphi$ with redundant premises. Instead, they will discard premises in Δ found to follow from others in Δ, until those that remain form an irredundant set. They might arrive by different routes at different irredundant subsets of the original set Δ; but those irredundant subsets of Δ will at least have the same logical strength, since each of them will logically imply all the others' members.

We have seen that an argument can be strengthened by weakening its set of premises. If the original set of premises is irredundant, then discarding any of those premises will strengthen the argument. That is,

$$\left.\begin{array}{r}\Delta \text{ irredundant} \\ \Gamma \subset \Delta\end{array}\right\} \quad \Rightarrow \quad \Gamma : \varphi \text{ is stronger than } \Delta : \varphi$$

When a valid argument $\Delta : \varphi$ has a satisfiable set Δ of premises, considerable interest attaches to the question whether there are any *proper* subsets Γ of Δ for which $\Gamma : \varphi$ is still valid. (Note that this question can arise even when Δ is irredundant.) Most interesting among these proper subsets are the *inclusion-minimal* ones—those proper subsets Γ of Δ whose every proper subset does *not* logically imply φ. Such proper subsets Γ of Δ (which will obviously, themselves, be irredundant) yield the most ideal strengthenings $\Gamma : \varphi$ of the original argument $\Delta : \varphi$.

There is another way, however, to strengthen an argument, and that is to claim that its premises cannot all be true:

$$\Delta : \bot \text{ is stronger than } \Delta : \varphi$$

Note that when we say that an argument has been strengthened, we are not implying that the new (stronger) argument is itself valid. One can strengthen an argument so much that it becomes invalid—for example, by discarding sufficiently many premises, or by putting $\bot$ in place of the original conclusion when the original premises form a satisfiable set.

But when a strengthened version of a valid argument is itself valid, we have potential *epistemic gain* if we can prove it.

23.2 Epistemic gain

Our metalogical observations in this section should be accessible to the reader who has by now acquired a full understanding of the finer details of our proof system. This is a timely juncture at which to address the important issue of potential epistemic gain.

This primer has presented *relevant* formal logical systems for both constructive and classical deductive reasoning, in which the prospect of epistemic gain is prized. Ironically, this important feature of our relevant systems arises from examination of what is standardly thought to be a knock-down argument *for* the notorious, because *ir*relevant, First Lewis Paradox $A, \neg A : B$.

Purveyors of this argument invite one to consider a particular interpolant, the disjunction $A \vee B$. They point out that there is the obvious one-step proof

$$\frac{A}{A \vee B}\;;$$

and then that we also have the *Disjunctive Syllogism* $A \vee B, \neg A : B$, which in our system of constructive logic would enjoy the proof

$$\frac{A \vee B \qquad \dfrac{\neg A \qquad \overline{A}^{\,(1)}}{\bot} \qquad \overline{B}^{\,(1)}}{B}{\scriptstyle (1)}\;.$$

They then suggest that we should simply put these two proofs together (*via* the interpolant $A \vee B$) to get free passage from the premise-set $\{A, \neg A\}$ to the conclusion B.

That, however, is to fail to take seriously enough the prospect of epistemic gain. Such gain is to be had by applying the well-known *reduction procedure* for $\vee$. Every logical operator that is furnished with Introduction and Elimination rules has an associated reduction procedure. It is designed to get rid of unnecessary deductive detours through a sentence that is the conclusion of an Introduction at the end of one proof, but is the major premise of a corresponding Elimination at the beginning of another proof that is to be 'chained together' with the first proof. In our notation, the reduction to be performed in the case at hand would be

$$\left[\frac{A}{A \vee B} \quad , \quad \dfrac{A \vee B \qquad \dfrac{\neg A \qquad \overline{A}^{(1)}}{\bot} \qquad \overline{B}^{(1)}}{B}{\scriptstyle (1)} \right] ;$$

and this would—given *our* conception of how reductions are best put to work—deliver the *epistemically gainful* reduct

$$\frac{\neg A \quad A}{\bot} \; .$$

This rebuts the argument-by-interpolation for adoption of the First Lewis Paradox. Our system of inference rules and proof-formation ensures as much *transitivity of deduction* as can be demanded for the purposes of mathematical and scientific reasoning. We take care, however, not to avail ourselves uncritically of the orthodox treatment of such transitivity as *unrestricted*. That orthodox treatment would let the foregoing 'argument by interpolation' go through unchallenged, by use of the so-called 'Cut Rule' *in the system*:

$$\frac{\Delta : \varphi \quad \varphi, \Gamma : \psi}{\Delta, \Gamma : \psi} \; .$$

This, as just seen, would mire the user in the irrelevance of Lewis's First Paradox, leaving them frequently oblivious to the prospect of epistemic gain in 'chaining proofs together'.

One needs to stay alive to that prospect, but without losing anything essential in the way of overall logical power. We shall be proving in the intended sequel to this primer the following general result, of which the foregoing reduct is a very simple but illuminating instance. We call the following result the Cut Admissibility Theorem.

Metatheorem 3. *There is an effective method* [,] *that transforms any two proofs*

$$\begin{array}{ccc} \Delta & & \varphi, \Gamma \\ \Pi & & \Sigma \\ \varphi & & \theta \end{array}$$

(where $\varphi \notin \Gamma$ and Γ may be empty) into a proof (the 'reduct')

$$\begin{array}{c} \Omega \\ [\Pi, \Sigma] \\ \theta \end{array} \quad \textit{or} \quad \begin{array}{c} \Omega \\ [\Pi, \Sigma] \\ \bot \end{array} \ , \ \textit{for some } \Omega \subseteq \Delta \cup \Gamma.$$

The informally rigorous proof of this Metatheorem is both constructive and relevant (at the meta-level).[1] It proceeds by induction on the combined lengths of the proofs Π and Σ. It holds for constructive proofs, and it holds for classical proofs.

The Metatheorem has three immediate and important corollaries. In their statements, the single turnstiles stand for deducibility in each of the two systems—constructive and classical.

Corollary 2. CUT WITH POTENTIAL EPISTEMIC GAIN
If $\Delta \vdash \varphi$ *and* $\Gamma, \varphi \vdash \psi$, *then either* $\Delta, \Gamma \vdash \bot$ *or* $\Delta, \Gamma \vdash \psi$.

Corollary 3. CUT FOR ABSURDITY *(or* 'CUT *for* $\bot$'*)*
If $\Delta \vdash \varphi$ *and* $\Gamma, \varphi \vdash \bot$, *then* $\Delta, \Gamma \vdash \bot$.

Corollary 4. CUT ON CONSISTENT PREMISES
If $\Delta \nvdash \bot$ *and* $\Delta \vdash \theta$ *for every* θ *in* Γ *and* $\Gamma \vdash \psi$, *then* $\Delta \vdash \psi$.

These corollaries ensure that our system of *constructive* logic is adequate unto the methodological demands that can be placed on logics whose aim is to formalize the informally rigorous deductive reasoning that is to be found both in *constructive* mathematics and in the empirical sciences. Likewise, with the single turnstile read as classical deducibility, we are assured that our system of *classical* logic is adequate unto the methodological demands that can be placed on logics whose aim is to formalize the informally rigorous deductive reasoning that is to be found in *classical* mathematics

For those demands (or adequacy conditions), see Chapter 24. Proofs of the pertinent metatheorems will be furnished in the planned sequel to this primer.

---

[1]See Tennant [2012] and Tennant [2015a]. The result there was called 'Cut Elimination'. We now prefer 'Cut Admissibility', since our proofs do not contain any cuts that need to be eliminated.

23.3 Soundness and completeness

The SOUNDNESS claim above is to the effect that if one can prove an argument, then it is valid. More formally:

If $P_1, \ldots, P_n \vdash_S Q$, then $P_1, \ldots, P_n \models_L Q$.

We tacitly understand this to hold for all sentences $P_1, \ldots, P_n, Q$ in the language L. This can be read in two equivalent ways. Appealing to the definition of $\vdash_S$, we have

If there is an S-proof of Q from $P_1, \ldots, P_n$,

then $P_1, \ldots, P_n \models_L Q$.

And this in turn is equivalent to

For all Π, if Π is an S-proof of Q from $P_1, \ldots, P_n$,

then $P_1, \ldots, P_n \models_L Q$.

These are equivalent ways of claiming that the proof system S is *sound* for the language L.

We are using here the (metalinguistic) 'sortal variable' Π in place of an (unrestricted) individual variable x. This sortal variable is understood to range over just proofs (in the system S). But it would be easy to re-express claims involving the sortal variable Π as ones involving only an ordinary (metalinguistic) variable x ranging over things in general. One would use the translations

$$\exists x(x \text{ is an } S\text{-proof } \wedge \varphi(x)) \quad \text{for} \quad \exists\Pi\varphi(\Pi),$$

and

$$\forall x(x \text{ is an } S\text{-proof } \rightarrow \varphi(x)) \quad \text{for} \quad \forall\Pi\varphi(\Pi).$$

It is easy to prove the equivalence of any two sentences of the respective forms $\exists x\varphi(x) \rightarrow \psi$ and $\forall x(\varphi(x) \rightarrow \psi)$, when the bound variable x does not occur in ψ. We shall give two proofs, one to establish the deducibility claim

$$\exists x\varphi(x) \rightarrow \psi \vdash \forall x(\varphi(x) \rightarrow \psi),$$

and the other to establish its converse:

$$\forall x(\varphi(x) \to \psi) \vdash \exists x\varphi(x) \to \psi.$$

The respective formal proofs are composed by applying the introduction and elimination rules governing the three operators $\exists$, $\forall$ and $\to$ that appear in the sentences involved. The two proofs are

$$\dfrac{\dfrac{\dfrac{\dfrac{\overline{\varphi(a)}^{(1)}}{\exists x\varphi(x)} \quad \exists x\varphi(x) \to \psi}{\psi}}{\varphi(a) \to \psi}{\scriptstyle (1)}}{\forall x(\varphi(x) \to \psi)} \qquad \text{and} \qquad \dfrac{\dfrac{\overline{\exists x\varphi(x)}^{(2)} \quad \dfrac{\overline{\varphi a}^{(1)} \quad \dfrac{\forall x(\varphi(x)\to\psi)}{\varphi(a)\to\psi}}{\psi}}{\psi}{\scriptstyle (1)}}{\exists x\varphi(x) \to \psi}{\scriptstyle (2)}$$

23.4 Strong v. weak completeness, and compactness

One can *counterexemplify* an argument just in case it is *invalid.* More formally:

> There is a counterexample to $P_1, \ldots, P_n : Q$
>
> if and only if
>
> $P_1, \ldots, P_n \not\models_L Q.$

So no counterexemplifiable argument is valid. Hence—assuming SOUNDNESS holds—no counterexemplifiable argument is provable.

Everything said thus far leaves open the possibility that some L-valid arguments are unprovable in the (sound) system S. That is, the space of arguments might look like this (adopting the diagramming convention that each point in the diagram, like the one labeled α, represents an argument):

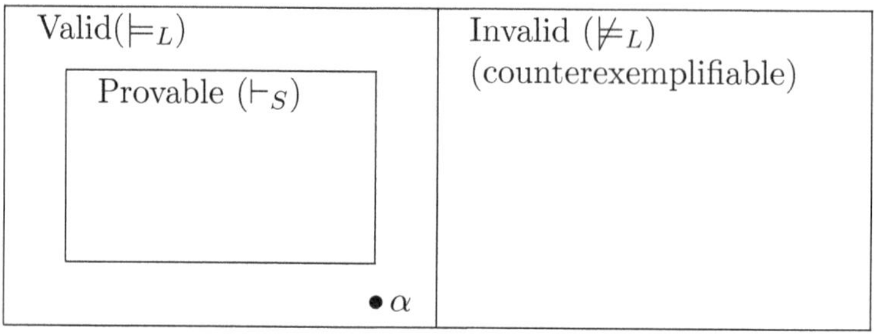

If there were any such (valid) arguments α, they would admit neither of proof (*ex hypothesi*) nor of counterexample (since they are *valid*). We need to rule out the possibility of such arguments α. That is, we need to ensure that every argument admits *either* of proof *or* of counterexample. We can express this by saying that the proof system S needs to be *complete*. That is, it needs to be the case that *every L-valid argument has a proof in the system S*:

STRONG COMPLETENESS
If $\Delta \models_L \varphi$, then $\Delta \vdash_S \varphi$.

Strong completeness would ensure that the border region of the left half of the diagram above was empty—that is, that there would be no argument like α. That region of the diagram would be uninhabited. The box labeled 'Provable' would expand to coincide with the box labeled 'Valid':

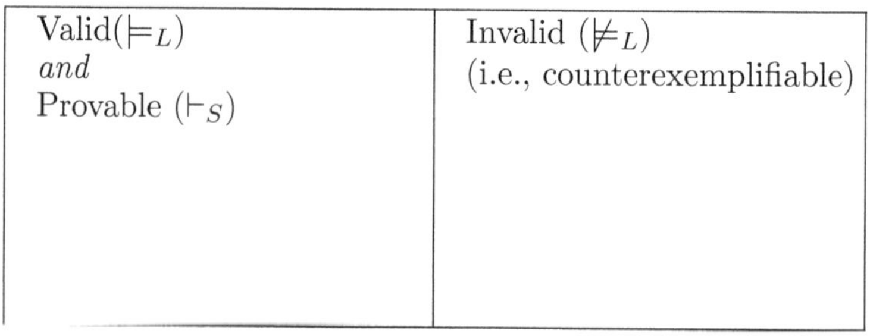

Note that STRONG COMPLETENESS is the most general possible *converse* of SOUNDNESS. The set Δ here may contain infinitely many sentences. If we were to restrict our attention to arguments with only finitely many premises, the statement of completeness would be weaker:

WEAK COMPLETENESS
If $P_1, \ldots, P_n \models_L Q$, then $P_1, \ldots, P_n \vdash_S Q$.

This is because for some languages L, some weakly complete proof systems S are not strongly complete.[2]

STRONG COMPLETENESS guarantees that logical consequence, like deducibility, is *compact* (for deducibility itself is compact):

COMPACTNESS OF LOGICAL CONSEQUENCE
If $\Delta \models \varphi$, then, for some finite subset Γ of Δ, we have $\Gamma \models \varphi$.

If a sound proof system S is weakly complete for a language L *whose logical consequence relation is compact*, then S is strongly complete for consequence in L:

$$\left.\begin{array}{r} S \text{ is weakly complete for } \models_L \\ \models_L \text{ is compact} \end{array}\right\} \rightarrow S \text{ is strongly complete for } \models_L$$

For, suppose that the two conditions on the left hold. Suppose $\Delta \models_L \varphi$. By compactness of $\models_L$, there are finitely members $P_1, \ldots, P_n$ of Δ such that $P_1, \ldots, P_n \models_L \varphi$. By weak completeness of S, we have $P_1, \ldots, P_n \vdash_S \varphi$. Hence $\Delta \vdash_S \varphi$.

It follows that any language whose proof system is weakly, but not strongly, complete has a non-compact relation of logical consequence.

Metalogic is that (mathematical) discipline devoted to establishing such results as that certain proof systems S are both sound and complete, for certain kinds of language L. As we explained earlier, the prefix 'meta' indicates a move 'one level up', as it were. At the lowest level, one is simply 'doing logic'. That is, one is negotiating (by means of proof) or blocking (by means of counterexample) logical passages *within* the object language L in question. At the metalevel ('one level up') one theorizes about how certain rules of inference guarantee the transmission of

[2]The first-order logic $L(Q_0)$ with the quantifier 'there exist *infinitely many* x such that ...' is not compact. Nor is the first-order logic $L(Q_1)$ with the quantifier 'there exist *uncountably many* x such that ...'. At the propositional level, examples of non-compact logics are the modal logic GL of Gödel and Löb $[\mathsf{K} + \Box(\Box p \rightarrow p) \rightarrow \Box p]$ and the modal logic Grz of Grzegorczyk $[\mathsf{K} + \Box(\Box(p \rightarrow \Box p) \rightarrow p]$. K is the system named after Kripke. It consists of just the Introduction and Elimination rules for $\Box$, and in its 'possible worlds' semantics no conditions at all are placed on the accessibility relation among worlds. Infinitary quantifiers and modal logics are not covered in this primer, but they will be covered in the intended sequel.

truth (soundness), and how the proof system S based on those rules is able to furnish any L-valid argument with a proof (completeness). Such theorizing will be undertaken in detail in the intended sequel to this primer.

23.5 Extremal cases of orthodox validity

One virtue of the orthodox definition of logical consequence is its simplicity. It has the drawback, however, of distorting pre-formal intuitions about validity, concerning cases at either one of two extremes.

The first extremal case is when $P_1, \ldots, P_n$ form an unsatisfiable set. Since then *no* interpretation makes $P_1, \ldots, P_n$ true, the universal generalization 'every interpretation that makes $P_1, \ldots, P_n$ true makes Q true' holds *vacuously*, for want of a counterexample. Hence, on the orthodox account, such an argument $P_1, \ldots, P_n : Q$ is counted as valid—even though there might not be any 'meaning connection' at all between the premises $P_1, \ldots, P_n$ and the conclusion Q. The former can be utterly irrelevant to the latter, as is the case with the notorious and counterintuitive argument

$$A, \neg A : B$$

—the First Lewis Paradox—an argument licensed as valid on the orthodox account of validity. But it is regarded as *invalid* (and *obviously* so) by those who insist on *relevance* of premises to conclusion as a precondition for the *genuine* validity of an argument. For such theorists, mere 'truth-preservation' is not enough for genuine validity. One needs *relevance* (of premises to conclusion) as well.

The second extremal case, dual to the first, is exemplified (in the case of classical logic) by the *Second Lewis Paradox*:[3]

$$A : B \vee \neg B.$$

The general case we have in mind here is one in which the conclusion Q is logically true, hence does not need the support of any of the premises $P_1, \ldots, P_n$, severally or jointly. To say that the argument $P_1, \ldots, P_n : Q$ is valid (for cases where Q itself is logically true) is to invite the more

[3]For constructive logic, in which the 'law of excluded middle' $B \vee \neg B$ does not hold, a similar paradox of relevance would be $A : B \rightarrow B$.

charitable reading that Q follows from *some (possibly empty) subset* of $\{P_1, \ldots, P_n\}$. The charitable reading, therefore, is that when one asserts the validity of an argument $\Delta : \varphi$, one is 'really' saying something like 'For some (possibly empty) subset Γ of Δ, the argument $\Gamma : \varphi$ is *genunely/relevantly/really* valid'. Or, to put it more pithily: a sequent is valid (in the orthodox sense) just in case it contains a *genuinely* valid subsequent (with the same conclusion). Here—that is, in a 'second Lewis' case—the 'subsetting down' that is envisaged is on the *premises'* side, namely on the left.

In a completely analogous way, the asserter of validity in a 'first Lewis' case might envisage 'subsetting down' on the *conclusion's* side, namely on the right. But the only way to 'subset down' on a *single-*conclusion argument $\Delta : \varphi$ is to consider the subsequent $\Delta : \bot$ (by *removing* the conclusion φ). So, the thought might go, when one says that the *irrelevant* sequent $A, \neg A : B$ is valid, one is saved by the fact that the *relevant* subsequent $A, \neg A : \bot$ is valid.

But the asserter of validity, in either the first extremal case or the second extremal case, should not be able to rely on such interpretive charity being extended. The ploy depends on being able to invoke some *(proper) subsequent* in which there is 'no funny business'—no lack of relevance between premises (on the left) and conclusion (on the right). So, the reforming thought can go, why not simply concentrate on the kind of validity displayed by the subsequents to which the orthodox logician has to appeal when invoking the 'subsetting down' considerations? Why not *restrict* the application of the honorific 'valid argument' to the arguments that really deserve it?

The problem with this proposal is that the weight of tradition is simply too great. It would be wiser, then, to introduce a new term—'perfect validity', say—to capture what we are after.[4] Ordinary validity can then be defined in the usual way; and it will turn out that every valid sequent has a perfectly valid subsequent.

These considerations lead one, however, to think more carefully about the exact form of strong completeness that one should require of one's logic. Why not simply require this?:

STRONG COMPLETENESS (EPISTEMICALLY GAINFUL FORM)

[4]See Tennant [1984].

If $\Delta \models_L Q$, then for some $\Delta' \subseteq \Delta$, either $\Delta' \vdash_S Q$ or $\Delta' \vdash_S \bot$

This form of strong completeness says that every valid sequent has a provable subsequent. It offers *potential epistemic gain* by possibly *strengthening* the argument when proving it. One could eschew some of the given premises, or even strengthen the conclusion Q to $\bot$.

This will be the form of strong completeness that our system of logic will enjoy. Given any (conventionally) valid argument $\Delta : Q$, we shall be able to construct a proof whose premises are in Δ, and whose conclusion is either Q or $\bot$. That enables one to deal forthrightly with the notorious Lewis sequent $A, \neg A : B$. The standard logician insists that it is indeed valid, on the orthodox construal of validity that we have just discussed. Very well then, we can respond: here is a proof of a *subsequent* of that sequent!:

$$\frac{\neg A \quad A}{\bot} .$$

23.6 Perfect validity

When an argument $P_1, \ldots, P_n : Q$ is said to be *perfectly valid*, all the premises $P_1, \ldots, P_n$ mentioned are taken to be 'doing some work' in helping to secure the conclusion Q. We shall also be assuming that Q is in some sense a prize to be had: it is not simply masking an underlying absurdity ($\bot$); *nor* is Q a logical truth. So, for perfect validity, the following will hold:

> No perfectly valid sequent has a perfectly valid proper subsequent.

Recall that an argument with no premises (or 'with the empty set of premises') amounts to the claim that its conclusion is *logically true*—that is, true under all possible interpretations of the non-logical vocabulary involved. So, for such an argument

$$\emptyset : Q$$

to be (perfectly) valid, Q itself must be logically true.

Arguments with at least one premise call for a different treatment.

Such an argument $P_1, \ldots, P_n : Q$ is *perfectly valid* just in case one of the following three mutually exclusive conditions holds:

1. (a) Q is $\bot$;
   (b) the set $\{P_1, \ldots, P_n\}$ is unsatisfiable; and
   (c) every proper subset of $\{P_1, \ldots, P_n\}$ is satisfiable.

2. (a) Q is not $\bot$;
   (b) Q is not logically true;
   (c) $\{P_1, \ldots, P_n\}$ is satisfiable;
   (d) every situation making every member of $\{P_1, \ldots, P_n\}$ true makes Q true; and
   (e) condition (2d) does not hold of any proper subset of $\{P_1, \ldots, P_n\}$ in place of $\{P_1, \ldots, P_n\}$.[5]

3. (a) Q is not $\bot$;
   (b) $\{P_1, \ldots, P_n\}$ is unsatisfiable; and
   (c) $P_1, \ldots, P_n : Q$ is a substitution instance of an argument $P'_1, \ldots, P'_n : Q'$ that is valid by virtue of (2).

Examples of perfectly valid arguments of
type (1): There is only one thing, There are exactly two things : $\bot$;
type (2): $\varphi \wedge \psi : \varphi$;
type (3): $\varphi \wedge \neg\varphi : \varphi$, which is a substitution instance of $\varphi \wedge \psi : \varphi$.

Foils. On our definition of perfect validity, the following arguments, which are regarded as 'ordinarily' valid in standard logic, are not perfectly valid:

$$\varphi : \neg(\psi \wedge \neg\psi)$$

$$\varphi \wedge \neg\varphi : \psi$$

$$\varphi \wedge \neg\varphi : \neg(\psi \wedge \neg\psi)$$

The first argument is to the effect that any proposition logically implies a logical truth; the second is to the effect that every logically false proposition logically implies any proposition; and the third is to the effect

[5] Note that (2a) follows from (2c) and (2d). But we state (2a) explicitly in order to make it immediately evident that conditions (1), (2) and (3) are mutually exclusive. Note also that (1b) follows from (1a). But we state (1b) explicitly in order to highlight its contrast with (1c).

that every logically false proposition logically implies any logically true proposition. Virtually every beginner, and many a trained expert, finds it counterintuitive to call such arguments valid. For they lack the connection between premise(s) and conclusion that is intuitively needed in order to *constitute* the validity of the passage from the premise to the conclusion. But according to standard logic, such arguments, despite their lack of any such connection between their premises and their conclusion, are nevertheless (technically) valid. This is an over-simplification made only for (what is *alleged* to be) 'theoretical convenience'. Here, our intention has been to rectify this situation, by providing a more demanding definition of *perfect* validity, which coheres better with pre-formal intuitions.

Note that a sequent is valid (in the standard sense) just in case it has a perfectly valid subsequent.

Chapter 24

Adequacy Conditions on a System of Proof

It is time to take stock of what has been achieved, *methodologically*, by formulating our systems of logic for constructive and classical deductive reasoning.

24.1 Five Main Adequacy Conditions on a System of Proof

Consider a logical system $\mathcal{S}$ that is intended to provide formal proofs to regiment all the informally rigorous proofs in mathematics. It is what is colloquially called a 'no-brainer' that such a system $\mathcal{S}$ would have to satisfy the following five conditions:

1. $\mathcal{S}$-proofs are finite. Hence, the set of premises of any $\mathcal{S}$-proof is finite. The relation 'Π is an $\mathcal{S}$-proof of the sentence φ whose premises form the set Δ' is effectively decidable.

2. If there is an $\mathcal{S}$-proof of $\perp$ (absurdity) whose premises form the set Δ, then Δ has no model.

3. If Δ is a set of sentences that has no model, then there is an $\mathcal{S}$-proof of $\perp$ (absurdity) from premises in Δ.

4. If Δ has a model and there is an $\mathcal{S}$-proof of φ from premises in Δ, then every model of Δ makes φ true.

5. If Δ has a model and every model of Δ makes φ true, then there is an $\mathcal{S}$-proof of φ from premises in Δ.

Lemma 4. *(1), (2), (3), (4), and (5) do not collectively entail that if there is an $\mathcal{S}$-proof of $\bot$ whose premises form the set Δ, then there is an $\mathcal{S}$-proof of any sentence whatsoever from Δ.*

Proof. The system of classical first-order logic with identity that has been developed in this Primer meets requirements (1), (2), (3), (4), and (5). It furnishes a one-step (constructive) proof of $\bot$ from the set $\{A, \neg A\}$:

$$\frac{\neg A \quad A}{\bot}\ ;$$

but it furnishes no proof at all of 'arbitrary' B from $\{A, \neg A\}$. □

24.2 A Sixth Condition of Adequacy: Relevance

The following Condition of Adequacy is never mentioned in the orthodox literature on mathematical logic. We state it here because *the informally rigorous mathematical reasoning that a proof system is expected to regiment is always relevant.* The following is therefore a reasonable Condition of Adequacy to impose on any proof system $\mathcal{S}$ in which the regimenting is to take place.

6. The premise set Δ and the conclusion φ of any $\mathcal{S}$-proof are mutually relevant.

This notion of relevance is an informal one, even in the context of a system of formal proof. But it can be formally explicated as a syntactic relation $\mathcal{R}(\Delta, \varphi)$; and one can rigorously and constructively prove that *$\mathcal{R}(\Delta, \varphi)$ holds for every proof in the system of classical logic developed herein whose conclusion is φ and whose premises form the set Δ.*[1] (The result therefore holds also for every proof in the subsystem of *constructive* logic presented herein.)

[1] See Tennant [2015b].

That this italicized claim holds can be summed up in more expansive detail as follows. There are three mutually exclusive and collectively exhaustive possibilities for Δ and φ:

1. $\Delta \neq \emptyset$ and $\varphi = \bot$;

2. $\Delta \neq \emptyset$ and $\varphi \neq \bot$;

3. $\Delta = \emptyset$ and $\varphi \neq \bot$.

Note that the would-be fourth case ($\Delta = \emptyset$ and $\varphi = \bot$) is impossible. One cannot derive $\bot$ from no premises at all.

All one needs, in order to appreciate how 'tight' are the deductive connections forged by our foregoing rules of inference for classical logic, is the notion of positive v. negative occurrence of a subformula within a formula. Recall Definition 5 on p. 112:

> Any formula occurs positively within itself. ψ occurs negatively both in $\neg\psi$ and in $\psi \rightarrow \theta$. ψ and θ occur positively in $\psi \wedge \theta$ and in $\psi \vee \theta$. θ occurs positively in $\psi \rightarrow \theta$. ψ occurs positively both in $\forall x\psi$ and in $\exists x\psi$.

Armed now with these notions of positive and negative occurrences, we can explain what $\mathcal{R}(\Delta, \varphi)$ means in each of the three cases above.

In case (1), Δ is provably inconsistent. If Δ is a singleton $\{\theta\}$, then θ contains both a positive occurrence and a negative occurrence of some same atomic formula. (Simple example: $\neg\forall x(Fx \rightarrow Fx)$.) If, on the other hand, Δ contains more than one sentence, then Δ is 'bowtie connected': that is, every sentence in Δ can be connected to every other sentence in Δ by successive steps of the symmetric 'bowtie' relation $\psi \bowtie \theta$, which is defined as follows: some atomic formula occurs positively in one of ψ, θ but negatively in the other. (Simple example: $\forall xFx, \exists x\neg Fx$.)

Case (2) is the all-important central case, where Δ is a consistent set of axioms used in a proof of φ as a theorem of one's mathematical theory. Here, Δ partitions into one or more internally 'bowtie connected' cells; and each cell has a member θ that contains an atomic formula A occurring within it with the same 'parity' (positive or negative) that an occurrence of A enjoys in φ.

Case (3) is that of φ being a logical theorem; and φ will contain both a positive occurrence and a negative occurrence of some same atomic formula.

This effectively decidable condition $\mathcal{R}(\Delta, \varphi)$ can be implemented as a 'relevance filter' in automated proof search, to greatly increase its efficiency.

Exercise 64. *Look back at all the valid arguments that you have proved thus far. Satisfy yourself that in every one of those proofs the premise-set Δ and the conclusion φ stand in the relation $\mathcal{R}(\Delta, \varphi)$ that we have just explained by reference to the foregoing Cases (1), (2), and (3).*

24.3 Another Condition of Adequacy: Feasible Formalizability

There is much to be said in favor of one more condition. In all likelihood it (like Condition 6 above) has not been much considered by formal logicians, even when they address the matter of a logic's adequacy unto the deductive demands of mathematics and science.

7. Given any proof P in a mathematics journal or textbook, written by an expert mathematician, and acknowledged within the pertinent mathematical community as meeting their standards of informal rigor, the following task will be feasible for proponents of system $\mathcal{S}$: extract from the proof P its conclusion φ, along with the set Δ of its premises; then regiment P as a fully formal $\mathcal{S}$-proof of φ from Δ, in such a way as to merely supply missing detail, while preserving at the macro-level all the various 'lines of argument' that an expert mathematician can discern within P.

This is a much more exigent requirement on the relationship between informally rigorous proofs and their formal regimentations than any that has been broached in the logical literature thus far.

Let us focus now on Conditions (2)–(5).

(2) can be thought of as "soundness of $\mathcal{S}$ with respect to unsatisfiable

sets of sentences".

(3) can be thought of as "completeness of $\mathcal{S}$ with respect to unsatisfiable sets of sentences".

(4) can be thought of as "soundness of $\mathcal{S}$ with respect to satisfiable sets of sentences".

(5) can be thought of as "completeness of $\mathcal{S}$ with respect to satisfiable sets of sentences".

(3) entails that every logically false sentence φ can be refuted in $\mathcal{S}$; that is, there is an $\mathcal{S}$-proof of $\bot$ from (the singleton of) φ.

Observation 3. *The empty set (of sentences) is not merely satisfiable, that is, satisfied in* some *model; it is satisfied in* every *model. This is because every model makes every one of the empty set's members true—because there aren't any.*

(4) entails that any $\mathcal{S}$-theorem (i.e., sentence with an $\mathcal{S}$-proof whose premise-set is empty) is (by Observation 3) a logical truth, i.e. true in every model.

(5) entails that every logically true sentence φ can be proved outright in $\mathcal{S}$; that is, φ is the conclusion of an $\mathcal{S}$-proof whose premise-set is empty.

Note that Lemma 4 would be trivially true if one were to add condition (6) above to the list of five conditions that Lemma 4 mentions.

Lemma 4 tells us that (1), (2), (3), (4), and (5) do not force the logic to be what contemporary logicians call 'explosive'. We are morally certain that a strengthening of Lemma 4 can be achieved by incorporating (7); but at this stage this remains a conjecture.

It is worth stressing that we are engaging here in an *a prioristic* methodological investigation, insofar as it involves merely reflecting on, and analyzing, the clear and distinct *deductive needs* of both mathematicians and natural scientists. The aim is to formulate (as we have done above) *conditions of adequacy* on any logical system that is to satisfy those needs. Once we have our logic, *and have established metatheorems that show that it meets the conditions of adequacy*, we can proceed to develop our theories.

This is the proper order of foundational study.

It is the *opposite* of the recently touted 'abductive methodology' (see, for example, Williamson [2018] and the author's own critique of it, Tennant [2022]) according to which we are presumed to have prior access to theories in general (both mathematical and scientific) and then somehow—by means of criteria both vague and implausible—work backwards from those theories (closed, one might ask, under *what* relation of logical consequence or deducibility, exactly?) to some choice of logical system the preference for which is based on intuitive and unexplicated notions of simplicity, elegance, fit with evidence, and the like.

Chapter 25

Afterword

Ex Falso Quodlibet has been a persisting defect in formal logic from the ancients through the medievals. The following rogues' gallery of EFQ -purveying medieval logicians includes (left to right) Peter Abelard 1079–1142; Duns Scotus c. 1265/66–1308; John Buridan 1300–1361; Alexander Neckham, c. 1200; and Peter of Spain in the 1240s. (We have no image for William of Soissons, who was complicit with them, in the late 1100s.)

EFQ persisted even unto Frege, the founding father of modern (classical) first-order logic, and his most notable heirs and successors in the twentieth century. Frege's *Begriffsschrift* of 1879 was furnished with a completely formal system of proof, intended to be used in the rigorous formalization of classical mathematical reasoning. After the disaster of Russell's Paradox for Frege's class theory (this disaster resulting from Frege's failure to work in a *free* logic), Russell and Whitehead emerged as the next most influential custodians of logical tradition, still embodying this defect. Their monumental three-volume *Principia Mathematica* (1910–1913) was the 'Bible' of foundational investigations in mathematics for almost two decades; *and* it was committed to EFQ..

As those two decades drew to a close the world-famous mathematician-logician Hilbert (next to Frege in the twentieth-century rogues' gallery above) preserved the same canon of formal logic—with the wart EFQ and all—in the highly influential *Grundzüge der Theoretischen Logik*, co-authored with Ackermann, and published in 1928. This book was so canonical that it was the source of the completeness problem for first-order logic, which Gödel solved positively in his *Doktorarbeit* of 1929. This was published as Gödel [1930]. The book was also the source, for Alonzo Church, of the *Entscheidungsproblem* for first-order theoremhood, in what Church called 'the engere Funktionenkalkül of Hilbert and Ackermann'. See Church [1936].

Even Brouwer, the revolutionary intuitionist who famously rejected the Law of Excluded Middle,[1] indifferently tolerated EFQ.[2] EFQ even found its way into Gentzen's 1934-5 breakthrough formal renderings of deductive proofs as natural deductions—which Hilbert and Ackerman hailed in their second, revised edition of their *Grundzüge* in 1938.[3]

EFQ has ever since continued to besmirch all subsequent revampings of the systems conventionally known as Intuitionistic Logic and Classical Logic, whenever they were used or theorized about by the later famous

---

[1]Brouwer [1908]. For an English translation see Atten and Sundholm [2016].

[2]Or perhaps we should say that his student *Heyting* did, with his formalization of what he took to be Brouwer's logical canon. See Heyting [1930].

[3]Hilbert and Ackermann [1938]. The English translation is Hilbert and Ackermann [1950].

figures in the gallery above. We must mention, however, that one (and only one) of the giants of twentieth-century mathematical logic in the gallery above ventured into print with some principled misgivings about EFQ. The reader will presently learn who that figure was

Exercise 65. *Identify all the twentieth-century logicians pictured above.*

The systems of Intuitionistic Logic and (orthodox) Classical Logic purported to offer *exactly* what is needed (no more, and no less), inferentially, for the full formalization (i.e., rigorous regimentation) of the deductive reasoning that is to be found in constructive and classical mathematics respectively. But these two systems *overprovide* in that regard, by serving up as would-be licit proofs ones that are *irrelevant*, and therefore not needed at all for the formalization of mathematical reasoning.

The extraordinary thing about EFQ is that it is *extremely counter-intuitive* yet has been preserved for so long within the tradition. The usual pedagogical rationalizations of EFQ for the beginning student who expresses opposition to it, or even dismay and consternation about it, is that EFQ (allegedly) is an 'indispensable' part of the full canon of logical reasoning (*both* in constructive *and* in classical mathematics). The student is asked to accept that EFQ is there for 'rounding out', or 'systematically smoothing' what would otherwise be detracting wrinkles in the logical system as a whole.

These vaguely metaphorical and apologetic handwavings by experts, to the acolyte, have been going on for far too long. EFQ is *not* needed for the full formalization of deductive reasoning. It is *dispensable.* So why put up with it? Its presence in any system of logic commits the system to the 'relevance howler' of what is deservedly called Lewis's First **Paradox**:

$$\frac{A \quad \neg A}{B},$$

which Lewis brought to the attention of an indifferent audience as early as 1932.[4] The only logician at that time to try to get rid of EFQ was Johansson.[5] But his system of 'Minimal Logic' (later treated in Prawitz

[4] Lewis and Langford [1932].

[5] Johansson [1936].

[1965] alongside Intuitionistic and orthodox Classical Logic) failed to rid logic of the Lewis Paradox in its *negative* form

$$\frac{A \quad \neg A}{\neg B} \ .$$

We defy any would-be formalizer of mathematical reasoning to identify any passage in a mathematical text where the positive *or negative* form of the Lewis Paradox would be *needed* to make a proof go through. It simply isn't.

That it isn't has been the hidden lesson of all that has been expounded above.

Logicians, methodologists, and philosophers seem to have lost sight of how Tarski [1944], at pp. 167–8, anticipated eschewal of *Ex Falso Quodlibet* (EFQ):[6]

> I believe everybody agrees that one of the reasons which may compel us to reject an empirical theory is the proof of its inconsistency: a theory becomes untenable if we succeed in deriving from it two contradictory sentences. Now we can ask what are the usual motives for rejecting a theory on such grounds. Persons who are acquainted with modern logic are inclined to answer this question in the following way: A well-known logical law [EFQ—NT] shows that a theory which enables us to derive two contradictory sentences enables us also to derive every sentence; therefore, such a theory is trivial and deprived of any scientific interest. ...
> *I think that people who do not know modern logic are as little inclined to accept an inconsistent theory as those who are thoroughly familiar with it; and probably this applies even to those who regard (as some still do) the logical law on which the argument is based as a highly controversial issue, and almost as a paradox. I do not think that our attitude toward an inconsistent theory would change even if we decided for some reasons to weaken our system of logic so as to deprive*

[6]In connection with Tarski's mention of the proof of an empirical theory's inconsistency (either internal, or by conflicting with empirical observations), the reader should be reminded of our disproof display on p. 45.

> *ourselves of the possibility of deriving every sentence from any two contradictory sentences.* [Emphasis added]

This primer set out to introduce the beginner to the discipline of Symbolic Logic. It has departed significantly but, one hopes, barely detectably, from received tradition. It has done so in the very way that Tarski (approvingly) envisaged—that is, in 'weakening' the formal system of proof in the received tradition precisely 'so as to deprive the user of the possibility of deriving every sentence from any two contradictory sentences.' It has confined the beginner's attention (but not in so many words) to the relevant systems of *Core Logic* (for deductive reasoning that is constructive) and *Classical Core Logic* (for deductive reasoning that is non-constructive—or, as one also says, 'strictly classical'). Elsewhere we have called these two systems $\mathbb{C}$ and $\mathbb{C}^+$ respectively. But in this primer we have not used either of these labels. We have just spoken matter-of-factly of 'constructive logic' and 'classical logic'.

These two systems $\mathbb{C}$ and $\mathbb{C}^+$ were treated in depth in the research monograph *Core Logic* (Oxford University Press, 2017), whose intended audience was one of professional logicians, especially proof theorists. There is now a need, however, to introduce *beginners* to the inner workings of these two systems. This we have sought to do here with a minimum of fuss and without going out of our way to draw attention prematurely to their perhaps surprising adequacy unto the needs of deductive reasoning—surprising, because of the important differences between the two relevant systems $\mathbb{C}$ and $\mathbb{C}^+$, on the one hand, and the two *ir*relevant systems, in the received tradition, of Intuitionistic Logic and Classical Logic.

Researchers in Logic have already been steeped in certain aspects of the post-Fregean tradition that can make it difficult for them to appreciate the technical simplicity and the methodological adequacy of the Core systems. A beginner who approaches Symbolic Logic with no preconceptions at all, with a willingness to learn, and with an open mind, should be able to acquire both mastery of the technicalities and confidence in the systems' adequacy to fulfill the original foundational aims of the discipline of Logic.

The Introduction to *Core Logic* described it as a beautiful, elegant, and powerful system. It has been in Plato's Heaven all this time, waiting to be discovered. The same description applies to its classical extension $\mathbb{C}^+$. This book has set out to present the beginner with a gentle and accessible introduction to the two systems, without formally naming them. We have simply spoken of 'constructive logic' and 'classical logic' throughout. We hope their treatment here might have won some converts. If not, the failing is ours, not to have done the Core systems the justice they deserve.

Bibliography

G. B Airy. Account of some circumstances historically connected with the discovery of the planet exterior to Uranus. *Monthly Notices of the Royal Astronomical Society,*, VII(9):121–144, 1846.

J. E. J. Altham. *The Logic of Plurality.* Methuen, London, 1971.

J. E. J. Altham and Neil W. Tennant. Sortal Quantification. In Edward L. Keenan, editor, *Formal Semantics for Natural Language*, pages 46–58. Cambridge University Press, Cambridge, 1975.

M. van Atten and B. G. Sundholm. L. E. J. Brouwer's 'Unreliability of the Logical Principles': A New Translation, with an Introduction. *History And Philosophy Of Logic*, 38:24–47, 2016. doi: https://doi.org/10.1080/01445340.2016.1210986.

Eugenio Beltrami. Saggio di interpretazione della geometria non-euclidea. *Giornale di Matematiche*, 6:284–312, 1868.

Paul Benacerraf. God, the Devil and Gödel. *The Monist*, 51:9–32, 1967.

L. E. J. Brouwer. Over de onbetrouwbaarheid der logische principes. *Tijdschrift voor Wijsbegeerte*, 2:152–158, 1908.

John P. Burgess. Logic and Time. *The Journal of Symbolic Logicc*, 44 (4):566–582, 1979.

Alonzo Church. A note on the Entscheidungsproblem. *Journal of Symbolic Logic*, 1(1):40–41. *Correction, ibid.*, pp. 101–2, 1936.

René Descartes. *La Geometrie* (third appendix within *Discours de la méthode pour bien conduire sa raison, & chercher la verité dans les sciences*). Jan Maire, Leiden. Published anonymously, 1637.

Gottlob Frege. *Grundgesetze der Arithmetik. I. Band.* Georg Olms Verlagsbuchhandlung, Hildesheim, 1893; reprinted 1962.

Gottlob Frege. *Grundgesetze der Arithmetik. II. Band.* Georg Olms Verlagsbuchhandlung, Hildesheim, 1903; reprinted 1962.

Gerhard Gentzen. Untersuchungen über das logische Schliessen. *Mathematische Zeitschrift*, I, II:176–210, 405–431, 1934, 1935. Translated as 'Investigations into Logical Deduction', in *The Collected Papers of Gerhard Gentzen*, edited by M. E. Szabo, North-Holland, Amsterdam, 1969, pp. 68–131.

Edmund Gettier. Is Justified True Belief Knowledge? *Analysis*, 23(6): 121–123, 1963.

Kurt Gödel. Die Vollständigkeit der Axiome des logischen Funktionenkalküls. *Ergebnisse eines mathematischen Kolloquiums*, 37:349–360, 1930.

Kurt Gödel. Über formal unentscheidbare Sätze der Principia Mathematica und verwandter Systeme I. *Monatshefte für Mathematik und Physik*, 37:173–198, 1931.

Paul Henle. On the Fourth Figure of the Syllogism. *Philosophy of Science*, 16(2):94–104, 1949.

Arend Heyting. Die formalen Regeln der intuitionistischen Logik. *Sitzungsberichte der Preussischen Akademie der Wissenschaften. Physikalisch-mathematische Klasse. Deutsche Akademie der Wissenschaften zu Berlin, Mathematisch-Naturwissenschaftliche Klasse*, pages 42–56; 57–71; 158–169, 1930. English Translation in Mancosu, P. (Ed.) (1998). From Brouwer to Hilbert. The Debate on the Foundations of Mathematics in the 1920s. Oxford University Press.

D. Hilbert and W. Ackermann. *Grundzüge der Theoretischen Logik, 2. Auflage.* Springer, Berlin, 1938.

D. Hilbert and W. Ackermann. *Principles of Mathematical Logic.* Chelsea Publishing Co., New York, 1950.

Jaakko Hintikka. *Knowledge and Belief – An Introduction to the Logic of the Two Notions.* Cornell University Press, Ithaca, NY, 1964.

Ingebrigt Johansson. Der Minimalkalkül, ein reduzierter intuitionistischer Formalismus. *Compositio Mathematica*, 4:119–136, 1936. doi: https://doi.org/10.2307/2268823.

Philip N. Johnson-Laird. *Mental Models: Towards a Cognitive Science of Language, Inference, and Consciousness.* Harvard University Press and Cambridge University Press, Cambridge, Mass. and Cambridge, 1983.

John G. Kemeny. Models of Logical Systems. *Journal of Symbolic Logic*, 13(1):16–30, 1948.

Stephen Cole Kleene. *Introduction to Metamathematics.* Van Nostrand, Princeton, New Jersey, 1952.

Roman Kossak and Philip Ording. *Simplicity: Ideals of Practice in Mathematics and the Arts.* Springer International Publishing AG, Cham, Switzerland, 2017.

Saul Kripke. Semantical Considerations on Modal Logic. *Acta Philosophica Fennica*, 16:83–94, 1963.

Saul Kripke. *Naming and Necessity.* Harvard University Press, Cambridge, Mass., 1980.

Karel Lambert. Free Logic and the Concept of Existence. *Notre Dame Journal of Formal Logic*, 8(1–2):133–144, 1967.

George Leonard. *Mastery: The Keys to Success and Long-Term Fulfillment.* Plume (Penguin Publishing Group), New York, NY, 1992.

C. I. Lewis and C. H. Langford. *Symbolic Logic.* Century Company, New York, 1932.

David Lewis. *Counterfactuals.* Blackwell, Malden, Mass., 1973.

J. R. Lucas. Minds, Machines and Gödel. *Philosophy*, 36:120–124, 1961.

Gerald J. Massey. Tense Logic! Why Bother? *Noûs*, 3(1):17–32, 1969. URL `http://www.jstor.org/stable/2216154`.

R. Penrose. *The Emperor's New Mind: Concerning Computers, Minds, and the Laws of Physics.* Oxford University Press, Oxford, 1989.

Dag Prawitz. *Natural Deduction: A Proof-Theoretical Study*. Almqvist & Wiksell, Stockholm, 1965.

Willard Van Orman Quine. *Methods of Logic*. Holt, Rinehart and Winston, New York, 1950.

Nicholas Rescher. *Galen and the Syllogism: An Examination of the Thesis that Galen Originated the Fourth Figure of the Syllogism in Light of New Data from the Arabic*. University of Pittsburgh Press, Pittsburgh, PA, 1966.

Bertrand Russell. *The Principles of Mathematics*. Cambridge University Press, Cambridge, 1903.

Bertrand Russell. On Denoting. *Mind*, 14(56):479–493, 1905. doi: https://doi.org/10.1093/mind/XIV.4.479.

Peter Schroeder-Heister. A Natural Extension of Natural Deduction. *Journal of Symbolic Logic*, 49:1284–1300, 1984.

Henry Maurice Sheffer. A Set of Five Independent Postulates for Boolean Algebras, with Application to Logical Constants. *Transactions of the American Mathematical Society*, 14:481–488, 1913.

Timothy Smiley. Aristotle's Completeness Proof. *Ancient Philosophy*, 14(Special Issue No. 1):25–38, 1994.

Alfred Tarski. Der Wahrheitsbegriff in den formalisierten Sprachen. *Studia Philosophica. Commentarii Societatis Philosophicae Polonorum*, 1:267–405, 1935. First published, in Polish, in 1933.

Alfred Tarski. The Semantic Conception of Truth: and the Foundations of Semantics. *Philosophy and Phenomenological Research*, 4(3):341–376, 1944.

Alfred Tarski. The Concept of Truth in Formalized Languages. In J. H. Woodger, editor, *Logic, Semantics, Metamathematics*, pages 152–278. Clarendon Press, Oxford, 1956.

Neil Tennant. Recursive Semantics for Knowledge and Belief. *The Monist*, 60(3):419–430, 1977.

Neil Tennant. *Natural Logic*. Edinburgh University Press, 1978.

Neil Tennant. Entailment and Proofs. *Proceedings of the Aristotelian Society*, LXXIX:167–189, 1979a.

Neil Tennant. La Barre de Sheffer dans la Logique des Sequents et des Syllogismes. *Logique et Analyse*, 88:503–514, 1979b.

Neil Tennant. Perfect Validity, Entailment and Paraconsistency. *Studia Logica*, XLIII:179–198, 1984.

Neil Tennant. *Autologic*. Edinburgh University Press, Edinburgh, 1992.

Neil Tennant. Negation, Absurdity and Contrariety. In Dov Gabbay and Heinrich Wansing, editors, *What is Negation*, pages 199–222. Kluwer, Dordrecht, 1999.

Neil Tennant. Deductive versus Expressive Power: A Pre-Gödelian Predicament. *Journal of Philosophy*, XCVII(5):257–277, 2000a.

Neil Tennant. Anti-Realist Aporias. *Mind*, 109(436):825–854, 2000b.

Neil Tennant. On Turing Machines Knowing Their Own Gödel-Sentences. *Philosophia Mathematica*, 9:72–79, 2001.

Neil Tennant. Existence and Identity in Free Logic: A Problem for Inferentialism? *Mind*, 116 (461):1055–1078, 2007.

Neil Tennant. Inferential Semantics. In Jonathan Lear and Alex Oliver, editors, *The Force of Argument: Essays in Honor of Timothy Smiley*, pages 223–257. Routledge, London, 2010.

Neil Tennant. Cut for Core Logic. *Review of Symbolic Logic*, 5(3): 450–479, 2012.

Neil Tennant. Cut for Classical Core Logic. *Review of Symbolic Logic*, 8(4):236–256, 2015a.

Neil Tennant. The Relevance of Premises to Conclusions of Core Proofs. *Review of Symbolic Logic*, 8(4):743–784, 2015b.

Neil Tennant. A Logical Theory of Truthmakers and Falsitymakers. In Michael Glanzberg, editor, *Handbook on Truth*, pages 355–393. Oxford University Press, Oxford, 2018.

Neil Tennant. What is a Rule of Inference? *Review of Symbolic Logic*, 14 (2):307–346, 2021. doi: https://doi.org/10.1017/S1755020320000441.

Neil Tennant. On the Adequacy of a Substructural Logic for Mathematics and Science. *The Philosophical Quarterly*, 72(4):1002–1018, 2022. doi: https://doi.org/10.1093/pq/pqac001.

Neil Tennant. Frege's Class Theory and the Logic of Sets. In Kai Wehmeier and Thomas Piecha, editors, *Peter Schroeder-Heister on Proof-Theoretic Semantics,* in the series *Outstanding Contributions to Logic*, pages 85–134. Springer, Berlin, 2024.

Andrew Wiles. Modular elliptic curves and Fermat's Last Theorem. *Annals of Mathematics*, 141(3):443–551, 1995.

Timothy Williamson. Alternative Logics and Applied Mathematics. *Philosophical Issues*, 28:399–424, 2018.

Index

www.ingramcontent.com/pod-product-compliance
Ingram Content Group UK Ltd.
Pitfield, Milton Keynes, MK11 3LW, UK
UKHW020705041026
15590UKWH00001B/477

9 781848 905061